AIRCRAFT ELECTRICAL SYSTEMS, HYDRAULIC SYSTEMS, AND INSTRUMENTS

DRAKE'S AIRCRAFT MECHANIC SERIES

AIRCRAFT WOODWORK

AIRCRAFT WELDING

AIRCRAFT SHEET METAL

AIRCRAFT ENGINES

AIRCRAFT ELECTRICAL SYSTEMS,
HYDRAULIC SYSTEMS,
AND INSTRUMENTS

AIRCRAFT MAINTENANCE AND SERVICE

AIRCRAFT ENGINE MAINTENANCE AND SERVICE

AIRCRAFT ELECTRICAL SYSTEMS,
HYDRAULIC SYSTEMS,
AND INSTRUMENTS

by COLONEL ROLLEN H. DRAKE, B. S., M. A.

RESEARCH ENGINEER, PILOT AND GROUND SCHOOL INSTRUCTOR FORMERLY: INSTRUCTOR, VOCATIONAL AND AVIATION SUBJECTS, LOS ANGELES CITY SCHOOLS; CHIEF, AIR AGENCY UNIT AND CIVILIAN PILOT TRAINING SPECIALIST, CIVIL AERONAUTICS ADMINISTRATION; SUPPLY SPECIALIST, OFFICE OF THE QUARTERMASTER GENERAL

Originally published in 1949

ISBN: 1940001382

The Aviation Collection
by
Sportsman's Vintage Press
2015

This book is dedicated to

MY BELOVED WIFE

without whose inspiration and help

this book could not have been written

PREFACE

THIS book is planned to fill an ever-increasing need for a single volume written in nontechnical language covering aircraft electrical systems, aircraft hydraulic systems and aircraft instruments.

It would be impossible in any single volume to cover all phases of any one of these three fields. It is felt, however, that this volume covers rather completely the fundamentals of each of them.

The text is divided into three sections. Part I covers the fundamentals of electricity and aircraft electrical systems; Part II the fundamentals of hydraulics and aircraft hydraulic systems; and Part III the fundamentals of the principles underlying the construction and operation of aircraft and engine instruments.

This text should furnish all the theoretical knowledge necessary to qualify a mechanic for a specialist rating in these fields. It has been written as a classroom text but can also be used for reference and as a practical guide for self-instruction, or for mechanics in the field. It is intended for use in college vocational courses, trade schools, junior colleges, high schools, and in the rehabilitation programs and aviation ground schools. Perhaps the most outstanding feature of this book is the simple nontechnical language in which it is written. The author has avoided the use of formulas, graphs, confusing tables, obscure footnotes, and any other material which cannot be clearly understood by anyone. Particular attention has been given to the theory underlying the operation of the various electrical devices, hydraulic systems, and aircraft instruments.

The author wishes to express his grateful appreciation to the following who have so kindly furnished material which has been of assistance in the preparation of this book: the U. S. Office of Education; the Civil Aeronautics Administration; the Departments of Education of the various states, particularly New York, Pennsylvania, Utah, and Virginia; Douglas Aircraft Company, Inc.; The Glenn L. Martin Company; AC Spark Plug Division, General Motors Corporation; The Leece-Neville

Company; Electrical Engineering and Manufacturing Corporation; Jacobs Aircraft Engine Company; Eisemann Corporation; Scintilla Magneto Division, Bendix Aviation Corporation; The Electric Auto-Lite Company; Ranger Aircraft Engines Division, Fairchild Engine and Airplane Corporation; Pesco Products Co.; The B. F. Goodrich Company; Kollsmann Instrument Division, Square D Company; and Wright Aeronautical Corporation.

The author wants particularly to thank the following who not only furnished material for this text, but read and edited a large part of the manuscript: Eclipse-Pioneer Division, Bendix Aviation Corporation; Jack & Heintz, Inc.; Sperry Gyroscope Company, Inc.; and Minneapolis-Honeywell Regulator Company.

The author wishes also to thank his many friends who have so generously contributed their advice and assistance. He also wishes to express his particular gratitude to Earle R. Hough for his excellent work in the preparation of many of the drawings in this book, and to Mildred Pickrel and Alma Franklin for their untiring patience and cooperation in the preparation of the manuscript.

The frontispiece shows part of the main, hydraulic landing gear of a large airplane, and is reproduced by courtesy of Douglas Aircraft Company, Inc.

R. H. D.

TABLE OF CONTENTS

1 INTRODUCTION

In less than an average lifetime the airplane has developed from a fragile glider equipped with a 12-hp engine and built of bamboo poles, wire and cloth, to the monsters of today's skys which weigh scores of tons. The early airplanes were built as light as possible and made use

Fig. 1. Large aircraft are navigated largely by means of electrical equipment. (Courtesy The Glenn L. Martin Company)

of the lightest engine available. They flew neither high nor far. Each additional few minutes one of them remained in the air set new endurance records. Each few feet of additional altitude thrilled the pilot with his mastery of the air.

In the early days of aviation many avoidable accidents took the lives of pioneers in the field of aviation and gave the general public the impression that aviation was a dangerous and foolhardy sport to be indulged in by supermen, dare-devils and fools. The common causes of accidents at that time were overheating engines, running out of fuel or oil and getting lost and not having a suitable landing field at hand, or stalling and getting into the then-dangerous tailspin.

These early airplanes had no accessories such as electrical equipment, instruments, or aids to navigation. Among the first pieces of equipment to be added were the water temperature thermometer for the radiator,

Fig. 2. Electrical wiring and hydraulic tubing in fuselage of an airplane. (Courtesy of Douglas Aircraft Company, Inc.)

the oil pressure and temperature gauges and the fuel gauge. As the pilot began to take longer trips a compass was added. It was soon found that airspeed was important if the pilot wished to avoid dangerous stalls and spins, so a simple airspeed indicator was installed.

The mechanical equipment of the aircraft has followed closely the development of the airplane and the airplane engine. Mechanical equipment and accessories fall roughly into three divisions: the electrical system, the hydraulic system and the instruments.

The aircraft electrical system includes the engine ignition, the radio and communication equipment, the electrical accessories and electrical instruments.

The hydraulic system operates the landing gear, the flaps, and often the automatic pilot. It consists of pumps, storage tanks, pipes, valves and controls in addition to the actuating cylinders which bring about the movement of the parts.

Aircraft instruments are usually grouped as aircraft instruments and engine instruments. The aircraft instruments include flight instruments, navigation instruments, automatic pilots and other instruments

Fig. 3. A panel board used in testing electrical equipment. (Courtesy Douglas Aircraft Company, Inc.)

primarily concerned with the operation of the aircraft. The engine instruments include the instruments which are primarily concerned with the operation of the engine. These instruments include speed, temperature, and pressure indicators, fuel and oil gauges, fuel mixture indicators, and horsepower indicators.

The beginning of the complex electrical system used today seems to date from about the time of World War I. At this time it was decided to install battery ignition on some aircraft engines. This decision was brought about because the engines were hard to start. The magnetos were not very dependable and furnished a weak spark at the low speeds developed when the propellers were pulled through by hand. The battery furnished a hot spark even at these low speeds.

With the installation of the storage battery with its supply of electric current, a few lights were installed. These first lights were used to illuminate the few instruments then in use. These batteries had to be removed from the airplane for recharging. The battery was sometimes

Fig. 4. The main electrical system junction box on a large transport airplane. (Courtesy of Douglas Aircraft Company, Inc.)

found dead when the pilot wished to take off, so a generator driven by the engine was installed to keep the battery charged. With this extra supply of electric current, electrical instruments began to make their appearance, and small electric motors were put in to operate movable parts. The development of radio communication and radio aids to navigation placed further demands on the electrical system.

Pneumatics pertain to gases. All gases are highly elastic. This means that they can be compressed or caused to expand almost without limit. In this way, gases may be used to exert pressure or to create suctions. Pneumatic effects may be either positive or negative. When a gas is compressed, it exerts a pressure which may be said to be positive; when it is expanded, it creates a suction which may be said to be negative. The pneumatic system of the large airplane includes both pressure and suction. The pressure and suction are measured by pressure and suction gauges. For example, the manifold pressure is indicated by the manifold-pressure gauge which measures both pressure and suction.

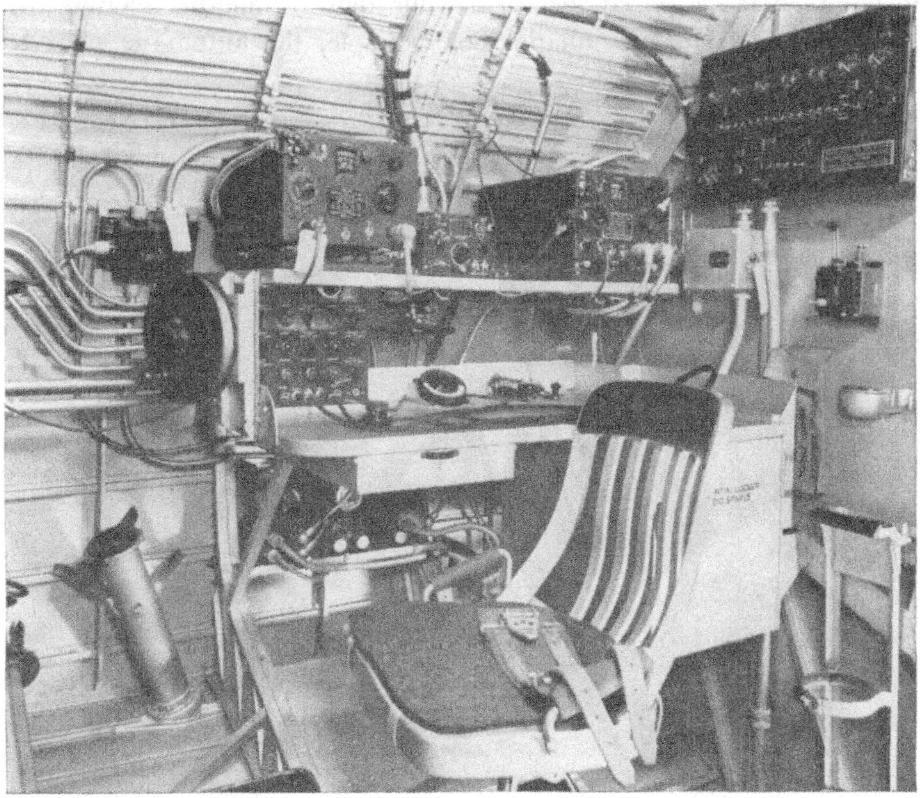

Fig. 5. Radio installation on a large twin engine flying boat. (Courtesy Consolidated Aircraft Corporation)

On some of the early aircraft compressed air was used to operate movable parts. However, because of the elasticity of gases the action brought about by this means is "spongy" and not precise. Hydraulic systems have largely replaced pneumatic systems for operating movable parts.

The hydraulic system depends upon the fact that it is practically impossible to compress liquids. When a closed container, such as a tank or system of pipes or tubes, is completely filled with a liquid, the liquid is said to be "confined." When pressure is applied to any point of a confined liquid, that pressure is carried all through the liquid without decreasing its force. This force is said to be produced by hydraulic pressure.

Electricity, pneumatics and hydraulics all play an important part in the functioning of aircraft instruments. It would be impossible to divide an aircraft instrument board in such a way that the instruments in one section are purely electrical while those in another section are purely hydraulic or pneumatic. For example, the automatic pilot is

Fig. 6. A hydraulic landing gear. (Courtesy Douglas Aircraft Company, Inc.)

largely dependent for its effectiveness upon delicate hydraulic, pneumatic, or electrical systems. The fuel mixture, which is so important to the proper performance of the aircraft engine, is accurately analyzed by means of instruments. The result of this analysis is transferred to the instrument board by an electric current. The position of the various parts of the airplane, such as landing gear and flaps, is indicated by an instrument on the instrument board which is operated by an electric current.

The close relationship between aircraft electrical systems, hydraulic systems, and instruments makes it necessary that the aircraft and engine mechanic be thoroughly familiar with all three. The position of the landing gear or hydraulically operated flaps is indicated on the instrument board by means of an electrical instrument. This operation involves the electrical system, the hydraulic system and the instrument board. While this text has separated these groups, it is impossible to operate the instruments separately, and it is essential that the aircraft mechanic have a complete understanding of each of them.

Fig. 7. A typical instrument layout for a two-motored airplane. (Courtesy Pioneer Instrument Company)

The first view of an instrument board of a large aircraft gives a person a feeling of helplessness. On some instrument boards, there are more than 100 gauges, switches, instruments, and gadgets. It seems a hopeless task not only to learn the name and use of each of these, but also to understand everything about each of them and be able to make adjustments and repairs. Each individual part is simple in itself, but each individual part has its own adjustment and function in the whole system.

Nowhere in the mechanical field is it so important that the "little thing wrong" be detected and remedied as in the field of aircraft maintenance and service. It is necessary that the aircraft mechanic have a thorough understanding of the theory and operation of all parts of the aircraft systems. He must understand the application of theory to the proper performance of the aircraft accessory and its relation to the aircraft as a whole. He must not only be able to determine when a part or instrument is not performing properly, but he must also be able to determine what is wrong and how it may be adjusted or repaired.

However, any person who is mechanically minded and mechanically handed can master any of the instruments or parts of the electrical or hydraulic systems used in the modern aircraft. It is only necessary, through experience and study, to be able to recognize each part and determine whether that part is in proper adjustment and, if not, the proper adjustment or repair to be made. There is only one way that it can be "right," and the fundamental law for the aircraft workman is to be sure that every part of the aircraft is right before turning the plane over to the pilot. This is the responsibility of the aircraft mechanic.

PART I AIRCRAFT ELECTRICAL SYSTEMS

II FUNDAMENTALS OF ELECTRICITY

As long ago as 600 B.C., some discoveries concerning electricity had been made by the Greeks. They found that, when amber was rubbed with a woolen material, it would attract certain light objects such as bits of pith or paper. Now we know that, if a hard-rubber fountain pen is rubbed on a piece of woolen cloth, it will attract bits of paper, and that a glass rod may be electrified by rubbing it with a piece of silk. Many experiments have shown that when any two different kinds of material are rubbed together or brought into close contact, they become somewhat electrified.

The name of electricity came from the word, electron, which is the Greek word for amber. A piece of material such as amber or glass, when rubbed with a proper substance, becomes electrified and is said

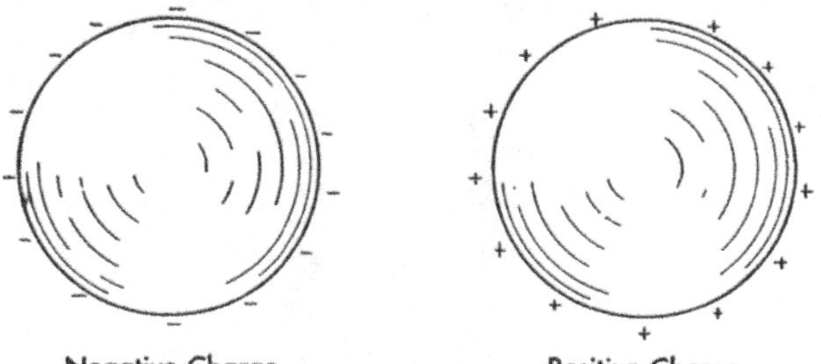

Fig. 8. The metal spheres carrying electric charges.

to have an electric charge. A metal sphere can be electrified by touching it with an object which has been electrified, such as a piece of amber which has been rubbed with a woolen cloth, a glass rod which has been rubbed with a piece of silk, or a piece of sealing wax which has been rubbed with cat's fur. If the sphere is suspended by a dry silk thread before being electrified, it will retain the charge of electricity for a considerable length of time. If, however, this metal sphere has been suspended by means of a metal wire or is touched by a metal object or by a moistened finger after having been electrified, the electric charge disappears from the sphere at once. This means that the electric charge has moved off the sphere along the metal wire. Electricity moving along

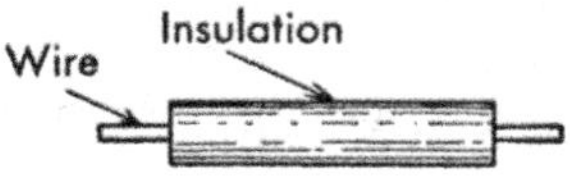

Fig. 9. Insulation on a piece of wire.

a wire is called an electric current. Any substance which allows electricity to move freely along it is called a "conductor." A substance, such as a dry silk thread, which does not allow the flow of electricity

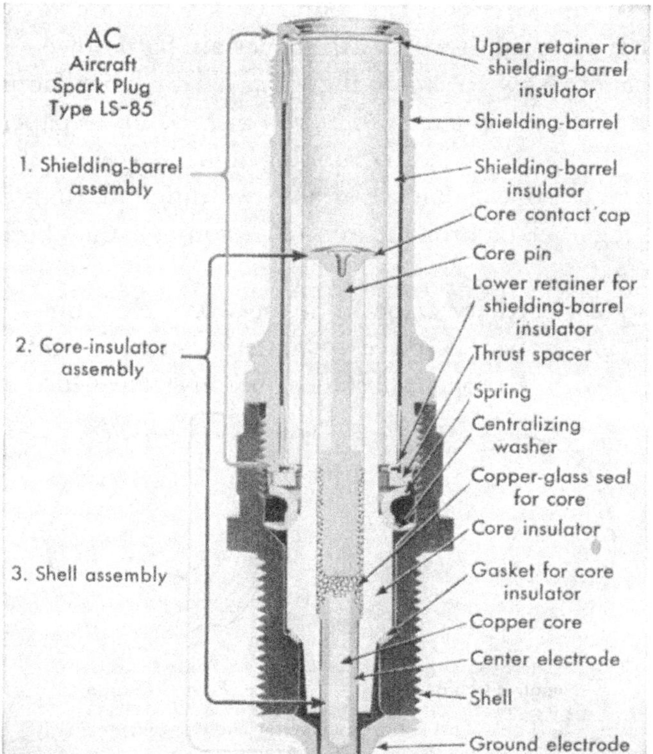

Fig. 10. A cut-away view of an aircraft's sparkplug showing insulation. (Courtesy AC Spark Plug Division, General Motors Corporation)

along it, is called a nonconductor or an insulator. Such substances are used to cover electric wires which might come into contact with other conductors, thereby losing current. Insulators are made of glass, rubber, porcelain, mica, and plastics.

A sharp line cannot be drawn between conductors and nonconductors. However, some substances are good conductors, some are poor conductors, and some are insulators or nonconductors. Most of the metals are good conductors. Graphite is a good conductor, as is water in which

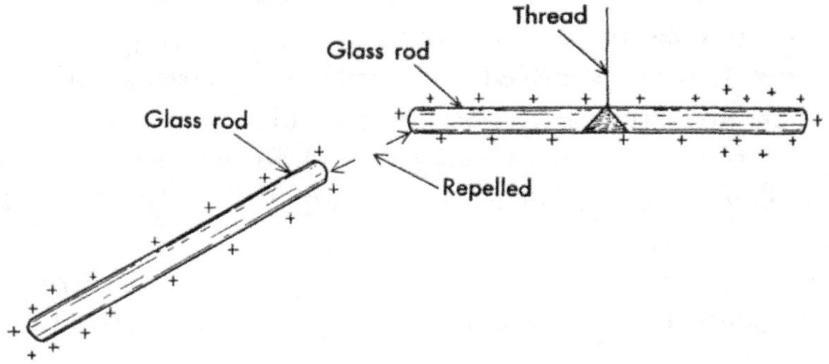

Fig. 11. A suspended electrified glass rod being repelled by a similar rod.

salts or acids have been dissolved. Dry wood, oil, and pure water are poor conductors. Such substances as glass, paraffin, plastics, rubber, and mica are good insulators. Substances which may be electrified by rubbing, fall into the class of insulators. A good conductor cannot be electrified by rubbing.

There are two kinds of electric charge. Experiments show that the kind of electric charge which is produced by rubbing a glass rod with a piece of silk is different from the electric charge produced when a rod

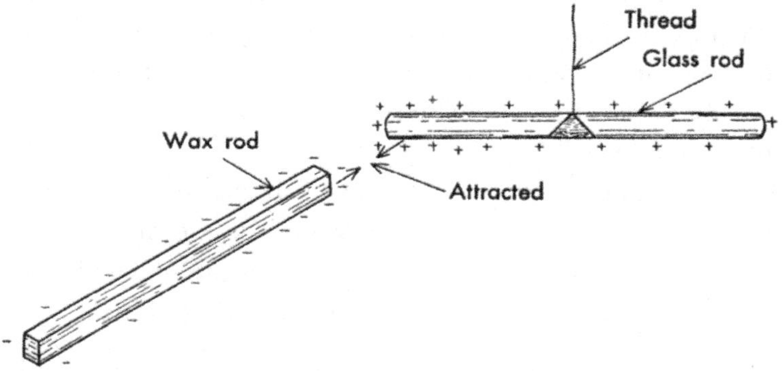

Fig. 12. A suspended glass rod being attracted by a wax rod carrying an opposite charge

of sealing wax is rubbed with cat's fur. If a glass rod which has been rubbed with a piece of silk is suspended by a silk thread so that it balances as on a pivot, another piece of glass rod which has been electrified in the same manner will, when brought close to the end of the suspended rod, repel the suspended rod. If, however, a piece of sealing wax which has been electrified by rubbing with cat's fur is brought close to the electrified glass rod, the glass rod will be attracted. This action is brought about by the two kinds of electric charge. For convenience, these two kinds of electric charge have been called "positive" and "negative."

Two objects electrified with positive electricity repel each other, as will two objects electrified with negative electricity. If, however, two objects are electrified, one with a positive charge and the other with a negative charge, the two objects will attract each other. A body carrying either a positive or negative charge will attract an uncharged body.

To find out which kind of electric charge a body is carrying, the charged body is brought close to a light object which carries a known

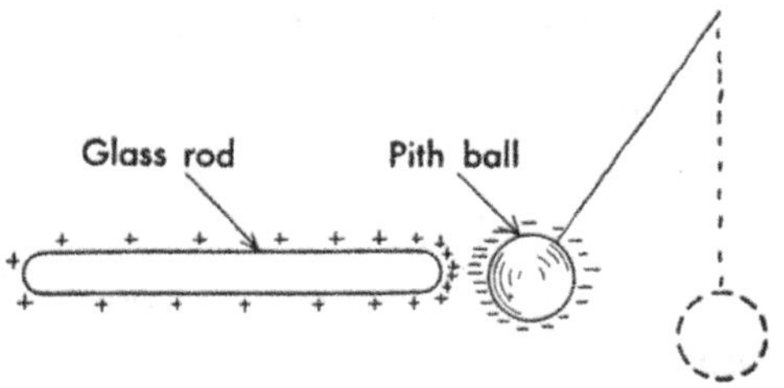

Fig. 13. A light ball having a negative charge is attracted by a rod having a plus charge.

charge of electricity, either positive or negative. If the light body is known to have a charge of positive electricity and is repelled by the first body, the first body is then known to carry a positive charge. If the first body attracts the light object, it is known to carry a negative charge of electricity.

Until recent times there was no clear understanding of what electricity was or what an electric current was. The atomic theory leads us to believe that all matter is made up of atoms or molecules which, in turn, are made up of two kinds of particles. These particles seem to be electricity. For example, the atom of hydrogen, which is the lightest of all substances, is thought to be made up of two particles, one of which is a particle of positive electricity called the "proton" and the other a negative particle of electricity called the "electron." The proton is approximately 1845 times as heavy as the electron.

Experiments have led us to believe that some of the electrons revolve about the protons in orbits much like those which the planets follow in their revolution about the sun. These particles travel in their orbits at tremendous speeds.

Each simple substance or element such as oxygen, iron or gold is made up of one kind of atom which always contains the same number of protons and electrons. In other words, each substance has its own

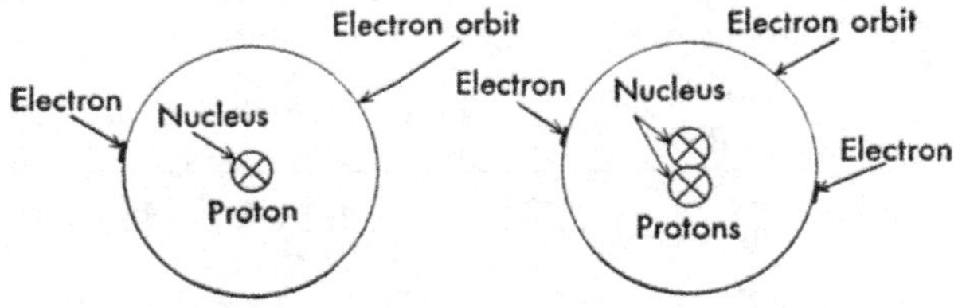

Fig. 14. Diagrams showing the theoretical make-up of the hydrogen and helium atoms.

particular kind of atom. The hydrogen atom contains one proton and one electron. The helium atom contains two protons and two electrons. The oxygen atom contains 16 electrons and 16 protons. Iron contains 56 particles of each. Sulphur contains 32 particles of each. Gold contains 197 particles of each. Nitrogen contains 14 particles of each. It has been generally accepted that there are 92 elements.

Some of the electrons revolve around the protons which are grouped in the center of the atom. This center group is called the nucleus of the atom. Not all electrons in an atom revolve about this nucleus. The electrons that do not are bound closely to the protons in the nucleus. One electron, for instance, revolves about the nucleus in the atom of hydrogen; two electrons revolve about the nucleus in helium; but in the rest of the atoms, approximately half of the electrons do not revolve.

Most of the electrons revolving about the nucleus are quite firmly fixed in their paths. In most of the atoms, however, there are a few (usually one, two, three, or five) electrons which travel in an elliptical path that carries them well outside the electron system. These may be quite easily displaced from the atom.

There are other electrons in most substances which do not seem to be fixed to any individual atom. These are known as "free electrons." These free electrons seem to wander from one atom to another without fixed paths. They can quite easily be moved through the substance by an external influence. Free electrons are more plentiful in conductors than in nonconductors.

If a supply of electrons is placed on one end of a piece of wire while electrons are removed from the opposite end of the wire, free electrons within the wire move from atom to atom resulting in a flow of electrons through the wire.

When a cupful of water is drawn from a faucet, water moves through the entire system. To keep the system full, a cup of water must be added at the opposite end of the water system to replace that which is drawn off at the faucet. The cup of water drawn from the faucet did not come

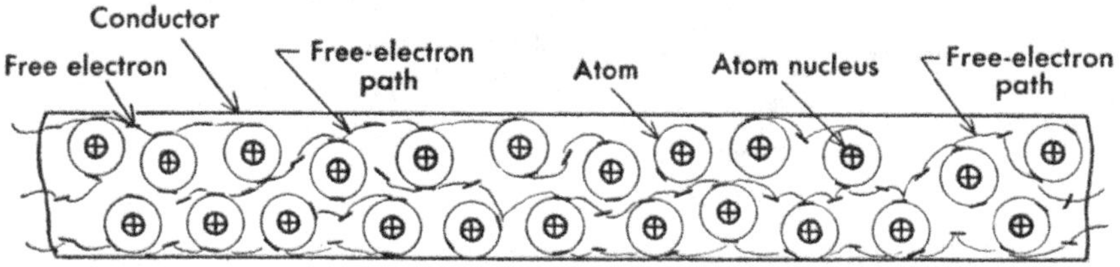

Fig. 15. A theoretical diagram illustrating how free electrons might travel through a conductor.

directly from the reservoir, but did cause a movement of water in all of the pipes from the faucet to the source of the water supply. Likewise, the electrons leaving one end of the wire are not the identical electrons being placed on the other end.

Each electron does not travel the length of the wire instantly but becomes a part of the stream of electrons moving through the whole length of the conductor. This movement of electrons through a conductor is called current electricity.

The kind of electricity which makes up a stationary charge on a body is called static electricity. When a glass rod is rubbed with a piece of silk, some of the electrons from the glass are rubbed off, leaving the rod with a positive charge. If a piece of sealing wax is rubbed with cat's fur, some of the electrons from the fur are rubbed off onto the wax, giving it an excess of negative particles, and it is said to have a negative charge. When a cat is stroked, large numbers of negative particles are rubbed off the fur by the hand and then jump back to the fur, forming minute electric sparks which can be seen in the dark. If the feet are scuffed over a dry woolen carpet and the finger is touched to a metal object or a conductor, an electric spark will jump from the finger to the conductor. This electric spark is made up of electrons which have been rubbed off the carpet and accumulated on the body. These electrons jump from the body to the conductor. An electric spark is made up of electrons passing through the air.

Lightning is the discharge of a tremendous number of electrons from a cloud to the earth or from the earth to a cloud. When a cloud is charged positively and passes over earth that has a negative charge, there is an attraction between the positive particles in the cloud and the negative particles in the earth. This attraction may be strong enough to cause the negative particles from the earth to discharge to the cloud, forming a lightning flash. If the earth is positively charged and the cloud is negatively charged, the flash occurs in the opposite direction. This kind of an electrical discharge is made up of static electricity.

Whenever a body contains more than the normal number of electrons, the body is negatively charged. If the body contains fewer than the normal number of electrons, it is said to have a positive charge.

Electricity is not visible, but its effects can be measured by means of the proper instruments. Electricity in an aircraft is used to produce heat, light, power and operate instruments and the communication system. Whenever an electric current flows through a conductor, it always produces heat. Most conductors designed to carry an electric current have such a low resistance that the heat is not noticeable. The wires in a heating element have high resistance and are heated red-hot by the electric current. A current of electricity always produces a magnetic effect about the conductor through which it is flowing. This magnetic effect may be simply a magnetic field, although a coil of wire carrying an electric current may show an effect similar to that shown by an ordinary magnet. Whenever electricity flows through a liquid, it produces chemical effects. This chemical effect is used in recharging ordinary storage batteries.

Electricity is able to do work. This ability is brought about by the effects of the current and not directly by the current itself. Electricity can be felt, of course, as anyone knows who has touched an electric wire and received a shock. Electricity is a form of energy. Light and heat are also forms of energy. In the electric light, the electric energy is changed to heat energy which heats the filament to such a high temperature that light is given off.

Energy, like electricity itself, cannot be seen. Its effects, however, may be measured and sometimes may be seen. The effects of energy may appear as heat, light, sound, and motion, all of which may be measured. Energy may appear as mechanical energy, electrical energy, magnetic energy, chemical energy, or heat energy.

The fundamental law of energy called "the law of conservation of energy" states that energy can neither be created nor destroyed. However, energy in any of its various forms can be changed to other forms of energy, but the amount obtained after the change is exactly the same as the amount of energy present before the change.

Energy has been defined as the ability to do work. Energy is also classified as kinetic energy and potential energy. Kinetic energy is energy which a body has because of its motion. If a ball is thrown, it takes considerable force to stop the ball instantly when it is caught. The force acting upon the hand or glove when stopping the ball was due to the motion of the ball, or its kinetic energy. Potential energy is the energy which anything has due to its position or its condition. For example, the snow on a mountaintop has stored up in it, because of its position, the power to turn a water wheel when it has melted and the water from it flows down the side of the mountain. A bomb filled with explosives has potential energy due to the condition of the substances making up the explosives contained within the bomb.

Work has been defined as the effect produced by energy acting through distance. Energy can be expended without performing work. For example, one may push against the side of a large building, expending a great deal of energy without moving the building. Since the energy has not been expended through distance, no work has been accomplished. If a brick is picked up, work has been done. If the brick weighs one pound and is lifted one foot, one foot-pound of work is said to have been done. The joule is another unit of measure for work and is used more often in connection with electrical energy. The joule is a unit of mechanical potential energy and is equal to 0.738 ft.-lb. If an object weighing 10 lb. is lifted to a height of 20 ft., 200 ft.-lb. of work have been performed on the object. There is now stored up in the object potential energy equal to the 200 ft.-lb. A joule equals one ampere of current at one volt flowing for one second. 200 ft.-lb. equal approximately 271 joules. If the weight is now dropped a distance of 20 ft., it will have 200 ft.-lb. of kinetic energy, due to its motion at the instant it strikes the earth. This energy will be expended in the form of heat.

Horsepower is a term used to express the rate of doing work. If 33,000 ft.-lb. of work are done in one minute, the rate of doing the work is 1 horsepower.

Electricity in motion is called "current electricity." Electricity which is in a state of rest, such as the kind of electricity produced by rubbing objects together, is called "static electricity." The attraction between positive and negative electricity causes electric currents to flow. The flow of electric currents may also be brought about by magnetic effects.

If one terminal, such as a battery terminal, is highly charged positively and the other terminal is highly charged negatively, the negative particles are attracted to the positive terminal. If a conductor, such as a piece of wire, is touched to both terminals at the same time, some of the negative particles on the negatively charged terminal will crowd onto the conductor, becoming free electrons. They can do this because the free electrons on the conductor are less concentrated than on the highly

Fig. 16. An inverted-V aircraft engine which develops approximately 500 horsepower. (Courtesy Ranger Aircraft Engines)

charged negative terminal. At the other end of the conductor, free electrons will pass off the end of the wire, being attracted by the large number of excess positive particles on the positive terminal. The free electrons scattered along through the conductor will move along from atom to atom to take the place of those which have moved off onto the positive terminal. This sets up a continuous flow of electrons through the conductor. This flow of electrons is an electric current.

The intensity of the charge on the terminals will determine the rate of flow of the current. The difference in intensity of charge between the two terminals is called their difference in potential, which is their difference in electrical pressure. Electrical potential is electrical pressure. The greater the difference in charge between the two terminals, the greater the electrical pressure will be and the greater the amount of current which will flow in a given length of time. The flow of current is fundamentally due to the attraction between the excess negative particles on the negative terminal and the excess positive particles on the positive terminal.

It will now be seen that the direction of flow of electric current is from negative to positive. The first experimenters with electricity thought that the direction of the flow was from positive to negative. Practically all of our electrical instruments and appliances are built on the latter assumption, and, conventionally, most people working with electricity assume that the direction of flow is from positive to negative. The positive terminal of any piece of electrical equipment is usually marked with a plus sign (+), and the negative terminal with a minus sign (−). Many terminals, such as those on storage batteries, have the positive terminal marked with red paint and the negative terminal marked with green paint.

There are certain common units in electrical measurement with which every person working with electricity should be familiar. Electrical pressure is measured in volts, as the pressure of water in a water pipe is measured in pounds of pressure per square inch. A volt is the unit which measures the difference in electrical pressure or in electrical potential.

When water is allowed to flow from a pipe, under pressure, the rate of flow is measured in some unit such as gallons per unit of time. The ampere is the unit used to describe the rate of flow of an electric current.

The rate at which water will flow through a pipe depends upon the pressure and the size of the pipe or the opening through which the water is flowing. The pipe and the faucet through which the water flows offer resistance to the flow of the water and determine how much will flow during a given time under a given pressure. Conductors carrying an electric current always offer resistance to the flow of the electric current. The ohm is the unit used to measure the resistance of a conductor.

There is a close relationship between the ohm, the ampere, and the volt. When the resistance of the conductor is one ohm and the electric pressure is one volt, one ampere of current will flow. The standard ohm is the resistance offered to the flow of electric current by a column of mercury 106.3 cm. long and having a cross section of 1 sq. mm.

When water is flowing from a pipe at the rate, for example, of 100 gal. per min., it is necessary that a definite interval of time be taken into consideration to determine the amount of water which has flowed. In this instance, 10 gal. of water will flow in 6 sec. At any given instant, no gallons of water flow, but the rate of flow is still 100 gal. per min.

When an electric current is flowing at the rate of 1 amp., it is again necessary to introduce the time element in order to measure the amount of electricity which has flowed. When 1 amp. of current is flowing, the amount of current which passes any point in the circuit in 1 sec. is called a coulomb. The coulomb is the unit by which quantities of electricity are measured. The coulomb is that amount of electricity which will deposit 0.001118 g. of silver from a chemical solution in 1 sec. When 10 amp. of current are flowing through a circuit, 10 coulombs of electricity pass any point in the circuit in 1 sec.

The ability of an electric current to do work is measured in terms of watts. A watt is the amount of power generated by 1 amp. flowing with a difference in potential of 1 v. The total number of watts may be determined by multiplying the amperes flowing through a circuit by the number of volts difference in potential. If the power is used over a period of one hour, the watts used will be given in terms of watthours. One ampere flowing with a voltage of 110 will produce 110 w. of power. In 1 hr., it will produce 110 whr. A kilowatthour is the number of watthours divided by 1000. One horsepower is equal to 746 w. A horsepower is equal to 33,000 ft.-lb. per min. or 550 ft.-lb. per sec.

Not all metals conduct electricity equally well. Silver, which is the best conductor of all the metals, is said to have an electrical conductivity of 100. All the other metals conduct electricity less readily. Copper is a good conductor of electricity, and 1000 ft. of No. 10 copper wire have a resistance of almost exactly 1 ohm. Ohm's law, which is one of the fundamental laws in electricity, simply states that "for any circuit or part of a circuit, the current in amperes is equal to the electromotive force expressed in volts, divided by the resistance in ohms." When E is used to express volts, I to express current in amperes, and R is equal to the resistance, Ohm's law may be stated in the following formulas.

$$\text{Current} = \frac{\text{Voltage}}{\text{Resistance}}$$

That is, amperes $= \frac{\text{volts}}{\text{ohms}}$. Using the letters or symbols, Ohm's law may be stated in the three formulas:

$$I = \frac{E}{R}; \quad R = \frac{E}{I}; \quad E = I \times R$$

In the following tables, the symbols commonly used in electrical wiring diagrams and blueprints are listed.

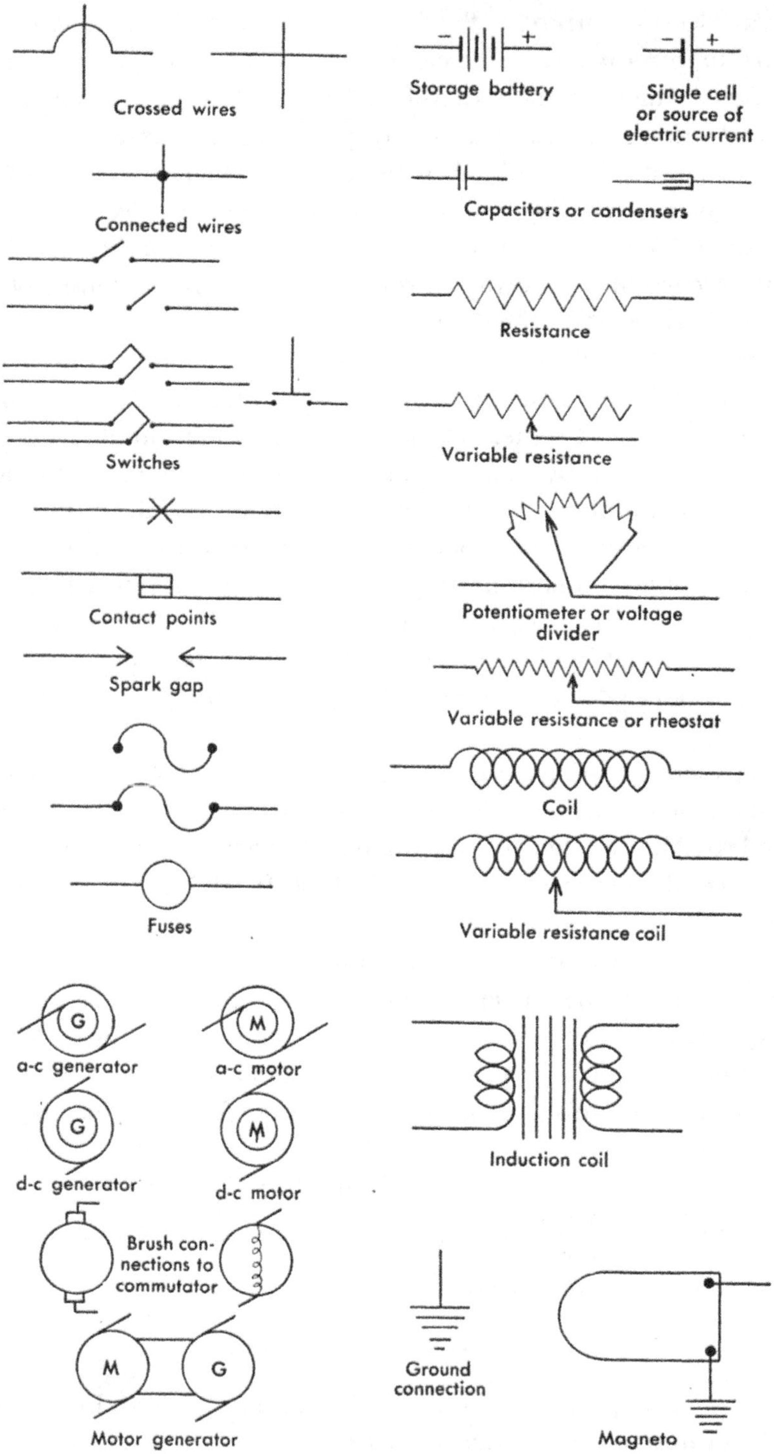

 Fig. 17. Conventional symbols used on electrical wiring diagrams.

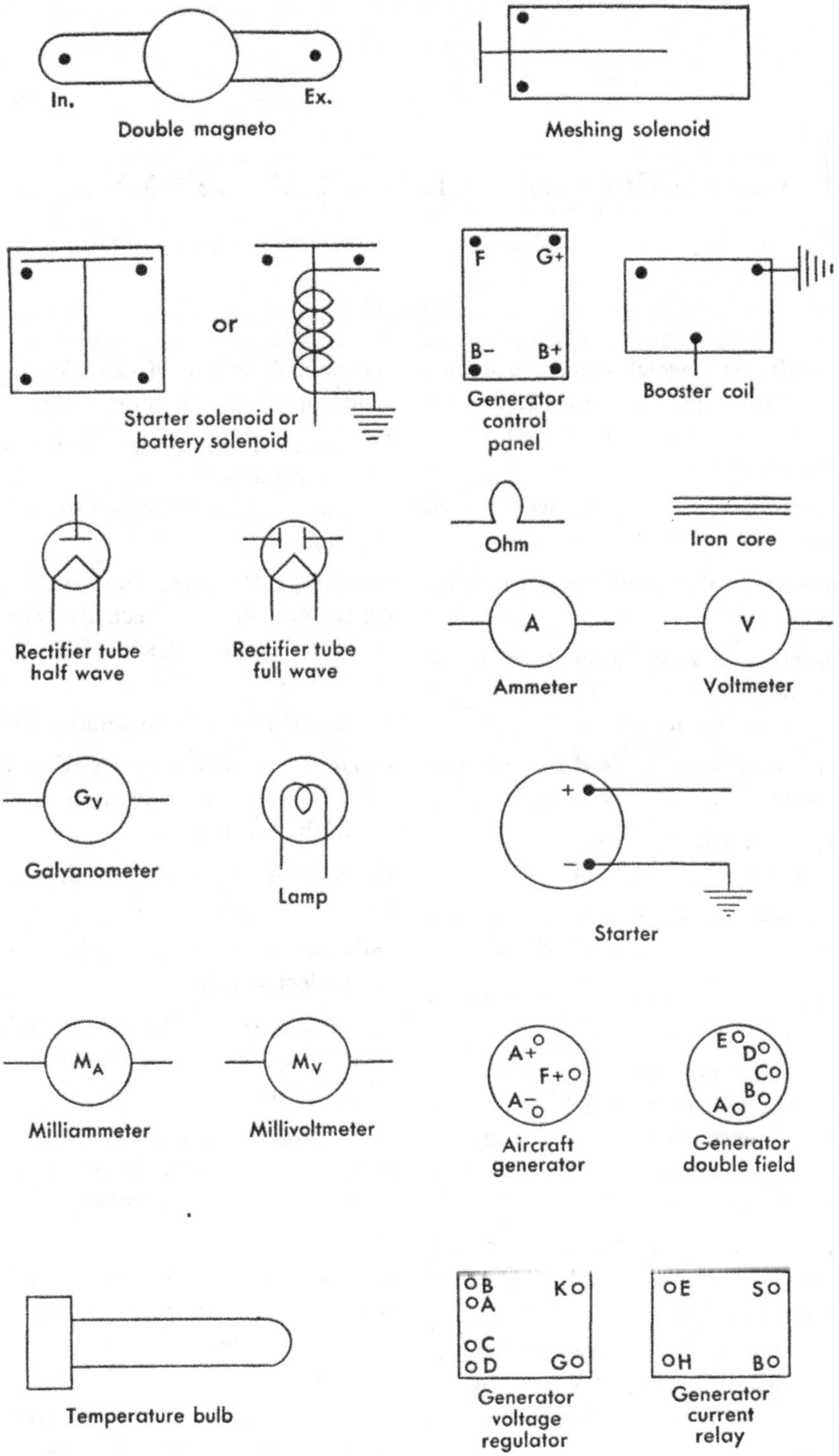

Fig. 18. Conventional symbols used on electrical wiring diagrams.

III GLOSSARY OF ELECTRICAL TERMS

aircraft. Any weight-carrying device or structure designed to be supported by the air, either by buoyance or dynamic action.

alternator. A generator which produces alternating electric current.

amalgam. Alloys of mercury with other metals.

ammeter. An instrument used to measure in amperes the rate of flow of an electric current.

ampere. The unit used to measure the rate of flow of an electric current.

ampere-hour. A current of one ampere flowing for one hour.

ampere-turn. One turn of wire in a coil carrying one ampere of current.

anode. The positive pole or plate of an electric cell.

armature. The revolving part of a motor or generator which carries the coils of wire through the magnetic field between the poles of the motor or generator.

armature reaction. The counter electromotive force in the windings of an armature.

atom. The smallest particle of a simple substance or element. An atom is believed to be made up of electrons and protons.

battery. A source of electric current made up of two or more electric cells.

brushes. The contact parts of a motor or generator which carry the electric current to or from the commutator or slip rings.

brushes, collecting. The brushes of a generator which collect the current generated from the slip rings or the commutator.

c.e.m.f. Counter electromotive force.

capacitance. The capacity of an electric circuit or a capacitator to store electricity temporarily.

capacitator. A device used to store quantities of electricity temporarily.

cathode. The negative pole or plate of an electric cell.

cell, capacity of. The number of ampere hours of current which the cell will produce.

cell, closed circuit. An electric cell designed to produce small quantities of electricity over extended periods of time.

cell, dry. A device for generating electricity by chemical means which is sealed to prevent loss of chemicals from the cell.

cell, open circuit. An electric cell designed to produce comparatively large quantities of electricity intermittently over short periods of time.

cell, primary. An electric cell, the activity of which cannot be restored when the materials of the cell have been used up.

cell, secondary (storage). An electric cell, the effectiveness of which may be renewed (recharged), such as a storage battery.

cell, wet. A device used to generate electric current by chemical means; the chemical is in the form of a liquid and the cell is not airtight.

circuit, closed. An electric circuit which is complete so that the current may flow.

circuit, electric. A continuous path over which an electric current may flow.

circuit, external. That part of an electric circuit outside the source of the current.

circuit, internal. That part of an electric circuit within the source of the current, such as inside the battery, cell or generator.

circuit, open. An electric circuit, some part of which is incomplete, such as an open switch, so that the current may not flow.

circuit, short. An accidental path over which an electric current may flow between parts of an electric circuit.

cobalt. A metallic element having magnetic properties.

coil, booster. A coil used to increase the voltage of an electric current, usually in an ignition system.

coil, field. A coil used to produce a magnetic field in generators and motors.

coil, induction. A coil used to induce an electric current in a second circuit. An induction coil is made up of a primary coil and a secondary coil.

coil, primary. The coil in a transformer or induction coil carrying the original electric current.

coil, resistance. A coil introduced into an electric circuit to increase the resistance of the circuit.

coil, secondary. The coil in a transformer or induction coil which carries the induced current.

commutator. A device used in a direct-current generator to allow the alternate surges of electric current within the armature to leave the generator in one direction only.

condenser. A device made up of conductors separated by nonconductors and designed to store quantities of electricity temporarily.

conductor. A substance through which an electric current may flow.

converter. An electric device which changes electrical energy to mechanical energy and then reconverts mechanical energy into electrical energy.

copper. A metallic element largely used in electrical equipment.

core. A piece of soft iron placed within coils carrying electric current. The core may be made up of bundles of soft iron wire or sheets of soft iron.

core, laminated. A core made up of sheets of soft iron insulated from each other.

coulomb. A unit used to measure quantities of electricity. One coulomb of electricity will deposit from a chemical solution 0.001118 g. of silver in 1 sec.

current, alternating (ac). A current of electricity which reverses its direction of flow many times per second. A 60-cycle current reverses its flow 60 times per second.

current, direct (dc). A current of electricity which flows in one direction only.

current, eddy. A small local electric current set up within a core of a transformer or armature.

current, electric. The flow of electrons along a conductor.

current limitator. A device used to regulate the current output.

depolarizer. A substance placed in an electric cell to prevent or decrease polarization.

dyne. A unit of force that will give one gram an acceleration of one centimeter per second.

electricity, negative. By common scientific consent, it has been agreed to call "negative" the kind of electrification which appears on sealing wax or vulcanite when rubbed with flannel.

electricity, positive. By common scientific consent, it has been agreed to call "positive" the kind of electrification which appears on a glass rod when rubbed with silk.

electricity, static. Electricity not in motion.

electrolyte. A solution through which an electric current can flow.

electromagnet. A temporary magnet which obtains its magnetism from the effect of an electric current.

electromotive force. The tendency of an electric current to flow. This is measured in volts.

electromotive force, counter. The electromotive force set up in a conductor when an electric current is flowing in the conductor which tends to resist the flow of current.

electron. A negative particle of electricity.

e.m.f. Electromotive force.

energy. Power efficiently and forcibly exerted. The capacity to do work.

energy, kinetic. The energy which a body has because of its motion.

energy, potential. The energy a body has because of its position or state of strain.

erg. The absolute unit of work in the c.g.s. (centimeter-gram-second) system. It is equal to a unit force (1 dyne) acting through a unit distance (1 cm.).

excitor. An electric device, usually a small d-c generator, used to energize the field coil of an alternator.

farad. A measure of capacitance or capacity. One farad of capacity will store one coulomb of electricity.

flux. An electric field made up of magnetic lines of force.

flux, magnetic. A magnetic field set up by a magnetic substance.

frequency. The cycles of an alternating current, or the number of fluctuations per second of a direct current.

fuse. A safety device, usually of metal having a low melting point, used to prevent overloading an electric circuit.

galvanometer. A device similar to a voltmeter used to detect the flow of electric currents.

galvanometer, d'Arsonval. A sensitive instrument of the voltmeter type used to detect the flow of electric currents, but not usually calibrated in terms of volts.

generators. An electrical device used to produce electric current.

graphite (crystalline carbon). A nonmetallic element which is a good conductor of electricity. It is also used as a lubricant and in collector brushes.

henry. A unit of inductance. A circuit has an inductance of 1 h. when a current, changing at the rate of 1 amp. per sec., induces an electromotive force of 1 v. in the circuit.

horsepower. The power necessary to do work at the rate of 33,000 ft.-lb. per min. One horsepower equals 746 w.

hydraulics. Pertaining to liquids in motion.

hydrostatics. Pertaining to liquids at rest.

hysteresis. A kind of molecular friction caused by reversal in the position of the minute molecular magnets which iron is thought to possess.

induction. Induction is the imparting of electrical or magnetic energy to an object or a circuit by means of magnetic lines of force, during which there is no physical contact.

induction, mutual. The induction of electromotive force in a coil as the result of a change in the magnetism of another coil which is close to it.

induction, self. The induction of electromagnetic force in a conductor due to changes in the strength of the electromagnetic field of the conductor. These changes are usually due to changes in current flow.

inverter. An electrical device for changing the characteristics of an electric current.

iron. A metallic element possessing pronounced magnetic properties.

joule. An electrical unit of energy equal to a watt-second. It is found by multiplying amperes by volts by seconds.

kilowatt (1000 watts). One kilowatt equals 1.34 hp.

laminated. Built up of layers of material; each layer is called a lamination.

lead sulphate. A chemical compound formed on the plates of a storage battery as the battery is discharged.

leyden jar. A condenser in the form of a glass jar coated inside and out with tinfoil.

lodestone. A naturally occurring magnetic substance; the iron ore, magnetite.

magnet. A piece of material, such as steel or magnetite, which will attract particles of magnetic material. Magnetite is a natural magnet, while pieces of steel which have been magnetized are artificial magnets.

magnet, compound. A magnet made up of two or more magnets.

magnetic. The property of a substance which causes it to attract small particles of such substances as iron, steel, cobalt, or other magnetic substances.

magnetic field. The space through which magnetic lines of force pass.

magnetic flux. Magnetic lines of force.

magnetic line of force. A line which indicates at its every point the direction in which a free north-seeking pole would tend to move.

magnetic poles (north and south). The portions of a magnet to which magnetic particles are attracted are called its poles. These are the magnetic poles of the magnet. One pole is called the north-seeking or north pole of the magnet and the other is called the south-seeking or south pole.

magnetism. The property of being magnetic.

magnetism, induced. Magnetism which is brought about in any substance by the effect of a magnetic field.

magnetism, residual. The magnetism which remains in a magnetic substance, such as soft iron, after it has been temporarily magnetized.

magneto. A form of generator used largely in ignition systems in which the magnets rotate while the field coils remain stationary.

molecule. A particle of any substance made up of two or more atoms.

motor. An electrical device used to change electrical energy into mechanical energy.

motor action. The action of the field in a motor which brings about the rotation of the armature.

neutral plane. A plane at right angles to the lines of force of the field of a motor or generator where no current is induced in the armature coils.

nickel. A metallic element. Nickel possesses magnetic qualities.

nitrogen. A nonmetallic gaseous element forming approximately 78 per cent of the earth's atmosphere.

nonconductor. A substance through which an electric current may not flow.

nucleus. Thought to be the center portion of an atom about which electrons revolve.

ohm. A unit used to measure the resistance to the flow of an electric current.

ohm, a standard. A standard ohm is the resistance of a column of mercury 106.3 cm. long, having a cross section of 1 sq. mm. at 0° C.

oxygen. A gaseous nonmetallic element forming approximately 21 per cent of the earth's atmosphere.

phase. One of a series of alternate fluctuations of an alternating current.

plate. That part of the cathode and anode submerged in the electrolyte in an electric cell, or one of the plates in a condenser.

pneumatics. Pertaining to air or other gases and their reaction.

polarity. Pertaining to the magnetic poles of a magnet or a coil.

polarization. A condition within a cell where bubbles of gas collect on the anode of the cell decreasing the current output.

pole pieces. The parts of an electric generator or motor around which the field coils are wound.

pole, unit. A pole of such strength that if it is placed a distance of 1 cm. in the air away from a like pole, the poles will repel each other with a force of 1 dyne.

polyphase. A term used to denote an alternating current made up of more than one series of cycles.

proton. A positive particle of electricity.

p.s.i. Pounds per square inch.

radar. A device designed to make use of reflected radio waves to give a visual indication of remote objects.

radio. A device by which signals may be transmitted by or received from the ether by means of radio waves.

radio shielding. A method of preventing interference with radio operation by miscellaneous radio waves, particularly those set up by the ignition system of an airplane, by enclosing parts of the electrical system in metal shields or hollow metal cables.

relay. A device by which an electric circuit may be opened or closed by remote control. The relay contact points are controlled by means of an electromagnet. The action may be automatic.

reluctance. The resistance which a substance offers to magnetic flux.

resistance coil. A coil of wire used to introduce resistance into an electric circuit.

resistance, electric. The opposition which is offered by an electrical conductor to the flow of current.

resistance load. A load in an electric circuit which consists largely of resistance to obtain the heating effect from the electric current, for example, an electric stove.

resistor. Any form of resistance introduced into an electric circuit to regulate or prevent the free flow of an electric current.

retentivity. The degree with which a substance retains its magnetism after having been magnetized.

reverse-current cutout. A device to prevent a current from reversing its direction, for example, from a storage battery back through the generator.

rheostat. A variable resistor.

rings, collector (slip rings). Rings on an a-c generator (alternator) which collect the surges of induced current from the armature. The collector brushes ride upon these rings and are connected with the external circuit.

ripple. The slight variations in the strength of a direct current from a direct-current generator.

r.p.m. Revolutions per minute.

sal ammoniac. Ammonium chloride, a chemical compound used in the preparation of dry cells.

segments, commutator. The bars of copper which are connected with the coils of the armature over which the collector brushes slide in a d-c generator or motor. These segments form the commutator.

self-excitor. A generator or motor which excites its own field coil.

self-induction. The induction of an electromotive force in a circuit or a conductor caused by changes in the current flowing through that same circuit or conductor.

slip rings. (*See* collector rings.)

solenoid. An artificial magnet formed by a conductor in the form of a coil carrying an electric current. To increase the magnetic effect, a soft iron core is usually placed within the coil.

solenoid switch. The switch operated by means of a solenoid.

sulphated. A condition in a storage battery where the plates have become covered with an excessive amount of lead sulphate.

sulphation. The changing of the material in the plates of a storage cell into lead sulphate.

sulphur. A nonmetallic element.

switch. A device used to open or close an electric circuit.

terminal. The point at which the connection is made between the electric circuit and the source of the electric current, such as battery terminals or generator terminals.

thermocouple. A temperature-measuring device formed by closely joining the ends of two wires of different composition, such as iron and constantin. When this joint is heated, an electric current is set up in the wires which may be measured by a sensitive voltmeter, which is usually calibrated in degrees.

transformer. A device formed by coils of wire wrapped around a mutual core of soft iron used to induce an electric current in one coil by passing a fluctuating current of electricity through the other coil. A transformer is usually used to change the voltage of an electric current.

vibrator. An electric device used to interrupt the flow of electric current at very short intervals, usually many times a second.

volt. A unit used to measure the electromotive force or electrical pressure.

voltage. Electromotive force, or electrical pressure of an electric current.

voltage regulator. An electrical device to control the voltage output from a generator.

voltmeter. An instrument used to determine the voltage of an electric current.

voltohmmeter. An instrument used to measure the resistance of an electric circuit, or the voltage of an electric current.

watt. A unit for measuring electrical energy which is found by multiplying amperes by volts.

watt-hour. A unit of electrical energy which is equal to the energy used at the rate of 1 w. for 1 hr.

wattmeter. An instrument used to measure the rate at which electrical energy is being used.

wound, compound-. A method of connecting the field coils by a combination of series and shunt winding.

wound, series-. A method of winding the field coils in a motor or generator in which all of the current from the armature passes through the coil windings.

wound, shunt-. A method of winding the field coils in a motor or generator in which only a part of the current in the armature passes through the coil windings.

zinc. A metallic element often used as one plate in wet and dry cells.

IV MAGNETISM AND MAGNETIC EFFECTS

It has been known for hundreds of years that a certain kind of rock would attract small pieces of iron. These small pieces of iron or steel could be picked up by pieces of this kind of "rock" which is an iron ore called "magnetite" and, when it possesses polarity, is called "lodestone." Lodestone is a natural magnet. A piece of this ore, when suspended in a state of balance by a fine silk thread, will arrange itself with one of its axes in a north-and-south direction. This material acts in the same way

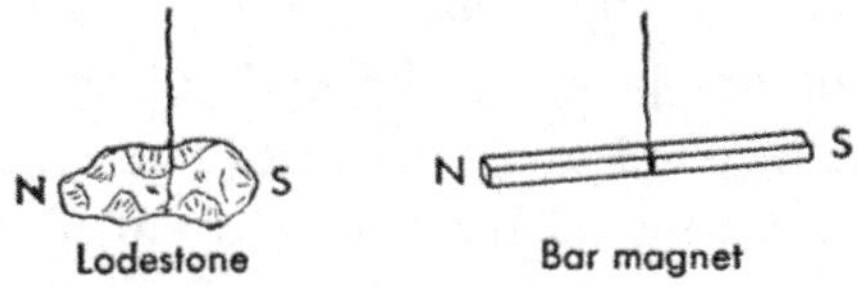

Fig. 19. A lodestone and a bar magnet suspended to form simple compasses.

that a bar magnet acts in an ordinary compass. The first compass was probably a piece of this material suspended in this manner. It was found many years ago that a piece of iron or steel, when rubbed with a piece of lodestone or natural magnet, became magnetized and took on the properties of the magnetic material. These were the first artificial magnets.

Most common magnets are made of a special kind of steel. The magnet will attract bits of iron or steel and is itself said to be magnetized. A substance such as iron or steel which is attracted by a magnet is said to be magnetic. Nickel and cobalt are also magnetic, but most metals and other substances show no signs of this property. A piece of iron or steel, when brought under the influence of an electric field, becomes magnetic. A magnet does not attract bits of paper or small pieces of pith as do amber and glass rods which have been electrified. Substances which have to be electrified in order to attract these nonmagnetic particles are not magnets and they are not magnetized. The substances have been given an electric charge.

Steel has been found to be the best material from which to make magnets. Iron is more easily magnetized than steel but loses its magnetism much more readily. Steel is said to have greater retentivity than does iron.

Electromagnets are made by placing a core of soft iron within a coil of wire carrying an electric current. Whenever the wire is carrying

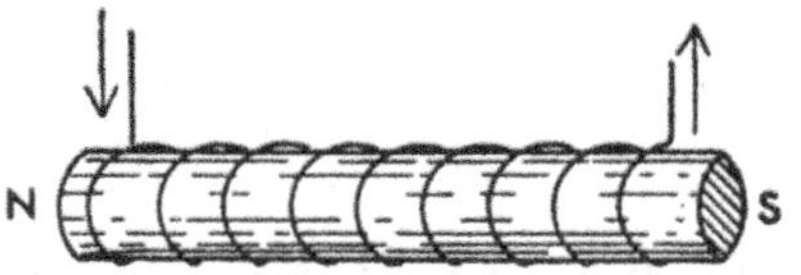

Fig. 20. A soft iron core surrounded by a coil of wire to form a simple electromagnet.

the current, the iron core becomes a very strong magnet. The electromagnet loses its magnetism instantly when the current ceases to flow in the coil.

The fact that one end of the magnet or lodestone pointed toward the North Pole, and the other end pointed toward the South Pole probably gave the name, pole, to the ends of the piece of lodestone and the magnet which acted as a compass. The end of the magnet or lodestone which points toward the north is called the north pole or north-seeking pole; the other end is called the south pole or south-seeking pole.

The earth itself acts as though it were a great magnet, having both north and south magnetic poles. These poles are called the North Magnetic Pole and the South Magnetic Pole. The magnetic poles are not located at the same place as the geographical North and South Poles. Through common usage, the magnetic pole located nearest the North Pole is called the North Magnetic Pole, and the north-seeking end of a magnet is customarily called the north pole of the magnet. It must, however, be understood that if, as we know, unlike poles attract each other, either the North Magnetic Pole is in itself a south pole or the north-seeking pole of a magnet is in reality a south pole. Through common usage, however, the north-seeking end of a compass or magnet is called the north pole of the magnet or the north end. The north-seeking end of a compass or magnet is conventionally marked with a plus sign or an N, and the south end is marked with a minus sign or S.

The magnetism which a bar of iron or steel takes on seems to be due to its molecules or atoms. It is thought that each atom or molecule is in itself a tiny magnet, and that the sun, earth, moon, and other planets act as magnets, each having a north and a south pole. When the atoms and molecules are arranged at random within a piece of material, it shows no magnetic properties. But when the piece of iron or steel has been stroked with a natural magnet or has been placed in a magnetic

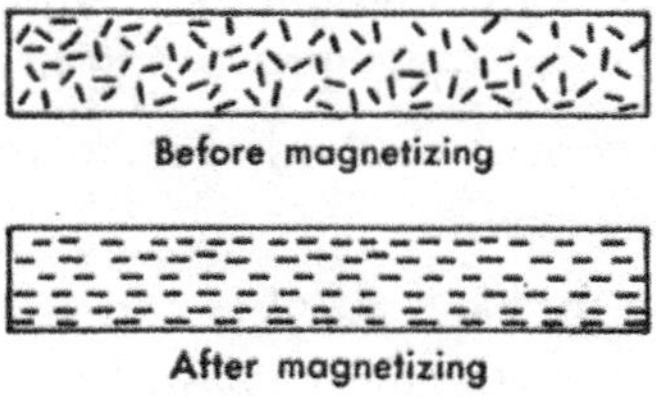

Fig. 21. Diagrams to show the theoretical arrangement of atoms or molecules before and after magnetizing a piece of steel.

field in such a manner that it becomes magnetized, it is believed that the atoms and molecules have rearranged themselves so that their north poles all point in the same direction, while their south poles all point in the opposite direction. This theory is proved to some extent by the fact that if a long magnet is broken in the middle, each portion becomes a magnet having a north pole and a south pole. If these pieces are again broken, each piece becomes again a complete magnet, having a north and a south pole. The north and south poles of the pieces are at the ends of the pieces which pointed toward the north and the south pole of the original magnet.

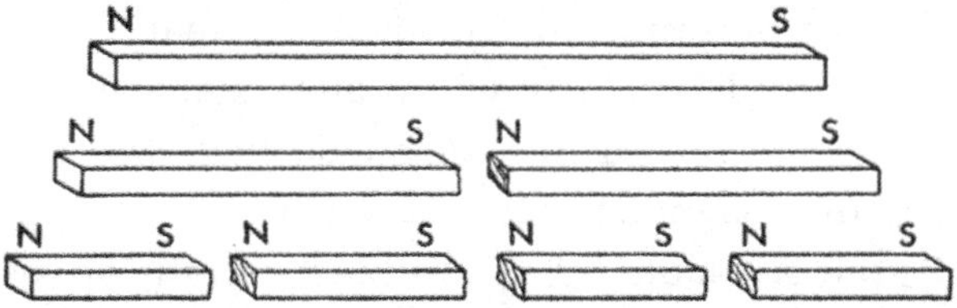

Fig. 22. A bar magnet which has been broken to show how each piece forms a complete magnet.

Since pieces of iron or steel may be lifted by a magnet, the size of the piece depending upon the strength of the magnet, the magnet must exert force upon the piece being lifted. It is possible to cause a piece of iron or steel to move over the surface of a piece of glass by drawing a magnet under the glass. The piece of magnetic material is caused to move without the magnet's coming in contact with it. Magnetic force seems to act through almost all substances. The magnetic force from the magnet, however, does not act effectively through a magnetic substance such as a sheet of iron. The attraction between the magnet and the magnetic substance is mutual for a magnet can be attracted to a piece of iron. The strength of the magnet is not measured by its lifting power. Other conditions will vary the lifting power of the magnet. Some of these conditions are the kind of magnetic substance being lifted, the shape of the body being lifted, and the method in which the load is applied to the magnet. The shape of the magnet also affects its lifting power. The accepted method of measuring a magnet's strength is by its action upon another magnet.

If a magnet is placed under a sheet of paper or a thin sheet of glass, and iron filings are sprinkled on the paper or glass, the filings, when the glass or paper is tapped, arrange themselves in a definite pattern, as shown in Figure 23. This pattern seems to be made up of lines which extend from one end of the magnet out through space and then back

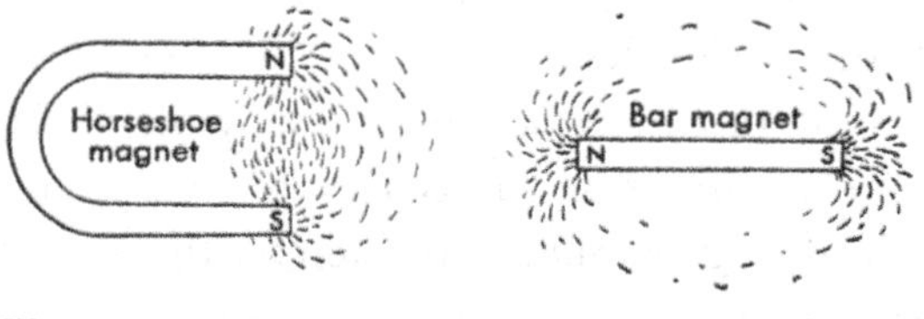

Fig. 23. The lines of force about a horseshoe and a bar magnet shown by the arrangement of iron filings.

into the other end of the magnet. These lines are called magnetic lines of force, and the space surrounding the magnet is called the magnetic field of the magnet or simply the field of the magnet. The magnetic field may be considered as a force which always acts in a definite direction. A magnetic pole placed in a magnetic field will move along the lines of force. If the magnetic pole is a north pole, it will move in one direction, while if it is a south pole, it will move in the opposite direction. To find the direction of a field, the polarity of the pole placed in the field must be known. The positive direction of a magnetic field at any point is defined as the direction in which a free north pole, if placed in the field, would tend to move. A compass needle placed in a magnetic field arranges itself along the lines of force. A line of force may be defined as a line whose direction at every point through which it passes is the direction in which a free magnetic north pole would move. Lines of force leave the magnet at the north pole and enter the magnet at the south pole. This is the accepted conventional direction of a magnetic field.

While the field of a magnet may be indicated by a few lines, there are actually thousands of lines of force for each magnet. It is assumed that the magnetic lines of force travel inside the magnet from south to north. It can be assumed that the lines of force are continuous and cannot be broken. The strength of a magnetic field is greatest at the poles of the magnet.

The magnetic lines of force seem to travel through various substances with varying degrees of readiness. Since all magnets do not have the same strength, the fields do not have the same number of lines of force.

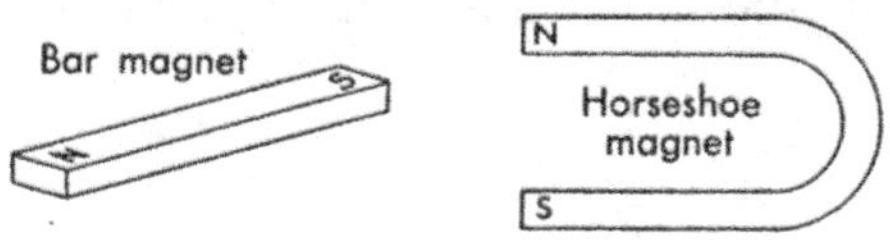

Fig. 24. A bar magnet and a horseshoe magnet.

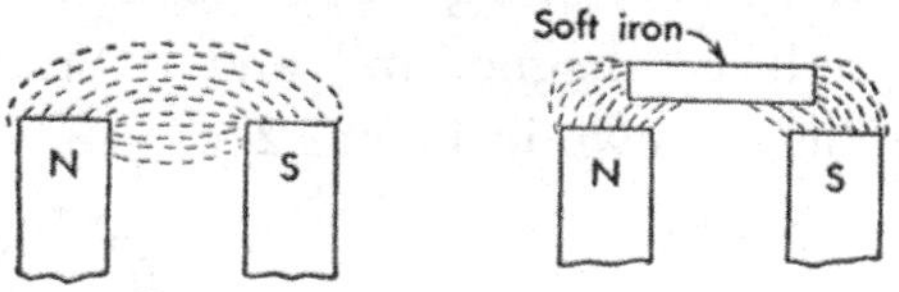

Fig. 25. The lines of force between the poles of a magnet concentrate themselves through a piece of soft iron.

The number of lines of force for any given magnet may be increased by decreasing the distance these lines of force must travel outside the magnet. A horseshoe magnet, where the north and south poles are quite close together, is stronger than a bar magnet of the same size and state of magnetism. The resistance which any substance offers to the lines of force is called "reluctance." Air offers considerable resistance to the lines of force, as do copper and other nonmagnetic materials. The strength of the field may be increased by placing a magnetic substance, such as soft iron, in the field. The lines of force concentrate themselves to pass through this soft iron which is a path of least resistance and this concentration produces a stronger field, as shown in Figure 25. When the magnet is in the form of a bar, a return path of soft iron may be arranged, as shown in Figure 26, to increase the effectiveness of the magnet. A

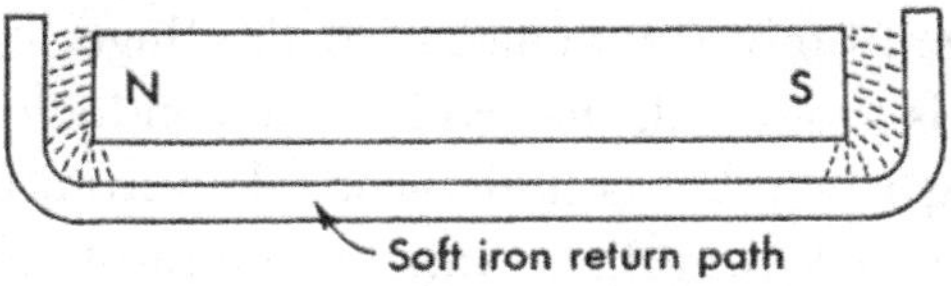

Fig. 26. The lines of force of a bar magnet pass through a soft iron return path.

shield may be made of soft iron to protect other objects from the magnetic lines of force. This is done by placing the object in the center of a cylinder or circle formed of soft iron. The magnetic lines of force pass into and through the shield, which is the path of least resistance.

If a needle or small nail is picked up by a magnet, the needle or nail itself may act as a magnet when removed. It has itself been magnetized and retains some of the magnetism. This type of magnetism is called "residual magnetism." The amount of residual magnetism which a substance has depends upon the strength of the magnetic field in which it was placed and the retentive qualities of the substance. If a magnetic substance such as soft iron is brought into a magnetic field, it itself becomes a magnet, even though it is not touching the magnet or whatever is developing the magnetic field. This type of magnetism is called "induced magnetism" and is brought about by magnetic induction. This is shown in Figure 27.

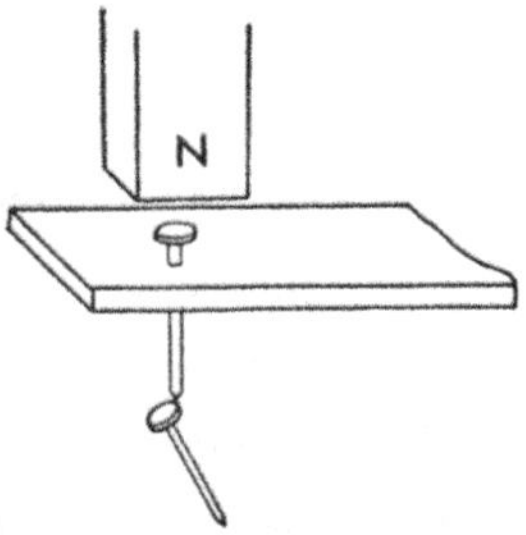

Fig. 27. The small nail is held in position by induced magnetism in the large nail.

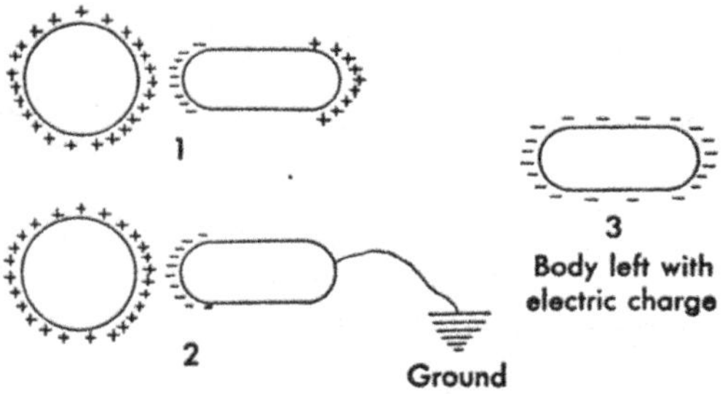

Fig. 28. An electric charge may be placed on a body by induction.

If a body carrying a charge of electricity is brought close to an insulated conductor (Figure 28), it can be shown that the part of the conductor nearest the charged body shows a charge of electricity opposite to the kind carried on the charged body. It can also be shown that the opposite end of the insulated conductor takes on a charge of electricity which is of the same kind as carried on the charged body. If, while the charged body is held close to the conductor, the part of the conductor nearest the charged body is touched with the finger or with any other conductor which is grounded, that end of the conductor loses its electric charge. If the finger or other conductor is taken from the charged body which is then removed from the vicinity of the insulated body, the insulated body retains a charge of electricity that is of the opposite kind to that carried on the charged body. This charge of electricity is called an "induced charge." This charge is not induced magnetism, but is induced electricity.

One of the most important uses made of induced electricity is shown by the action of a condenser. The simplest form of condenser is the

Leyden jar shown in Figure 29. The Leyden jar is simply a glass jar coated inside and out with tin foil or other conducting material. A metal rod is inserted through an insulator (which acts as a stopper for the top of the jar) and connected by means of a metal chain or other conductor to the tin foil lining the jar. If the outer tin foil is connected to a ground and the brass rod is brought in close contact to a source of electricity such as a static machine, a comparatively large charge is induced on the tin foil lining the jar. If the charged body is removed and the brass rod is touched with the finger, a strong spark of electricity will jump from the rod to the finger. Or, if the rod is connected by means of a conductor to the tin foil on the outside of the jar, the spark will jump. The size of the spark depends upon the size of the jar and the amount of tin foil in it. The ordinary condenser used in the ignition system of an engine acts upon this principle.

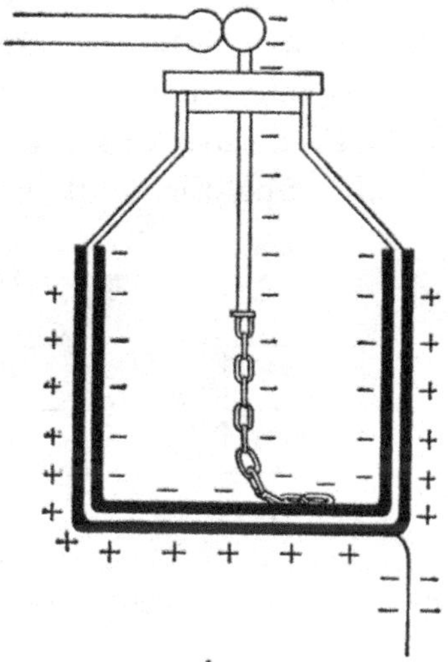

Fig. 29. A Leyden jar.

A magnet may be made by stroking a piece of steel with a magnet as shown in Figure 30. Figure 21 shows an arrangement of the molecules or atoms before and after stroking with the magnet. The end of the magnet stroked by the north pole of the magnet becomes a south pole, and the other end becomes a north pole.

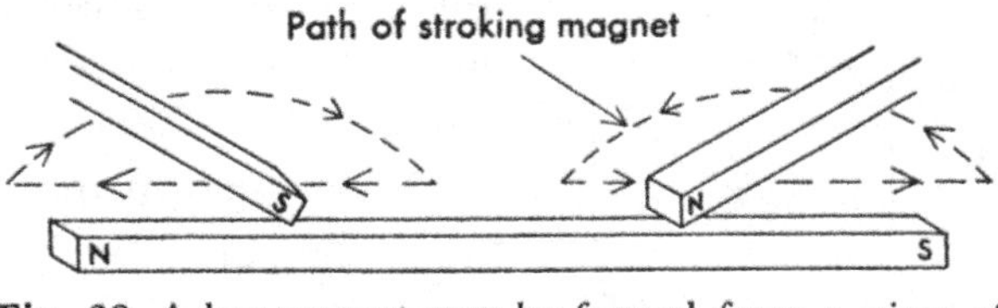

Fig. 30. A bar magnet may be formed from a piece of steel by stroking it with a bar magnet.

When soft iron is placed between two magnetic poles, an additional number of lines of force pass through the iron. Fewer of the lines of force between poles pass through the air outside the iron, as shown in Figure 25. The readiness with which lines of force pass through a substance is called its permeability. This permeability is the ratio of the number of lines of force which pass through a given space when it is occupied by a substance to the number of lines of force passing through that space when it is occupied by air. For nonmagnetic substances, the ratio is approximately one. It is therefore seen that nonmagnetic substances have about the same permeability as air.

Compound magnets may be used when it is desired to create a strong field. In constructing a compound magnet, the north poles are placed together and the south poles are placed together. With this arrangement, the field passes between the poles across the air gap, as shown in Figure 31. If a compound magnet is wrongly built up by putting the north pole into contact with the south pole of each end, as shown in Figure 32, the magnetic field takes a circular form and does not cross the air gap between the poles. This arrangement destroys the effect of the magnets.

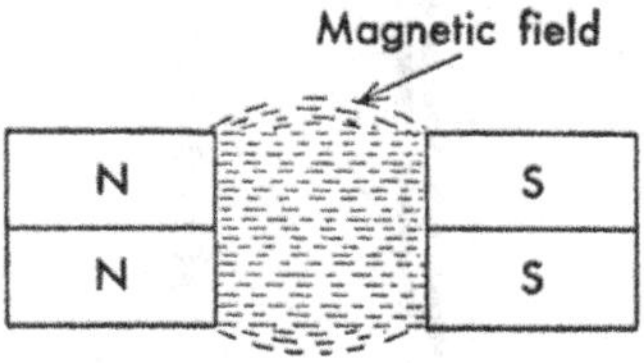

Fig. 31. The magnetic field between the poles of a properly arranged compound magnet.

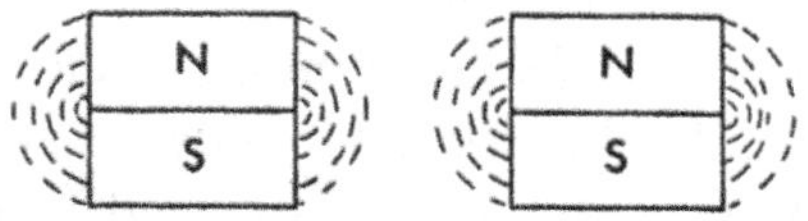

Fig. 32. The magnetic field about the poles of an improperly arranged compound magnet.

If the ends of a horseshoe magnet are connected with a soft iron bar or bent into the form of a ring or square, as shown in Figure 33, a ring magnet is made. The ring magnet has a definite field direction, but no poles. If a piece is cut out of the ring, a north pole and south pole are immediately established. Such magnets are used as a core in repeating coils and transformer coils.

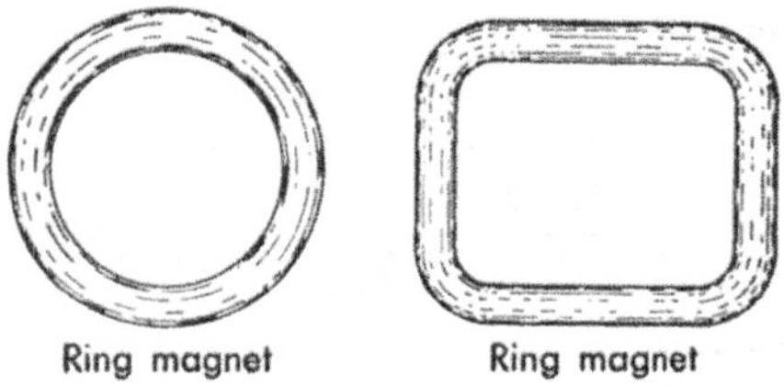

Fig. 33. Two forms of the ring magnet.

The magnetism may be removed from a magnet by heating or by excessive vibration such as that brought about by pounding. This is particularly true if a bar magnet is placed in an east-and-west direction and then pounded with a hammer. Hitting the magnet seems to allow the molecules to move into random positions. A piece of steel may be magnetized by holding it parallel to the magnetic lines of the earth and then pounding one end with a hammer. Airplanes having welded steel fuselages become quite magnetic if they land continually in the same direction on a north-and-south runway.

Electromagnetism, as compared with natural magnetism, is a magnetic field set up around a conductor by the passage of an electric current through it. Whenever an electric current passes through a conductor, a magnetic field is set up around the conductor. The lines of magnetic force are arranged around the conductor in the form of concentric

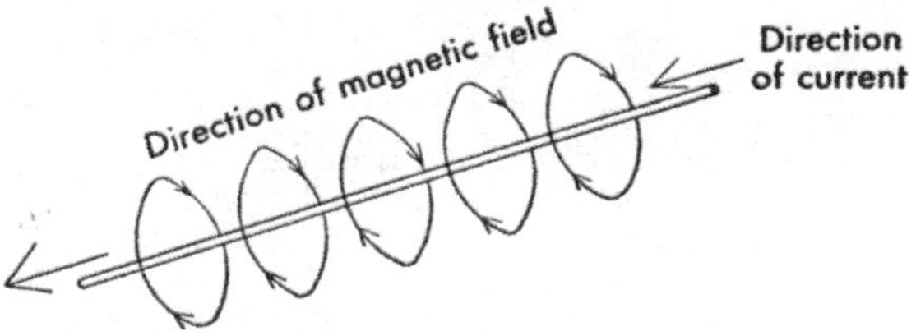

Fig. 34. The direction of the magnetic field formed about a conductor carrying an electric current.

circles. The greater the distance from the conductor, the weaker the magnetic field becomes. If the direction of flow of the electric current in the conductor is reversed, the direction of the magnetic lines of force surrounding the conductor is also reversed. This magnetic field may be tested by holding a small compass close to the conductor. The compass needle will tend to arrange itself at right angles to the conductor, that is, along the magnetic lines of force. If, while the compass is in the magnetic field, the direction of the current through the conductor is reversed, the compass needle will reverse its direction.

A common method of determining the direction of the magnetic field about a conductor is called the righthand rule. If the fingers of the right hand are placed around the conductor with the thumb pointing in the direction the current is flowing, the fingers will point in the direction of the lines of force in the magnetic field around the conductor. If a conductor is bent in the form of a loop, as shown in Figure 35, all the magnetic lines of force will pass through the loop, continuing in their circular path around the conductor. If several turns of the conductor are arranged so that they form a coil, as shown in Figure 36, the coil will act like a bar magnet. It will have a north and a south pole. Most

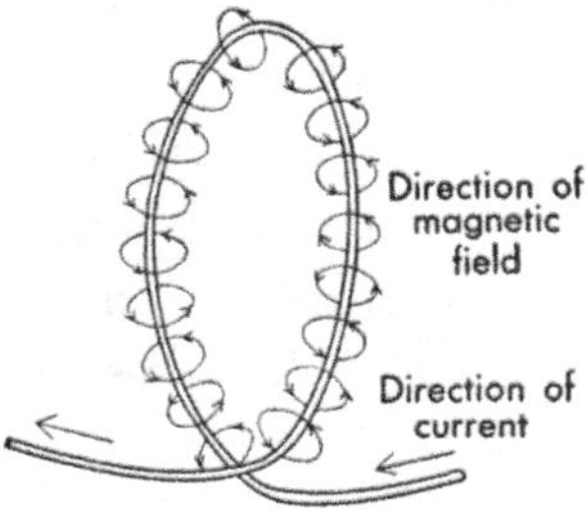

Fig. 35. The direction of the magnetic field formed about a conductor in the form of a loop carrying an electric current.

of the lines of force will pass through the coil from the south pole to the north pole, out through space, and back to the south pole. This type of coil is called a solenoid. If the fingers of the right hand are placed around the coil in the direction of the current flow, the thumb will point to the north pole of the coil. The magnetizing force set up by a

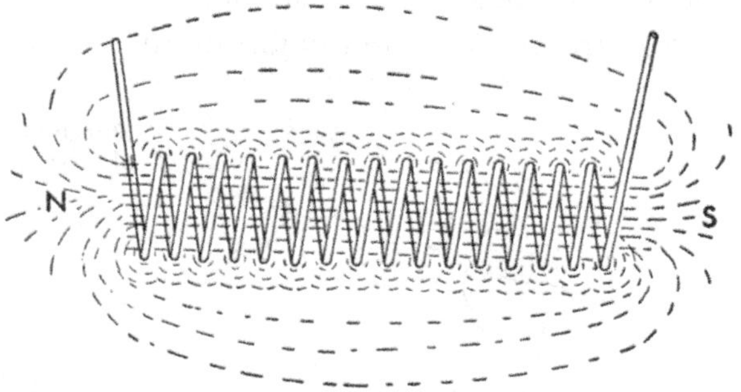

Fig. 36. The magnetic field about a conductor carrying an electric current when the conductor is wound into a coil. This type of coil is called a "solenoid coil."

coil or solenoid is known as the magnetomotive force. In any coil, the magnetomotive force is in proportion to the product of the number of turns of wire and the amount of current flowing. When the current is in amperes, the unit is known as the ampere turn. A coil having 10 turns with 1 amp. of current flowing would develop a magnetomotive force equal to 10 ampere turns. If a core of soft iron is placed in the solenoid coil, the iron core becomes highly magnetic and acts as a very strong magnet. This is true because the permeability of iron is approximately 2000 times as great as the permeability of air. The magnetic lines of force are often called "flux" or "magnetic flux."

A piece of iron may be pictured as containing billions of tiny magnets which ordinarily are arranged in random patterns, many of them forming little loops. These little loops and the irregular arrangement of these tiny magnets neutralize each other so that no magnetic field is set up around the iron. If the piece of iron is brought into the influence of a magnetic field, these tiny magnets arrange themselves along the magnetic lines of force in much the same way that the compass needle arranges itself along the lines of force in a magnetic field. When this happens, all of the north poles of the tiny magnets point in the same direction, and all the south poles point in the opposite direction. Therefore, the magnetic field of each particle is combined with and strengthens the magnetic field of all the other particles. The strength of the magnetic field depends upon whether or not all of the particles have been lined up. When they are lined up, the piece of iron becomes a strong magnet. The particles of soft iron, however, do not remain lined up when the magnetic field is removed. They swing back to their mixed-up positions, and we say the iron has "lost its magnetism." A few of the particles, comparatively, may not swing back to their original positions, and the iron will continue to show some magnetic properties due to residual magnetism.

If the bar of metal, instead of being soft iron, is hard steel and the magnetic field is strong enough to swing the particles into line, the steel will retain its magnetism and become a permanent magnet. When a core is placed in a solenoid, the polarity of the magnet thus formed is the same as the polarity of the coil. The core maintains its magnetic properties as long as the current continues to flow and, as long as it flows in the same direction, the polarity of the magnet will not change. If the direction of current flow is rapidly reversed, the polarity of the core changes with each change of current direction.

The iron core also becomes heated when the direction of the current is rapidly changed, and energy is given off in the form of heat. This heat is believed to be caused by hysteresis and eddy currents. Hysteresis is a kind of molecular friction caused by a reversal in position of the tiny molecular magnets which are thought to make up the iron. Eddy currents are currents produced by an electromotive force induced in metal when the metal is moved in a magnetic field or when a magnetic field in the metal changes its strength. In order to reduce the heating effect and the losses due to eddy currents in electromagnets, the cores are sometimes made up of bundles of soft iron wire or of sheets of soft iron insulated from each other. When the cores are made up in this way, they are called laminated cores, and this construction retards the flow of eddy currents from one part of the core to the other.

The wet cell was the first source of experimental electric current. Today, however, electric currents are largely generated because batteries able to produce enough current for a large building or a small city would be enormous and, therefore, impractical. The electric generator depends upon induction for the production of electric current.

Electromagnetic induction may be defined as the electromotive force which is induced in any conductor that cuts or is cut by magnetic lines of force. Induced currents may be set up in a conductor in three ways: (1) the conductor may be made to cut the magnetic lines of force;

(2) the magnetic lines of force may cut the conductor; or (3) the intensity of the magnetic field may be varied in strength. Electric currents may be set up in one conductor which is close to another conductor carrying a fluctuating current of electricity. This effect is known as mutual induction. Electromotive force may be induced in a conductor by a change in the strength of the current flowing in the conductor itself.

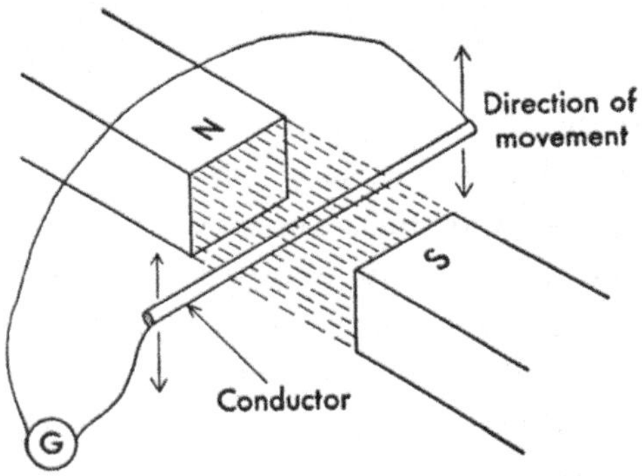

Fig. 37. A current of electricity may be induced in a conductor by passing it back and forth through an electric field between the poles of a magnet.

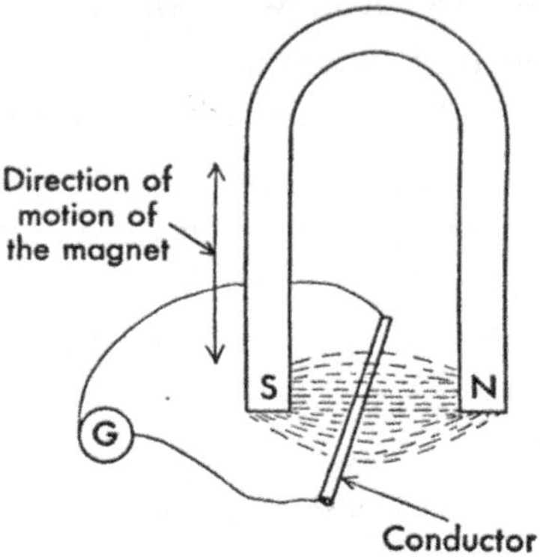

Fig. 38. A current of electricity may be induced in a conductor by moving the electric field between the poles of a magnet back and forth past the conductor.

This type of induction is known as self-induction. When a conductor is moved across magnetic lines of force in one direction, the induced current flows through the conductor in one direction, and it reverses itself in the conductor when the conductor cuts the magnetic lines of force in the opposite direction.

Figure 37 shows the generation of an induced current in a conductor cutting across the magnetic lines of force between the poles of two magnets. A galvanometer is connected in the circuit to detect the flow of electric current and indicate its direction. As the conductor moves across the lines of force in one direction, the needle of the galvanometer is deflected to the right, but when it cuts the lines of force in the opposite direction, the needle is deflected to the left.

Figure 38 shows how an electric current may be induced in a wire by moving a permanent magnet back and forth past the wire. A fundamental law of physics states that "every action or force in nature has an opposite and equal reaction or force." Since there is a force set up in the conductor which cuts across the lines of force, there must also be an opposite force which tends to resist the passage of the conductor through the magnetic field. This can be shown to be true for, if a wire which is free to move carries an electric current while in a magnetic field, the wire will be displaced from the field in a direction opposite to that which would be required to induce a corresponding current in the wire. This action is called "motor action" and is made use of in the moving coils of electric motors.

Mutual induction is the production of electromotive force in a circuit, resulting from the circuit being cut by a changing magnetic field from a neighboring circuit. Figure 39 shows how a current may be induced in a circuit by the magnetic field set up by another circuit.

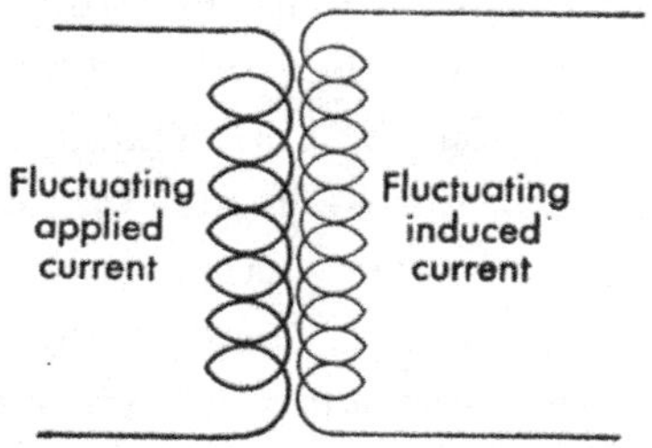

Fig. 39. An electric current may be induced in a coil by placing it in the magnetic field of another coil carrying a fluctuating electric current.

The current in the first circuit is interrupted by means of a switch or vibrating points, thus causing the magnetic field about the primary coil to expand and contract. As the field expands and contracts, the magnetic lines of force cut the coil of the secondary circuit, inducing a current in the secondary circuit. The current induced in the secondary circuit alternates in direction as the field expands and contracts. This kind of arrangement is called an induction coil. The induced current varies in strength, depending upon the number of turns of wire and the strength of the field set up.

Self-induction is the setting up of an induced current in a conductor or circuit by changes in the current flowing through that same conductor or circuit. As the current starts to flow through the conductor, a magnetic field is built up within and around the conductor. This field may be considered to cut the conductor as it builds up or collapses. During the time that the magnetic field is being built up, an opposing electromotive force is built up in the conductor. This electromotive force always acts in a direction which tends to oppose the flow of current in the conductor. When the conductor is in the form of a coil, the field built up around any single turn of the coil will cut some or all of the other turns of the coil. This will induce in the other turns an electromotive force which also opposes the current flowing in the coil. When the current in the coil dies down, an electromotive force is induced in the coil which opposes the dying-down of the electric current. A coil will have much more self-induction than a straight conductor.

The effect of self-induction may be increased by the use of an iron core in the coil to decrease the magnetic resistance or reluctance. Self-induction is very similar to inertia, which is the force with which a body resists being placed in motion, or being stopped if it is already in motion. Since this force is opposed to the applied electromotive force, it is sometimes called back electromotive force or counter electromotive force. Counter electromotive force may be abbreviated c.e.m.f.

The principle of induction, both self- and mutual, is given by Lenz's law. This law states that "whenever a conductor cuts across a magnetic field, or whenever the magnetic field surrounding a conductor changes, a voltage is induced in the conductor in such a direction that, if the circuit is closed, a current will flow and produce a flux or magnetic field to oppose the change in the magnetic field."

It has been reported that a new alloy has been developed for the manufacture of permanent magnets. It is stated that magnets made of this material have a lifting power many times that of magnets made from cobalt steel. This alloy is made of aluminum, nickel, and cobalt and has been given the name, "Alnico."

V SOURCES OF ELECTRIC CURRENT

Until near the beginning of the last century, most experiments were carried on with static electricity or "electricity at rest." It was only when current electricity began to be used that electricity really became useful to man. All of the electrical equipment in common use today depends upon current electricity.

It is not necessary to know exactly what electricity is if one has a clear understanding of what it will do. Current electricity may be thought of as electricity moving along a conductor. It has been found that the "flow" of electricity is along the outside of the conductor and not through the conductor. A thin metal pipe will carry just as much electric current as a solid rod having the same diameter. It is possible, however, to draw a comparison between the flow of electric current along a conductor and the flow of water through a pipe. The current can be made to flow in large or in small amounts. The electric pressure can be raised or lowered. The tendency to flow and the rate of flow can be measured.

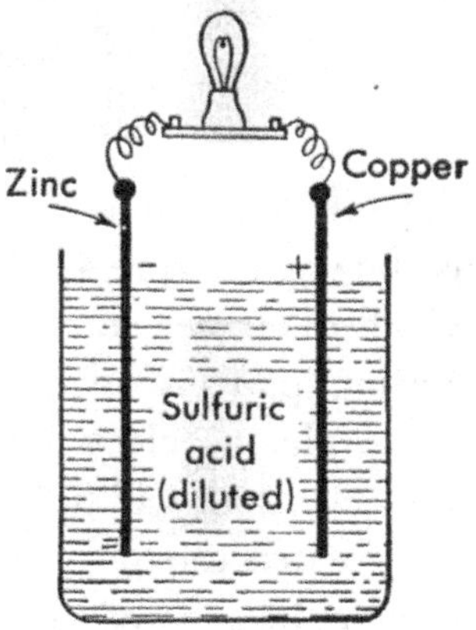

Fig. 40. A wet cell.

The first source of current electricity was the simple wet electric cell. If a strip of copper and a strip of zinc are placed in an ordinary water glass, as shown in Figure 40, and dilute sulphuric acid is added, a current of electrons will flow between the plates if they are connected by a conductor. It can also be shown that the copper plate becomes charged positively and the zinc plate becomes charged negatively. The chemical action causes an excess of electrons to be deposited upon the zinc plate. This is a primary wet cell. There are two kinds of primary cells, the wet cell and the so-called dry cell. Cells are commonly called batteries.

In the simple wet cell, the zinc plate is one charged body and the copper plate is the other charged body. If they are connected by a conductor forming a path along which electrons may pass, this path, including the cell itself, is called an electric circuit. The connecting conductor between the plates is called the external circuit, while the part within the cell itself, namely, the plates and the liquid, is called the internal circuit. The source of an electric current is always a part of a circuit.

If the conductor is cut outside the liquid, the circuit is broken. This can be done by means of a simple switch. When the switch is open, the circuit is open. When the switch is closed, the circuit is closed. If a very large conductor is used, such as a heavy wire, there is little resistance to the flow, and a short circuit is produced.

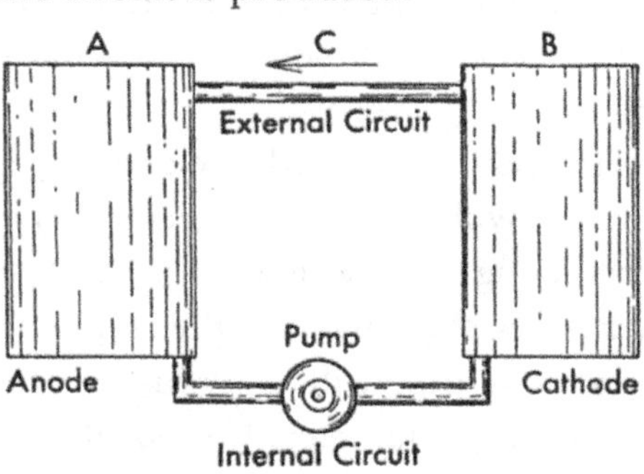

Fig. 41. A diagram to show how electric particles might be forced from one pole to the other in a simple cell using water to represent the electric current.

A simple cell may be compared with a pump which is pumping water, as shown in Figure 41. If the tank, A, is closed at the top, the force with which the water comes out at C, into tank, B, depends upon how rapidly the pump, P, is operated. In a simple cell, the chemical action is the pump placing electrons upon the zinc plate. The larger the plate, the more electrons will flow. The faster the pump acts in pumping the water, the greater the pressure of the water flowing through the pipe. The greater the area of the plates in the cell, the greater will be the flow of current.

For many years, the only source of electric current was the wet cell. This cell could not be moved readily because of the liquid which was necessary to its action.

The desire to have some sort of portable cell led to the development of the so-called dry cell. This cell, which is the type used in flashlights and for ringing doorbells, is constructed of a cup-shaped zinc container.

The zinc container acts as the negative plate. A carbon rod acts as the positive plate and replaces the copper plate used in the wet cell. The zinc container is lined with a layer of absorbent material such as blotting paper, and the space between this and the carbon rod, as shown in Figure 42, is filled with a paste made of granulated carbon and manganese dioxide. This mixture is saturated with a solution of sal ammoniac in water. Before placing this mixture in the container, the blotting paper is usually saturated with a solution of sal ammoniac and zinc chloride. The container is filled to within about ¾ in. of the top and sealed with a sealing compound. The container is usually lacquered or varnished on

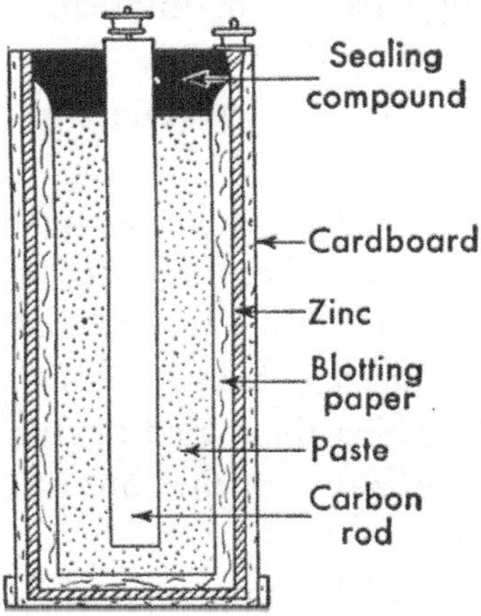

Fig. 42. A dry cell.

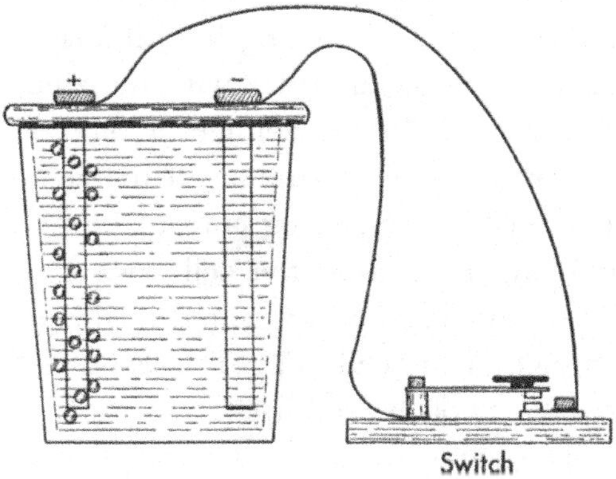

Fig. 43. A diagram, of a simple wet cell, to show how gas bubbles cause polarization.

the outside and then placed in a close-fitting cardboard container. The liquids in the cell take the place of the diluted sulphuric acid used in the wet cell.

Terminals are fastened to the top of the carbon rod and to one edge of the zinc container. These terminals are the positive and negative poles. The carbon terminal which is in the center of the top of the cell is the positive connection, and the terminal attached to the zinc container is the negative connection. This type of cell develops a current of about 1 v. It cannot be used for extended periods of time. It develops full voltage for a short time and then the voltage begins to fall. If the flow of current is stopped, the cell will build up to its full voltage again.

When a primary cell, such as the dry cell or simple wet cell, is furnishing a current of electricity, hydrogen gas is formed on the positive electrode. The gas collects in tiny bubbles which do not rise to the surface as rapidly as they form. Some of these bubbles collect on the electrode, preventing the electrolyte from making good contact with the plate. This causes a loss in the amount of current produced by the cell. This action is called "polarization." Polarization may cause a large drop in the voltage of the cell when it is used continuously for a comparatively short time. If the current is shut off, the cell soon builds up to its full strength again.

Some cells have an oxidizing agent placed in the cell to absorb the hydrogen. These agents are called "depolarizers." In a simple electric cell, if the zinc electrode is chemically pure, there is no chemical action when the circuit is open. If the electrode is not pure, a small amount of local action takes place. This is due to small particles of impurities embedded in the zinc. This action can be reduced by thoroughly cleaning the zinc and coating it with a small quantity of mercury. The mercury forms an amalgam with the zinc.

The open-circuit voltage of the cell is measured by placing a voltmeter across the terminals of the cell when it is not delivering current. If the cell is delivering current to an external circuit while the voltage is being measured, the cell always shows a reduced current reading. The voltage, when measured in this manner, is called the closed-circuit voltage of the cell. The difference in the voltage is due to the internal resistance of the cell itself.

The capacity of a cell is rated in ampere-hours. An ampere-hour is a current of 1 amp. flowing for 1 hr. For example, if a cell can furnish 1 amp. of current continuously for 100 hr., the cell has a 100 amp.-hr. rating.

The voltage of primary cells depends upon the materials from which the plates are made and the type of electrolyte used in the cell. A zinc copper cell has a voltage of approximately 1 v. If a cell uses a carbon plate and a zinc plate with a chromic acid electrolyte, it will develop approximately 2 v. The voltage of all primary cells having plates made of the same material and using the same electrolyte is always the same, regardless of the size of the plates. With larger plates, however, the cell is able to deliver a greater amount of current without being overloaded.

The capacity of a cell is influenced by:

1. The internal resistance of the cell.
2. The area of the electrodes exposed to the electrolyte. The larger the electrodes, the less the internal resistance of the cell will be.
3. The distance between the electrodes in the electrolyte. The closer the electrodes are together, the less will be the internal resistance.

4. The size of the cell. A large cell will furnish more current than a small cell, over a longer period of time.

Some cells are designed to supply a large current for a short time at irregular intervals. Other types are designed to supply a small current almost continuously. The type designed to supply a large current for a short time is called an open-circuit cell. The other type is called a closed-circuit cell.

The plus and minus terminals are not usually marked as such on dry cells, but are usually marked on storage cells, either by coloring the terminal or marking it with a plus or minus sign. The positive terminal is marked with a plus sign or red paint and the negative terminal with a minus sign or green paint.

When a primary cell is worn out, either by the dissolving of the plates in the wet cell or the exhaustion of the chemicals in the dry cell, it must be thrown away or have certain elements of the cell replaced.

The storage cell or the secondary cell, as compared with the primary cell, can have its usefulness restored by recharging. When a storage cell or, as it is more commonly known, a storage battery is furnishing current, it is discharging. When the cell is receiving current, it is said to be charging. Fundamentally, there is much similarity between the primary cell and the secondary cell or storage cell. Each converts chemical energy into electrical energy when delivering current. In the storage or secondary cell, this action may be reversed by the application of an electric current to the cell. This current is applied in the opposite direction to the current which the cell furnishes, and it simply reverses the chemical action which takes place when the cell is discharging. It is always necessary to put more electrical energy into the cell than can be taken from it, because of losses due to resistance and the development of heat.

A storage battery does not actually store electricity. The electric current furnished by a storage battery is the result of chemical action. As in the simple cell, the action of the acid on the metal plates produces an electric current. A simple lead storage battery or storage cell can be made by placing two sheets of lead in a glass jar and covering the lead sheets with dilute sulphuric acid.

To charge the battery, the two plates are connected with the plus and minus terminals of a d-c generator having a voltage output slightly higher than that of the battery (1 or 2 v. higher), and an amperage output about equal to the normal discharge rate of the battery. (Every battery should be charged in accordance with the manufacturer's recommendations.) Oxygen is formed at the anode or positive plate, and hydrogen at the cathode or negative plate. After passing the current through the cell for 15 to 20 min., the generator may be disconnected. A voltmeter across the terminals will show a difference in potential of about 2 v. A simple cell of this type will ring an electric bell. The cathode or positive plate has had deposited upon it a coat of brown lead peroxide (PbO_2). The negative plate shows the usual color of lead (Pb). The charging of the cell brought about by passing the electric current through the cell produces a chemical reaction whereby lead peroxide and sulphuric acid are produced. To bring about this change, electrical energy is used. When the cell discharges, it gives off electrical energy. Lead sulphate and water are formed plus the electrical energy given off as shown in the following equation:

Fig. 44. An aircraft storage battery. (Courtesy Delco-Remy Division, General Motors Corporation)

$$PbO_2 + Pb + 2H_2SO_4 \underset{\text{Discharge} \longrightarrow}{\overset{\longleftarrow \text{Charge}}{=}} 2PbSO_4 + 2H_2O + \text{electrical energy}$$

When the cell is being charged, lead peroxide is formed on the positive plate. Sulphuric acid is also formed, which remains in the solution. Water is broken up in this action: the hydrogen combines with the sulphate radical to form the sulphuric acid, and the oxygen combines with the lead to form the lead peroxide. In discharging, the sulphuric acid breaks up: the sulphate radical combines with lead to form lead sulphate, and the hydrogen and oxygen combine to form water or escape as free gases.

In building up the commercial lead storage cell or storage battery, the negative plates are made up of pure spongy lead. The positive plates are in the form of a lead grill in which the spaces are filled with lead peroxide. During the charging process, the positive plate takes on the brown color of lead peroxide, while the negative plate remains the gray color of pure lead. During discharge, both plates gradually become covered with light-colored lead sulphate which, when pure, is white. Sulphuric acid is much heavier than water and has a specific gravity of

1.835. A mixture of sulphuric acid and water having a specific gravity of 1.300 is added to the battery when it is to be placed in service and the cell will be ready for immediate use. During discharge, the specific gravity becomes less. As the specific gravity decreases to less than 1.200, the battery is in a low state of charge and must be recharged.

To determine the state of charge of a storage battery, a hydrometer is used. The hydrometer is included in a syringe arrangement. A fully charged battery should have a specific gravity of approximately 1.300, while a medium charge shows a specific gravity of approximately 1.250. The hydrometer is usually marked with a scale having graduations from 1.10 to 1.300.

Storage batteries are made up of a number of storage cells connected in series. Each cell has a number of positive and negative plates. This number may vary from about 12 to about 24. The positive plates in each cell are connected with each other, as are the negative plates. The plates are separated by means of insulators which may be made of wood, rubber, or a composition material, which, being porous, do not prevent the passage of current through the solution from plate to plate. The insulators are usually ribbed from top to bottom to permit the free flow of the liquid and allow sediment to drop to the bottom of the cell. Each set of plates forming a cell is enclosed in a hard-rubber container having a ribbed bottom upon which the plates rest. The sediment settling to the bottom comes to rest between these ribs and does not "short" the bottom of the plates. The hard-rubber cover of the cell is sealed with a compound which is melted and poured into place. Each cell has a vent plug which may be removed to add water to the cell and a small vent which is open at all times to allow the escape of gases. Aircraft batteries have a special self-sealing vent which prevents leakage of the acid when the battery is on its side or in an inverted position.

A 6-v. storage battery is made up of three storage cells connected in series and contained in a single hard-rubber case. A 12-v. battery is made up of six cells connected in series and contained in a single hard-rubber case. The normal voltage of the lead storage cell is 2 v. and not 2.2 v. which is the actual voltage of each pair of plates. A slight fall in voltage is caused by the internal resistance of the cell.

A storage battery is rated in ampere-hours. The rating is the number of hours during which the battery will discharge 1 amp. of current from a state of full charge to a state of discharge. A 100-amp. battery would theoretically furnish 100 amp. of current for 1 hr. or 1 amp. of current for

100 hr. The rate of discharge, however, causes a variation in the amount of current which may be had from a given battery. A rapid rate of discharge causes heat and large losses within the battery itself. The more slowly the battery is discharged, the greater amount of current it will produce. Under actual operating conditions, a battery should never be

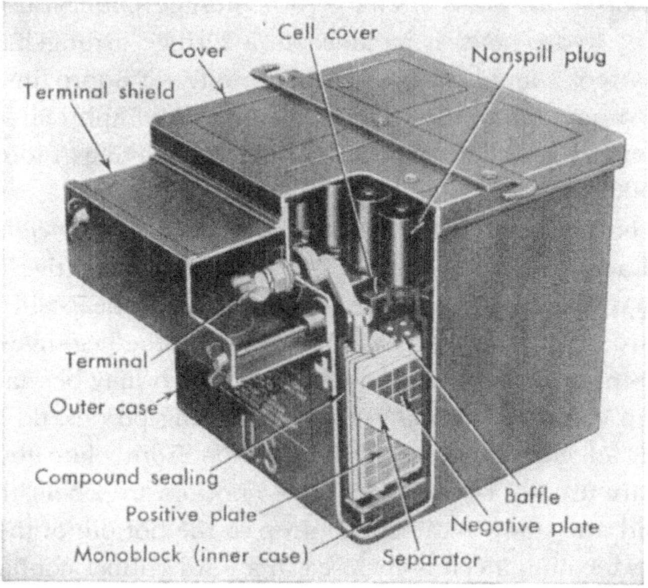

Fig. 45. A cutaway view showing the construction of an aircraft storage battery. (Courtesy Delco-Remy Division, General Motors Corporation)

completely discharged. Except when used for starting purposes, an excessive amount of current is not usually drawn from the battery. The battery in an aircraft electrical system is usually connected with a generator which, not only furnishes current to the battery, but also furnishes an excess amount which takes care of the lights and other electrical equipment. The battery furnishes current for the electrical equipment when the generators are not in operation.

The ampere-hour rating of the battery depends upon the total effective plate area. Cells having a large number of plates have a higher ampere-hour rating than cells having fewer plates. A number of identical cells connected in series have the same ampere-hour capacity as one individual cell, but the total voltage is equal to the sum of the voltages of the cells connected. When the cells are connected in parallel, the voltage is the same as that of any one cell, but the ampere-hour capacity is equal to the total of the ampere-hours of the cells connected in this manner.

The current used to charge the battery must be slightly in excess of the voltage of the battery. For example, a 12-v. battery should be charged with a current of slightly more than 14-v. If the charging voltage is just equal to that of the battery, no current will flow and the battery will not be charged. The charging current must always flow in the same direction and must, therefore, be a direct current. The positive pole of the charging apparatus should be connected to the positive pole of the battery. A battery in a low state of charge may be charged at a high starting rate but, as the battery approaches full charge, the charging rate should be decreased. Some gas is always given off during the charging process. If the liquid in the battery seems to boil violently, the charging rate should be reduced. If only alternating current is available for charging, a Tungar rectifier may be used. Batteries are not usually charged at a rate higher than approximately 5 to 7 amp. A battery may be considered fully charged when no increase in specific gravity is shown by the hydrometer over a 30-min. period. Overcharging may damage the battery. Batteries having large plates may be charged at higher rates than batteries having small plates.

A storage battery should never be allowed to reach such a state of discharge that most of the active material in the plates has become lead sulphate. If the battery is not charged before this condition is reached, the cell will be damaged. When this condition exists in a storage cell, the cell is said to have been sulphated. The condition itself is called sulphation.

If the liquid in the battery is allowed to become low or is emptied out while the battery is in a low state of charge, exposure of the plates to the air makes it almost impossible to recharge the battery. When a battery has become sulphated and a large amount of the sulphate material drops to the bottom of the cell, it may form a layer thick enough to short the bottom of the plates. This is a common cause for failure of storage batteries. A discharged battery should never be placed in storage. The battery should be kept filled with electrolyte and charged from time to time. A discharged battery may freeze in cold weather.

In mixing electrolyte for a storage battery, never pour water into the acid. Always pour the acid slowly into the water, stirring continuously. When the acid is added to the water, the mixture becomes heated. Battery electrolyte should be mixed in a container which will not crack when heated. The mixture of hydrogen and oxygen gas generated by a storage battery on charge or discharge is highly explosive. Electrolyte should not be added to a storage cell, unless part of the electrolyte has been spilled.

Electric currents produced by generators are commonly divided into two types, alternating current and direct current.

A direct current is a current of electricity which flows continuously in one direction. A direct current has a constant polarity.

An alternating current is a current of electricity which is constantly reversing its direction of flow. An alternating current flows in one direction and then in the other, many times each second. During each cycle of the alternating current, the current builds up to a maximum, drops to zero, and then builds up to a maximum in the opposite direction and again drops to zero.

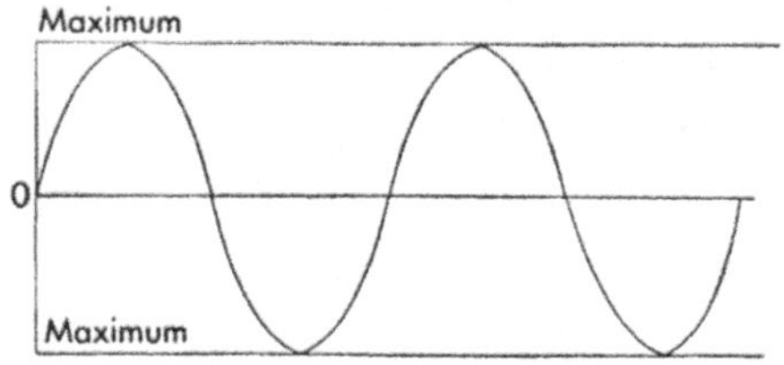

Fig. 46. A diagram to show the cycles of an alternating current, single phase.

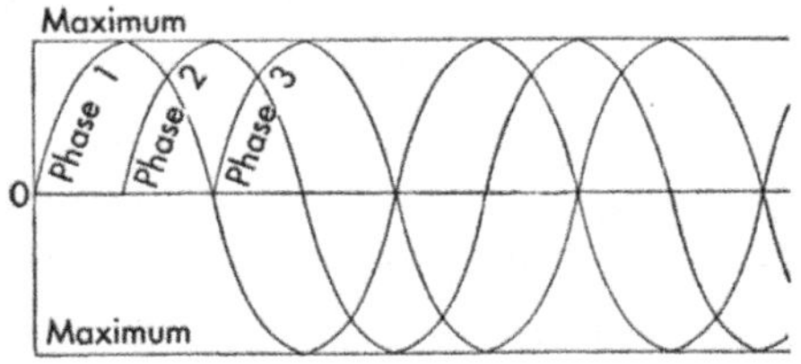

Fig. 47. A diagram to show the cycles of an alternating current polyphase.

The fundamental principle of a generator is that a current of electricity is induced in a conductor which is moved through a magnetic field, cutting the lines of force. A generator always contains a magnetic field. This field may be formed by the use of permanent magnets or by the use of electromagnets. Revolving in this magnetic field is the armature of the generator, which is made up of a large number of turns of wire wrapped lengthwise around an iron core. The armature is designed to revolve in the electric field formed by the magnets.

As shown in Figure 48, one side of the coil moves in one direction through the magnetic field, while the other side moves in the opposite direction. As the coil rotates through 180°, a current is induced in each loop of the coil and flows around the coil. As the loop starts on the second 180° of its rotation in the field, the current is induced in the opposite direction and flows around the coil in a direction opposite to its flow during the first 180° of rotation.

The alternations may be called positive alternation in one direction and negative alternation in the other direction. Each 360° of rotation of the armature causes a flow of current in one direction for 180°, beginning at zero and returning to zero, and in the opposite direction during the other 180° building up from zero and dying down to zero. One positive alternation and one negative alternation complete one cycle of alternating current. The current generated by each rotation of the armature completes one cycle. If the armature revolves at the rate of 60 times a second, a 60-cycle current is generated.

The ends of the coils of wire about the armature are connected with the slip rings. The terminals of the generator are connected to brushes, usually made of carbon, which slide around the slip rings of the generator. The leads from the generator are connected to these terminals,

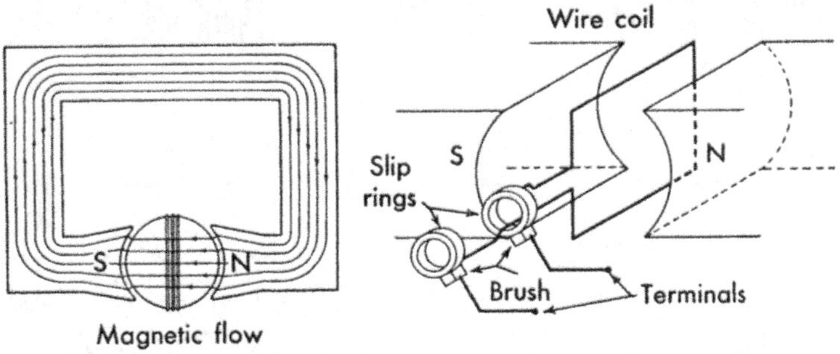

Fig. 48. A coil of wire rotated in an electric field has generated in it an electric current.

one lead being connected to the brush on one ring and the other lead connected to the brush on the other ring. The brushes make contact with two slip rings, each of which is connected with one end of the coil wound about the armature.

Starting with this coil at the zero position as it rotates through the magnetic field, a current flowing in one direction is induced in the coil. One side of the coil cuts the magnetic lines of force in one direction, while the other side of the coil cuts them in the opposite direction, inducing a current in the coil. This current flows through the slip rings to the line. As the coil reaches the zero position, the current dies down to zero. During the other half of the rotation, the current builds up to a maximum and then dies down to zero, but flows in the opposite direction. This is a single-phase current. The current builds up in one direction to a maximum, dies down to zero, and then builds up in the opposite direction to a maximum, again dying down to zero. This build-up and die-down in each direction completes one cycle. If the coil rotates 60 times a second, a 60-cycle current is produced.

The fundamental difference between the direct-current generator and the alternating-current generator is the commutator. A commutator is a device made up of copper segments insulated from each other, over which the brushes connecting with the external circuit slide. These segments are so connected with the leads from the generator that they allow the current induced in the armature to flow into the external circuit in

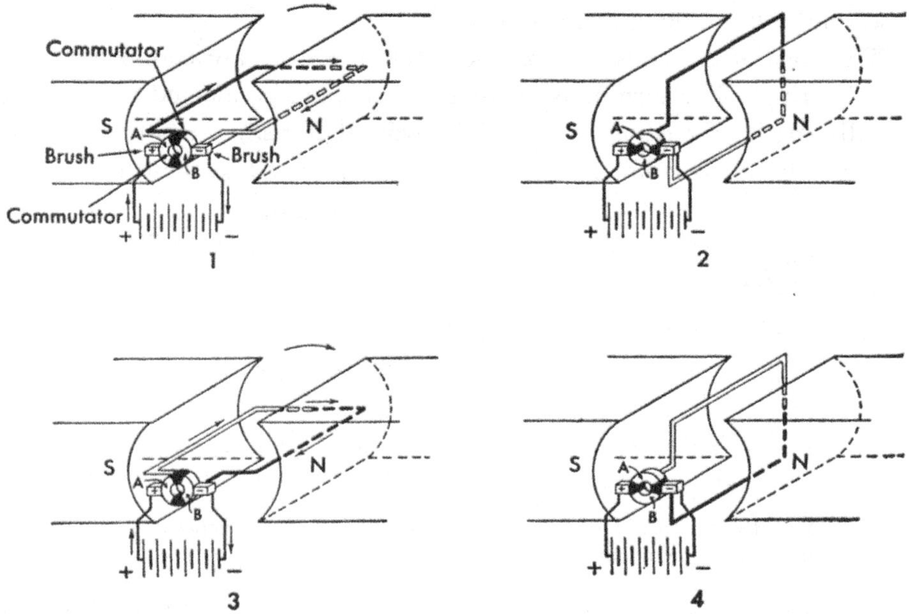

Fig. 49. A commutator is used to change alternating current surges from the armature of a generator to direct current.

one direction only. When a coil of the armature is cutting the magnetic field during the first 180° of its rotation, a pair of segments is so arranged that the induced current flows out into the external circuit through the brushes. As the armature starts on its second 180° of rotation, a pair of segments of the commutator connected with the opposite ends of the coil has been rotated to a position such that they have come into contact with the brushes. The current induced in the armature during the second half of the rotation then flows into the external circuit in the same direction as before. Although the current induced in the armature is reversed, it flows into the leads from the generator in the same direction as it did during the first half of the rotation. There are always two segments in the commutator for each separate coil in the armature.

In the d-c generator, the armature is rotated in the magnetic field the same as in the a-c generator. As the coils are rotated through the magnetic field, an ac is induced in the coils. Because of the commutator, the flow of current in the wires leading from the generator is continuous and forms a dc. As the coil approaches its zero point, the position when the coil is at right angles to the magnetic lines of force, no current is generated in the coil. This is the point where the brushes slide to the opposite pair of segments in the commutator.

If a single coil is used in the armature connected with a commutator, a pulsating direct current is generated. By increasing the number of coils in the armature this current may be smoothed out to an almost steady flow.

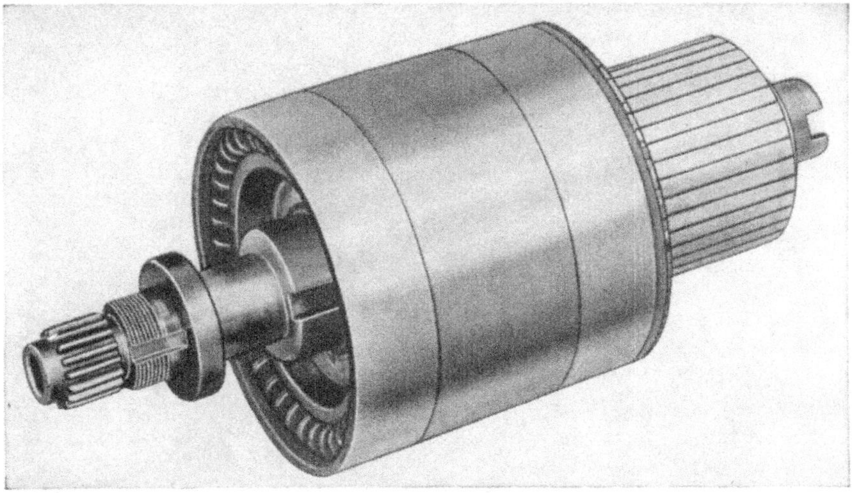

Fig. 50. A commutator and armature. (Courtesy of Jack & Heintz, Inc.)

A pulsating direct current is a direct current which changes its intensity at regular intervals. A pulsating direct current may be used to produce a pulsating magnetic field about a conductor. The value of the current may drop to zero and then build up, or may simply lose part of its value before building up to maximum again. Pulsating direct current is used in connection with induction coils.

The magnetic field around a conductor carrying a direct current of constant value remains constant. The magnetic field around a conductor carrying an alternating current fluctuates with each change in direction of the electric current. The field builds up in one direction, dies down to zero, and then builds up in the opposite direction. These fluctuations of the magnetic field occur at the same time as the fluctuations of the current itself.

Three things influence the electric current generated: (1) the strength of the field, (2) the speed of rotation, and (3) the number of turns of wire on the armature.

The strength of the field may be increased or decreased by using stronger or weaker magnets or electromagnets. The speed of rotation is varied by changing the rate at which the armature rotates. The number of coils is increased by increasing the number of turns of wire in each coil of the armature.

VI GENERATORS AND MOTORS

Generators and motors are two of the most important pieces of electrical machinery. They are both used extensively in aircraft electrical systems.

Both motors and generators may vary in size from those that produce only a small fraction of one horsepower to those furnishing hundreds of horsepower. Most motors and generators used in aircraft equipment are small, varying from a fraction of a horsepower to approximately 25 hp.

Fig. 51. An aircraft generator. (Courtesy Leece-Neville Company)

Generators. A generator is a machine which converts mechanical energy into electrical energy. Most aircraft generators are driven by the aircraft engine. A generator usually consists of an outer ring of iron called the yoke; an armature formed of coils of wire and a soft iron core; a set of collector rings, or commutator segments; and collecting brushes which collect the current from the armature. The pole pieces project from the yoke, and the field coils are wound on these pole pieces in such a way as to make them alternately north and south poles. The magnetic field, due to the alternate spacing of the poles, divides equally at each pole and, therefore, forms a magnetic field made up of the same number of horseshoe magnets as the number of poles. This is shown in Figure 53.

There must always be an even number of pole pieces. By winding the field coils on the pole pieces, the poles of the yoke become electromagnets. This arrangement greatly increases the field strength without increasing the size of the pole pieces. By using the pole pieces of the yoke to form horseshoe magnets, the air gap in the magnetic field is decreased to the minimum clearance necessary for the rotation of the armature. The smaller the air gap, the more efficient the electric field becomes. Therefore, the gap between the magnets and the armature should be as small as possible. The soft iron core of the armature is usually made of laminations to prevent eddy currents. The coils are made up of many

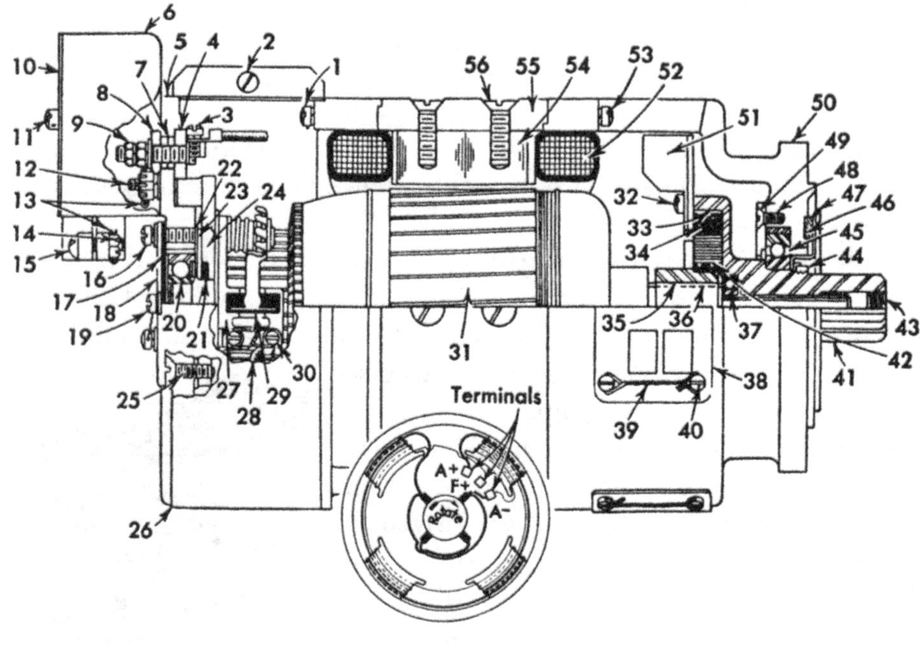

Fig. 52. Diagrammatic drawing to show the construction of an aircraft generator. (1) (2) (3) (9) (11) (12) (13) (14) (15) (16) (19) (25) (30) (32) (39) (40) (44) (48) (53) (56) Screws; guards; lock washers, nuts, etc. (4) insulator; (6) (50) housing; (7) bushing; (17) (22) gasket; (18) (23) (24) (49) retainer; (20) (45) bearing; (21) (46) felt washer; (26) band; (27) brush rigging; (28) lead assembly; (29) brush; (31) armature; (33) liner; (34) felt pad; (35) cam assembly; (36) straight keys; (37) bearing; (38) vent shield; (41) spline shaft; (42) spring assembly; (43) plug; (47) oil sling; (51) fan; (52) field coil; (54) pole piece; (55) field ring. (Courtesy Leece-Neville Company)

turns of copper wire. The soft iron core not only increases the effectiveness of the armature, but also acts as a support for the copper coils. The coils are placed in slots cut into the surface of the iron core and are fastened with wooden or fiber wedges to prevent their displacement by centrifugal force.

The electromotive force depends upon the speed of rotation of the armature. This speed is limited by the forces set up by centrifugal force. If the coil has to move only half as far to cut the same number of lines of force, the speed may be cut in half and the same power developed. An extra pair of poles has this effect.

Many generators have more than one set of poles. Some have as many as 36 poles to allow the speed necessary for a given output of power. The poles always have to be added in pairs.

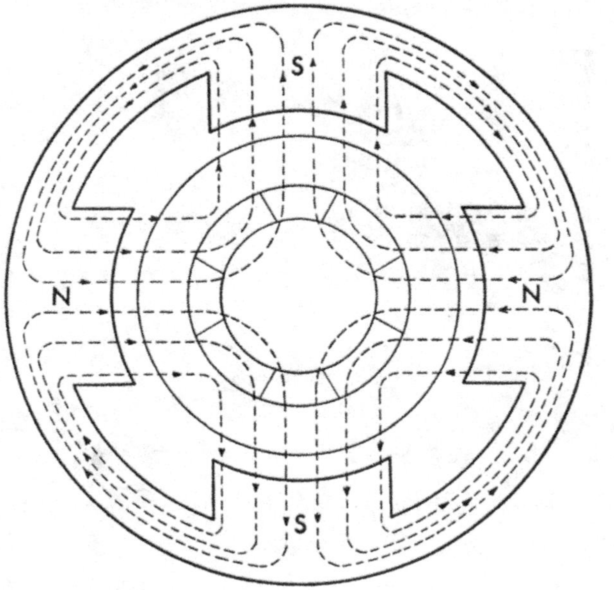

Fig. 53. A drawing to show a simple four-pole direct current generator and its magnetic circuits.

The a-c generator is similar to the d-c generator except that it is equipped with collector rings or slip rings which are placed around the shaft of the armature; the current drawn from these rings will be alternating current. Since an a-c generator cannot use the current supplied by the generator to excite the field windings, the field windings must carry a direct current. All a-c generators are separately excited by d-c from an outside source or are equipped with permanent field magnets.

An a-c generator is called an alternator. If a generator of this type has a single set of coils, all connected in a series, a single-phase current is generated, and the generator is called a single-phase alternator. It is possible to rearrange the coils or place more coils in the slots and wire them in two or three sets of coils spaced at such a distance on the armature that the angular distance between the sets of coils is 90° or 120°. The alternator will then produce two or three alternating voltages of the same strength, differing in phase from one another by 90° or 120°.

The coils are so arranged that the electromotive force in one coil reaches its maximum value a fixed time ahead of the electromotive force in the other coil or coils. This produces a multiple wave formed of two or more waves (see Figure 47, page 54). Such generators are known as polyphase, or two- or three-phase alternators.

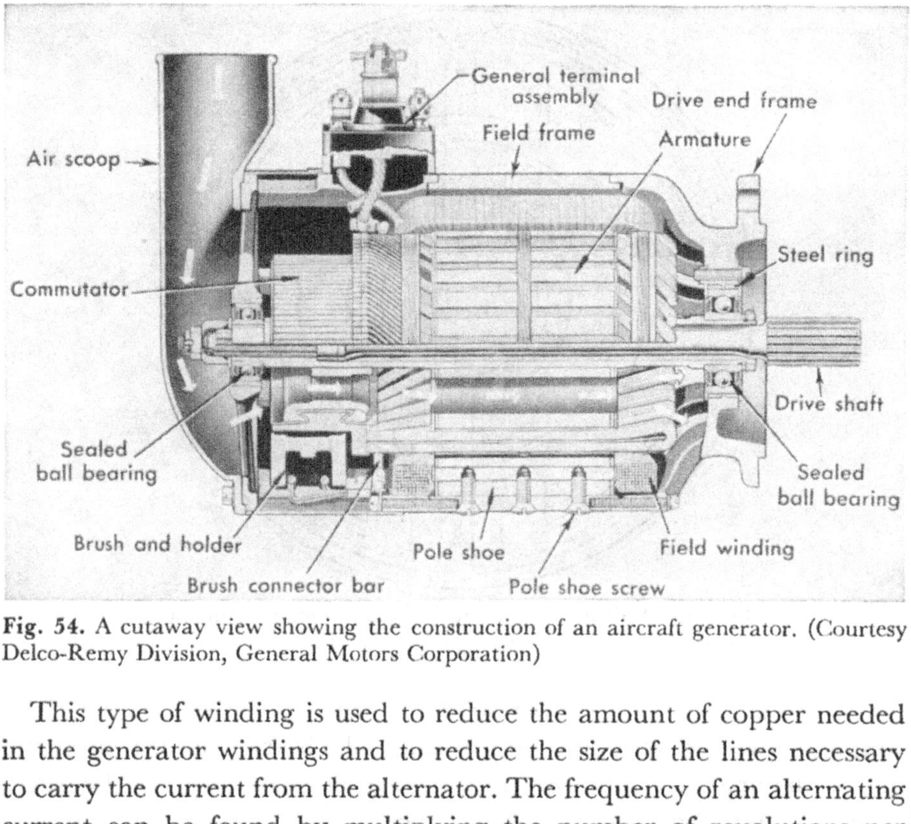

Fig. 54. A cutaway view showing the construction of an aircraft generator. (Courtesy Delco-Remy Division, General Motors Corporation)

This type of winding is used to reduce the amount of copper needed in the generator windings and to reduce the size of the lines necessary to carry the current from the alternator. The frequency of an alternating current can be found by multiplying the number of revolutions per second of the armature by the number of pairs of poles.

In some large generators, because of the strains placed upon the armature due to its rapid rotation, the armature coils and field coils are reversed. The armature coils are wound around a stationary yoke, while the field coils form the rotating part. In this type of generator, the stationary coils are called the stator and the rotating field is called the rotor. Figure 55 shows the construction of a d-c generator.

The main difference between a d-c generator and an alternator is the commutator. The commutator changes the alternating current surges from the armature into a current flowing in one direction only. The commutator is made up of a number of wedge-shaped copper segments fitted together but insulated from each other. Mica is the common insulating material used in the building of the commutator. The two ends of each coil of the armature winding are fastened firmly to adjacent commutator segments. Each commutator segment carries one terminal of each two adjacent coils. One terminal is the forward end of one coil,

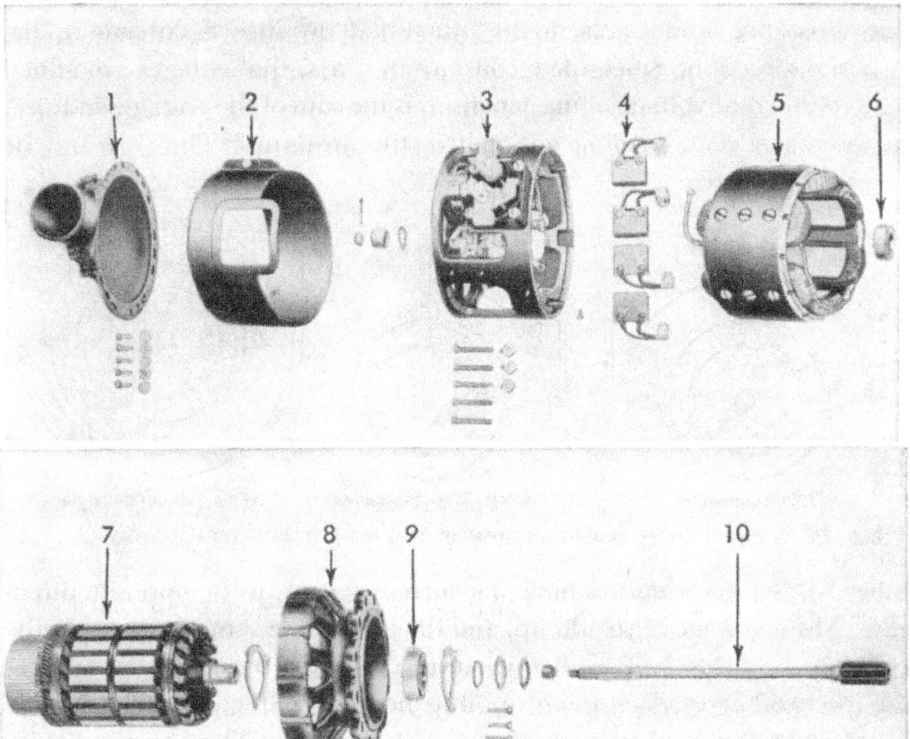

Fig. 55. An exploded view of an aircraft generator. (1) Air scoops; (2) cover bands; (3) commutator end frame assembly; (4) brushes; (5) field frame assembly; (6) ball bearing; (7) armature; (8) drive end frame; (9) ball bearing; (10) drive shaft. (Courtesy Delco-Remy Division, General Motors Corporation)

and the other the rear end of the adjoining coil. Each coil has two terminals, and each commutator segment makes contact with two coil terminals. Figure 56 shows the commutator arrangement for a d-c generator. There must be a commutator segment for each coil in the armature. Of course, if there is only one coil, there must be two commutator segments.

The direction of the electromotive force is shown by the arrows in Figure 57. The coils are represented by a few turns of wire. In both sides of the armature winding, current induced in the coil flows toward the first segment and away from the segment which is directly opposite. This is true of each opposite pair of coils and segments. It should be noted that the coils on one half of the armature produce a current moving in one direction, while coils in the other half produce a current in the opposite direction. Since these coils are in series, the voltages are added up and the total voltage of the generator is the sum of the voltages induced in the set of coils forming one half of the armature. The coils on the

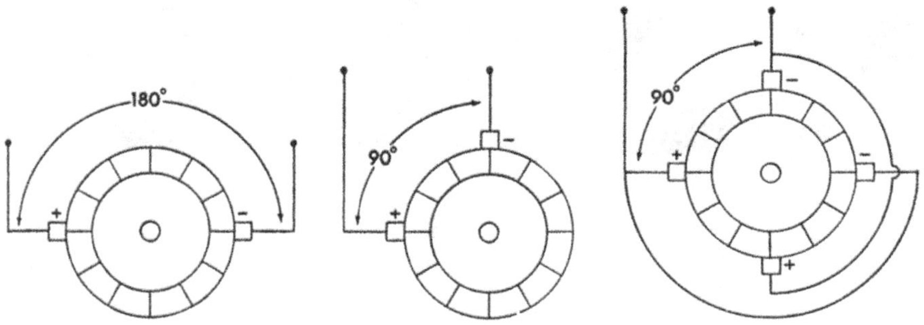

Fig. 56. A commutator and brush arrangement for direct current generator.

other half of the armature have the current set up in the opposite direction. These voltages also add up, and the total is the same as those for the coils on the other half of the armature. Being in parallel, the voltages for the two halves do not add up, but the total voltage of the generator is equal to the total voltage of one half the coils. The brushes are so arranged that they come into contact with opposite segments of the commutator. The upper brush will collect the current flowing from the coils in one direction, while the opposite brush will carry the current from the coils which are moving in the opposite direction. The upper brush is the positive terminal for the external circuit. The lower brush is the negative terminal for the external circuit and receives the current from the external circuit.

It must be understood that the current from any type of generator does not simply flow out from the generator and disappear some place. There must always be a complete circuit, and the current which flows outward from one brush of the generator must pass through a closed external circuit back into the generator through the other brush and through the coils of the armature. A cord may be arranged around a pulley system to represent a closed circuit which includes the coils of an armature. If the cord is pulled steadily around, it may be thought of as representing a direct current. If the cord is jerked rapidly back and forth, it may be thought of as representing an alternating current.

Every electric current, in order to flow, must have a negative and a positive terminal connected by a closed circuit. In the battery, the circuit

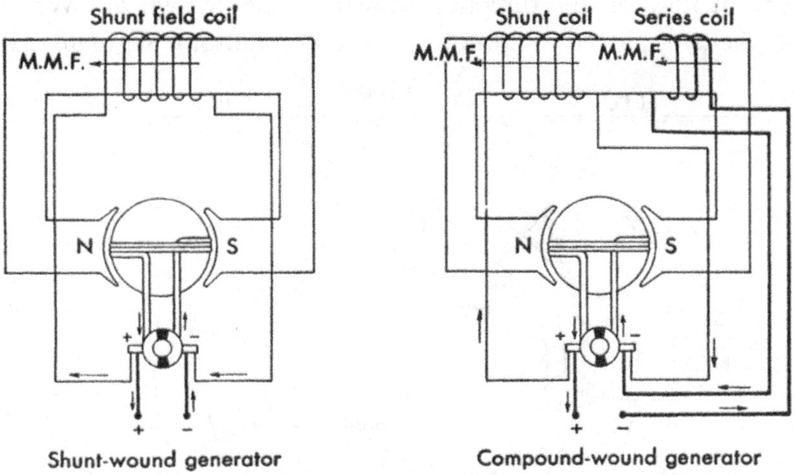

Fig. 57. Simple two-pole generators showing commutator arrangement and field windings.

is completed by means of a conductor from one terminal to the other, and the electrolyte within the battery makes the connection between these terminals and completes the circuit.

A line drawn perpendicular to the magnetic lines of force between the poles of a generator may represent the neutral plane. The neutral plane is the plane in which no electromotive force is produced by the motion of the armature coil in the magnetic field. The brushes of a generator are set in the neutral plane. When a generator is delivering current to an external circuit, a current flowing through the coils of the armature produces a magnetic field which will have a definite effect upon the size and direction of the magnetic field produced by the field coils. This effect is called the "armature reaction" of the generator.

The neutral plane of a generator is at right angles to the lines of force of the field only when there is no current flowing through the armature coils. The neutral plane is shifted forward in the direction of rotation as soon as current flows in the armature. For this reason, the generator brushes are not set exactly in the perpendicular plane between the poles. The brushes of a d-c generator are set just beyond the shifted neutral plane. This position of the brushes prevents sparking or arcing between the brushes and the commutator. Armature coils are highly inductive. Whenever current is flowing in a circuit and the circuit is broken, sparking is apt to take place, particularly in an inductive circuit.

Before the armature coil reaches the shifted neutral plane, as shown by the letter *X*, Figure 58, it is under the influence of the north pole, and the current flows in the direction shown by the arrowhead. When the first segment of the commutator and the commutator segment next to

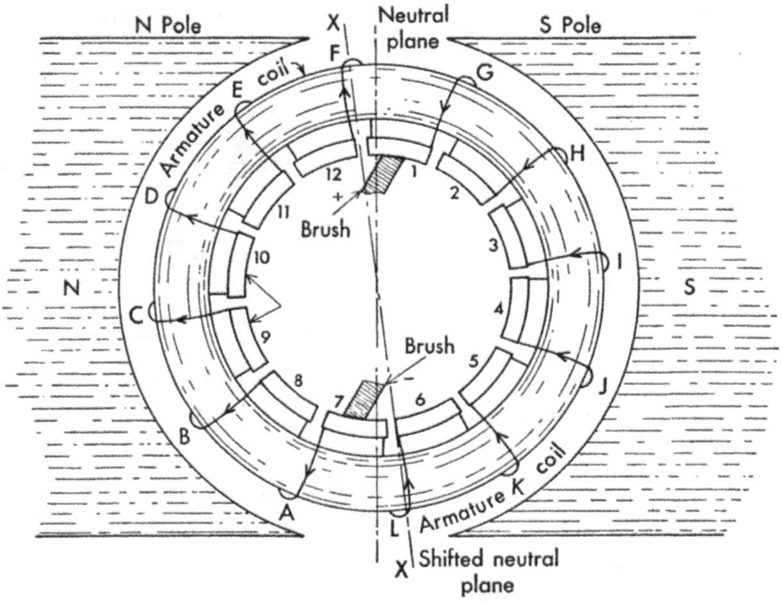

Fig. 58. A diagram to show the commutation of a d-c generator and the shifted neutral plane.

it are both in contact with the brush, coil *F* is short-circuited and a relatively heavy current flows through the circuit. This circuit then consists of coil *F*, two commutator segments, 1 and 12, and the brush, until just after the coil *F* passes through the shifted neutral plane. The size of this short circuit is reduced to a minimum by the use of carbon brushes. Carbon brushes are more resistant to current flow than are metallic brushes. When segment 1 leaves the brush, this short circuit will be broken and, unless the current in coil *F* has been reduced to zero, heavy sparking will occur at the trailing edge of the brush. Commutator brushes are so set that each segment of the commutator leaves the brush just after the coil has come under the influence of the south pole of the generator. At this position, the current in any coil is reduced to zero by the combining action of the self-induction of the coil and the action of the south pole upon it.

In a four-pole generator, there are four positions in which a coil cuts a number of lines of force. Such a generator has two neutral planes at right angles to each other. There must be two brushes for each neutral plane, and the four-pole generator has four brushes. Two of these brushes are positive, and the other two are negative. The two positive brushes are connected together by a low-resistance conductor, as are the two negative brushes. In this type of generator there are four paths through the armature, and the voltage of the generator is the sum of the electromotive force in one fourth of the windings. Although the electromotive

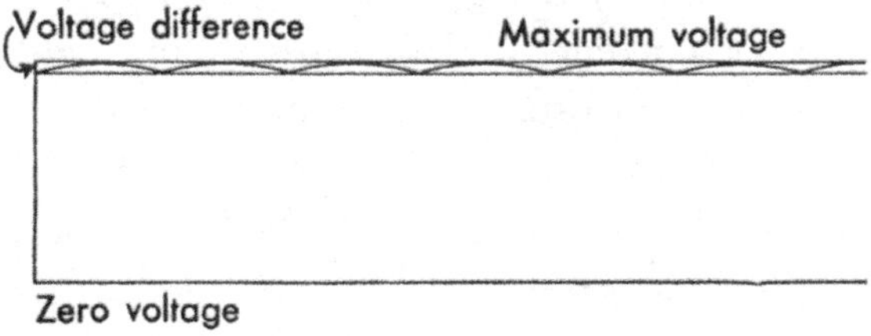

Fig. 59. A diagram to show "ripple" in a direct current.

force and current flow is in one direction, the intensity of the electromotive force and current varies somewhat from instant to instant. This is due to the fact that the coils cut the lines of force of the magnetic field at varying angles. This slight variation in the value of the electromotive force from a d-c generator is called "ripple" and is shown in Figure 59.

The larger the number of coils in the winding of the armature, the smaller will be the ripple. A d-c generator is normally self-excited. This is possible because of the residual magnetism remaining in the core of the electromagnet. When the generator is started and the armature starts turning, there is always a weak field produced by the small amount of magnetism retained in the poles of the electromagnets. The armature, in cutting this weak field, induces an electromotive force in the coils of the armature. This induced force drives a small current through the field coils which are connected in series, or in parallel, with the armature. This current increases the strength of the field magnets which, in turn, increases the amount of induced electromotive force in the armature coils, which again increases the strength of the electromagnets. The electromotive force of the machine does not increase indefinitely, as might appear from the above, as the maximum electromotive force of the generator is limited by the resistance of the field windings. The field-winding circuit usually has a rheostat installed in it to regulate the amount of current passing through the field windings.

At times, a d-c generator that has had the field coils reconnected after having been disconnected will not build up its voltage. It may be that the residual magnetism is opposing the field from the field current and the field connections should be reversed.

A d-c generator is generally known as a shunt-wound, series-wound, or compound-wound generator. These names simply tell how the electric current used to form the electromagnets is carried to the field coils.

In a shunt-wound generator, as shown in Figure 60, part of the current from the main leads of the generator is carried through the field coils. The amount of current flowing through the field coils may be regulated by means of a rheostat inserted in the field circuit. This type

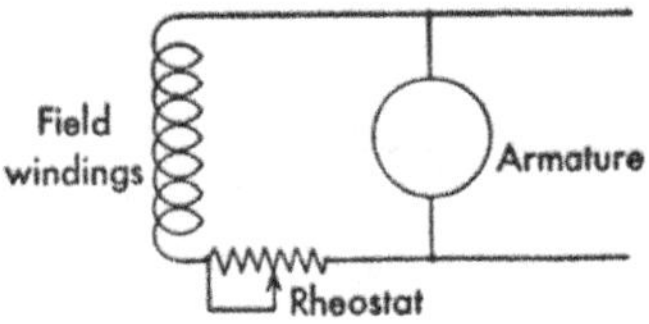

Fig. 60. A diagram to show the wiring of a shunt-wound generator.

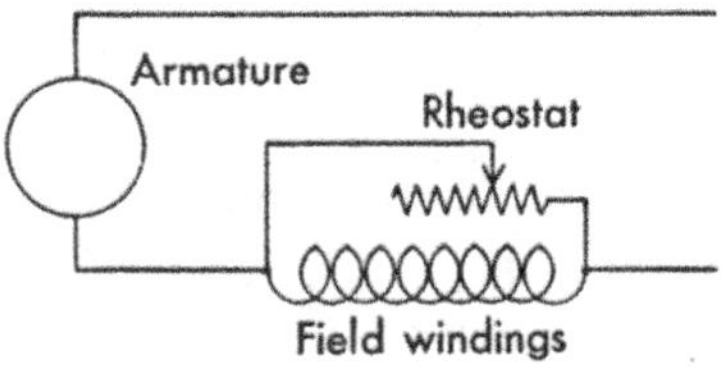

Fig. 61. A diagram to show the wiring of a series-wound generator.

of generator is not suitable for fluctuating loads if a constant voltage is required. Any increase in the current in the external circuit will increase the current flowing through the field coils, which, in turn, will cause an increase in the current output. If the current in the external circuit decreases, the current to the field coils will decrease, further decreasing the output. Of course, a balance will be reached in either case. Shunt-wound generators are used for such purposes as battery charging or where the current demand and load are quite constant.

In a series-wound generator, the field coils are in series with the external circuit. All of the current output from the generator passes through the field coils, as shown in Figure 61. Series generators have poor voltage regulation under a changing load. As the load from the generator in the external circuit decreases, the amount of current through the field decreases and weakens the field. This, in turn, will decrease the output of the generator. A rheostat is usually inserted in the field winding to regulate the strength of the field.

A compound-wound generator is a combination of the shunt-wound and the series-wound types. This method of winding is shown in Figure 62. In this generator, if the output voltage remains constant, the ampere turns of the shunt field are constant, but the ampere turns in the series field vary directly with the current passing through them to

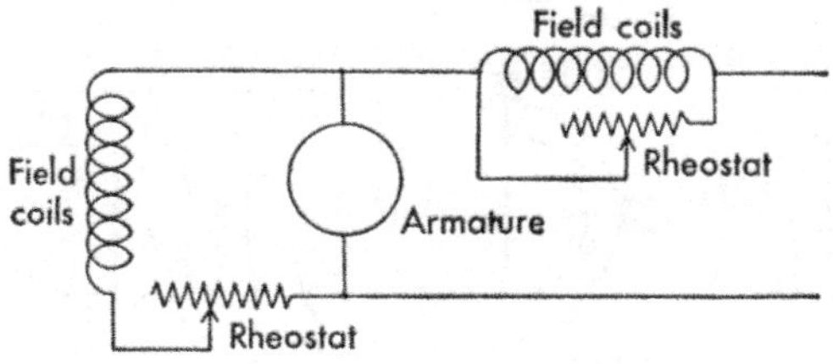

Fig. 62. A diagram to show the wiring of a compound-wound generator.

the external circuit. The field strength then increases, and the electromotive force of the generator increases with an increasing load. If the series field contains the proper number of turns of wire in its winding, the electromotive force of the generator may be made to rise enough to balance the increasing drop of potential in the armature. When the turns in the series field are just enough to keep the terminal voltage of the machine constant, the generator is said to be flat-compounded. If the terminal voltage rises with an increase in load, the generator is said to be over-compounded. If the terminal voltage decreases with an increased load, the generator is said to be under-compounded.

When it is desired to regulate the output of the generator for any reason, a rheostat is placed in series with the shunt field and another

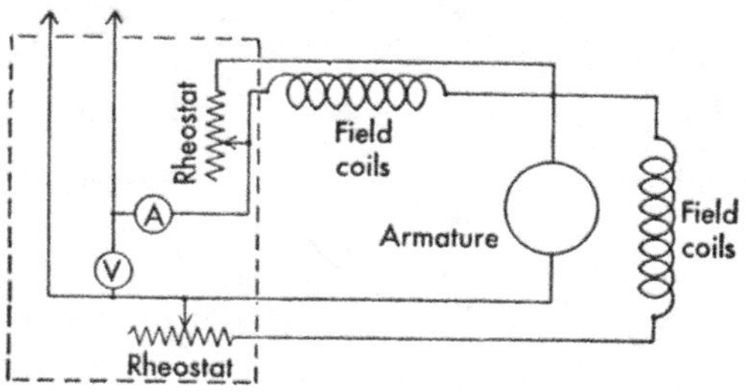

Fig. 63. A diagram showing how rheostats are placed in the field circuit of a generator.

rheostat is inserted in parallel with the series field, as shown in Figure 63. The current in the shunt field is regulated by means of its rheostat, increasing or decreasing the voltage output of the generator. The voltage may be controlled by the use of the rheostat in the series field which

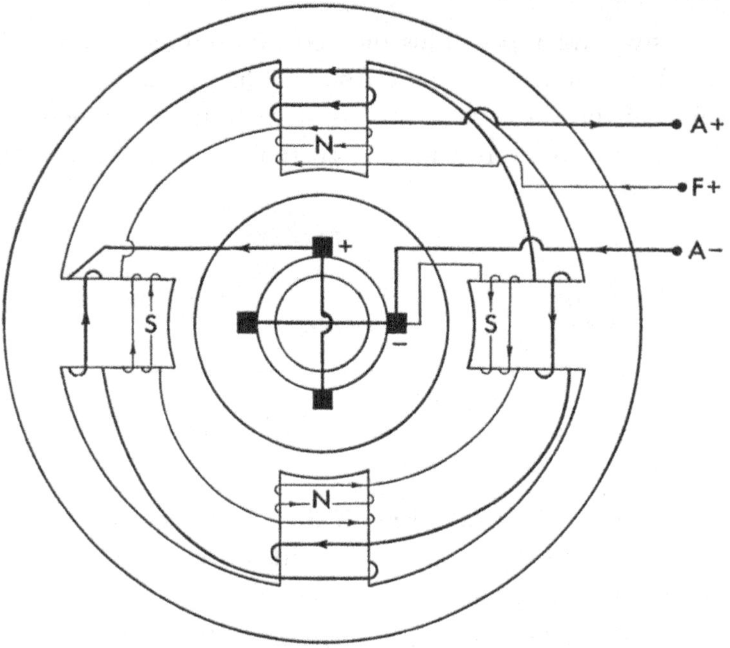

Compound-wound generator

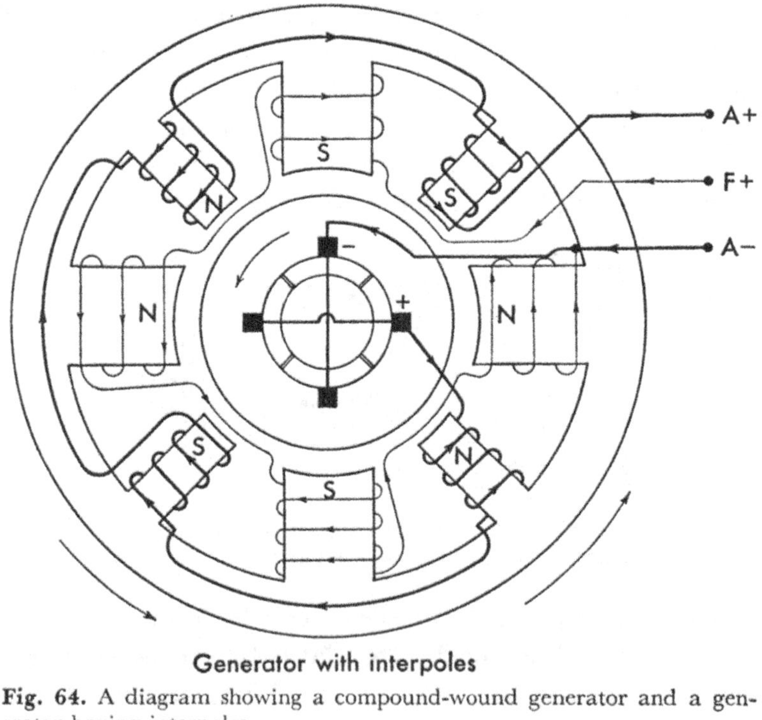

Generator with interpoles

Fig. 64. A diagram showing a compound-wound generator and a generator having interpoles.

controls the current by shunting a portion of the output from the generator around the series-field coils.

Motors. Motors may be considered as the opposite of generators. The generator converts mechanical power into electrical power and electric current, while the motor converts electrical power into mechanical power. Fundamentally, the motor is much like the generator. In fact, all d-c motors may be used as generators. The d-c motor has an armature with armature coils, field coils, and a commutator equipped with brushes, as does a d-c generator. In a generator, the armature is turned

Fig. 65. A landing gear retraction motor. (Courtesy Jack & Heintz, Inc.)

by mechanical means to generate electric current, while, in a motor, the armature is turned by means of an electric current to develop mechanical power.

The turning effect of an electric current is brought about by the magnetic field set up when the current is passed through the armature winding. The armature in the motor is mounted in an electric field similar to that in a generator. Figure 67 shows a single-turn coil mounted in a magnetic field. The magnetic lines of force between the poles of the magnet producing the field, if undisturbed, flow in the most direct line between the poles. If the loop of wire representing the armature is placed parallel to the lines of force and an electric current is passed through the coil, the coil tends to arrange itself at right angles to the lines of force. This turning force is brought about by the lines of force in the field being distorted by the field set up around the wires in the coil, as shown in Figure 68. These distorted or bent lines of force try to flow in a straight

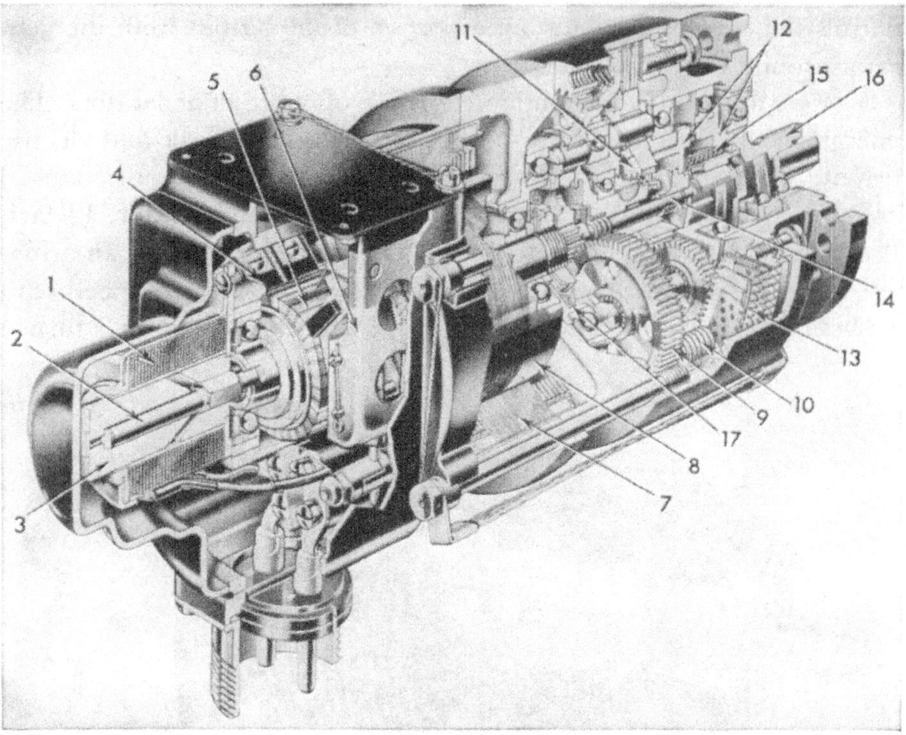

Fig. 66. A sectional view of a landing gear retraction motor: (1) Solenoid; (2) Push rod; (3) Solenoid core, movable; (4) Brush; (5) Commutator; (6) Brush holder; (7) Stator; (8) Armature; (9) First stage planetary gearing assembly; (10) Clutch springs; (11) Second stage planetary gearing assembly; (12) Engaging jaws; (13) Clutch packs; (14) Push rod bumper; (15) Jaw spring; (16) Output drive shaft assembly. (Courtesy Jack & Heintz, Inc.)

path. The bending of the lines of force develops a force which presses upward on one side of the coil and downward on the other.

The lefthand rule is used to find out which way the armature of a motor will rotate. The left hand is placed in such a position that the

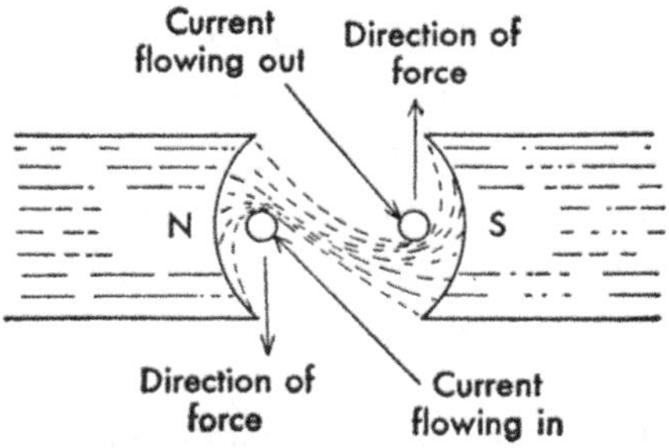

Fig. 67. A diagram to show the distortion of a magnetic field by loops of wire carrying an electric current.

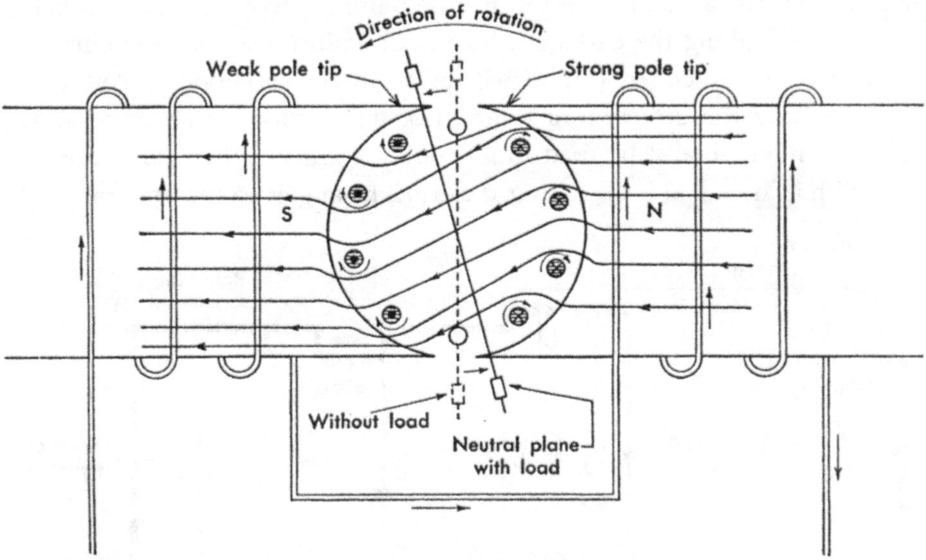

Fig. 68. The lines of force in the field are distorted by the coils of the armature. This causes a shift in the neutral plane.

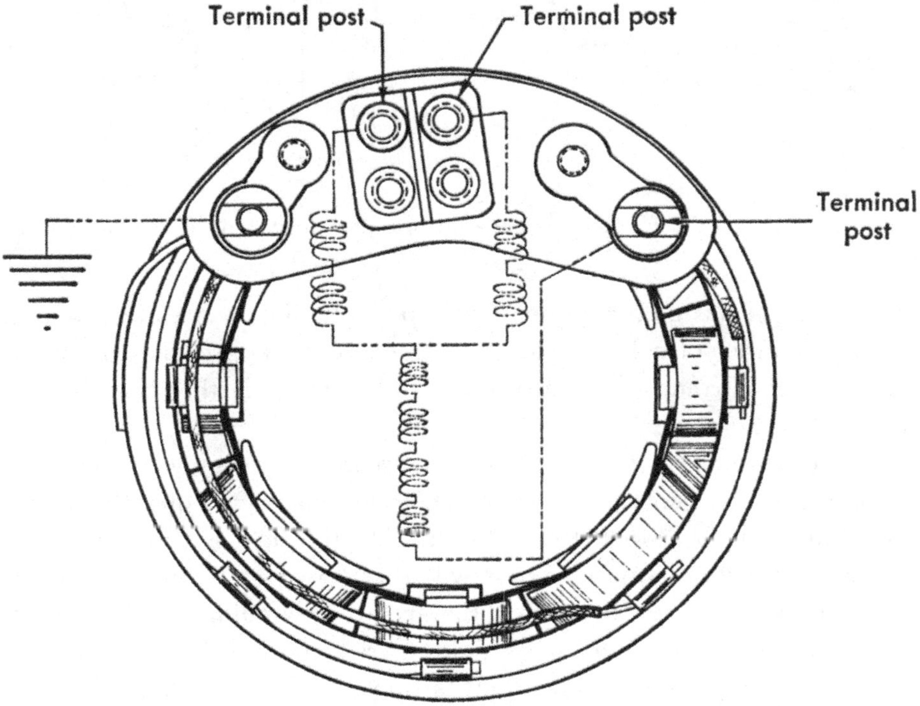

Fig. 69. Field circuits of a retraction motor. (Courtesy Jack & Heintz, Inc.)

lines of force in the field point toward the palm of the hand. If the fingers are extended along the conductor in the direction in which the current is flowing, the extended thumb will point in the direction of rotation. To determine the direction of rotation of the armature in a generator, the right hand should be used in a similar manner. When the plane of the coil has arranged itself so that it is at right angles to the lines of force

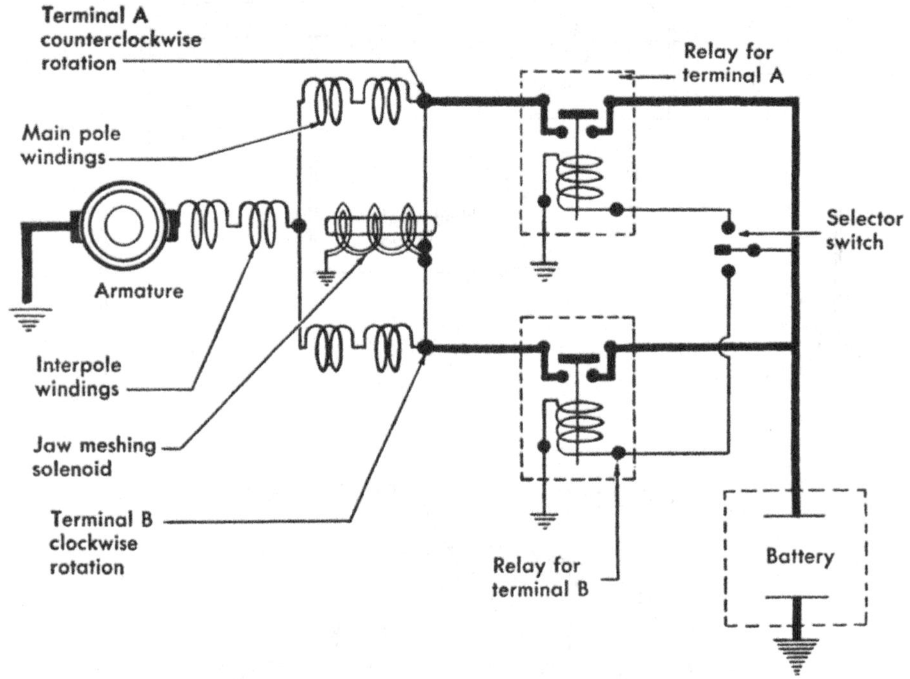

Fig. 70. A diagram to show the wiring of a landing gear motor unit. (Courtesy Jack & Heintz, Inc.)

in the field, it has no further tendency to rotate. The armature is made up of a number of coils of wire and, as each coil approaches the zero position, the current is switched to the coil which is farther back in the plane of rotation.

A shunt-wound motor is constructed in much the same manner as the shunt-wound generator. The field coils are connected directly across the power supply, and the current through the field is constant. The power of the shunt-wound motor will therefore vary with the size of the current being passed through the armature winding. As the armature rotates through the magnetic field, there is induced in the armature itself an electromotive force which is opposed to the current flowing through the armature. This is called the counter electromotive force (c.e.m.f.). When the armature is standing still, as in starting, there is no counter electromotive force present. To prevent overloading the armature in starting, a starting box should be installed. The starting box is simply a series of resistances which are cut out by the movement of the starting-box handle as the motor picks up speed. When the motor is running at top speed, all of the resistance is cut out. The speed of a shunt-wound motor may be regulated by a rheostat, either in the field winding or in series with the armature. A rheostat may be installed in each circuit. The shunt-wound motor will operate on direct current only.

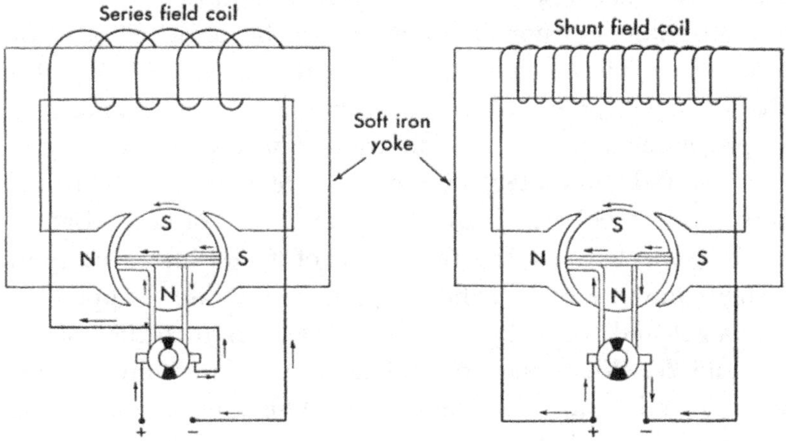

Fig. 71. A diagram of two-pole shunt-wound and series-wound motors.

The series-wound motor may be operated on either direct current or alternating current. The current through the armature also flows through the field coils. As the motor is operating on an alternating current, the brushes are set in a plane vertical to the field. When a direct current is used, the brushes are set back of the neutral plane to take care of the armature reaction. The series motor develops high power when starting. When a load is added and the armature slows down, there is a decrease in the counter electromotive force in the armature. This allows a greater current to flow through the armature, producing greater power. The speed of this motor is controlled by the load. The motor will race when unloaded. Starting boxes are generally used with large motors of this type.

The compound motor has a combination of series and shunt windings. There are two types of compound motors, the cumulative compound and the differential compound. In a cumulative compound, the coils are wound upon the core in the same direction. When wound in this manner, the magnetic fields tend to help each other, giving the name to the cumulative compound motor. This motor has good starting power, but the speed regulation is poor. The windings in the differential compound motor are wound in the opposite direction. With this winding, the series-field winding opposes the shunt-field winding. This motor will operate at almost constant speed under varying loads. When the motor tends to slow down due to load increase, the current in the series-field winding increases due to the decrease in counter electromotive force. A constant supply of current is maintained in the shunt-field winding, since it is connected directly across the line.

The multiphase a-c motors operate upon the principle of a rotating magnetic field. This rotating magnetic field is produced by multiphase current flowing through two or more groups of coils wound inwardly on the projecting poles of an iron yoke. The coils on each group of poles are wound alternately in the opposite direction. This winding produces opposite polarity in the alternate groups of poles. Each group of poles is connected to a separate phase of a multiphase current. Figure 72 shows how the rotating field is produced. Phase 1 supplies current to the poles A and A', and phase 2 supplies current to the coils on the poles B and B'. When the poles of A and A' are fully magnetized, the poles B and B' are not magnetized. This alternation of magnetic effect causes the rotation of the armature. The coils of the armature are forced around by the varying lines of force in the field.

One of the important types of multiphase a-c motors is the synchronous motor. It consists of a fixed element and a movable element. The fixed element is called the stator, and the movable element is called the rotor. They are different from d-c motors in that the field of the synchronous motor is usually the rotor and the armature is the stator. A multiphase alternating current is passed through the stator windings, producing in them a rotating magnetic field. A direct current is passed through the rotor coils. This sets up a magnetic field about the rotor coils. This motor is so designed that the two fields react upon each other in such a way that the rotor is pulled along by the rotating field of the stator. In this motor, since the a-c is of a fixed frequency, there can be only one speed of operation. This speed is known as the synchronous speed. A synchronous motor will not start itself. It is brought to speed by some other source of power before the direct current is passed through the rotor. It will then synchronize itself with the rotating field, and the load may be applied.

The induction motor, like the synchronous motor, has a stator and a rotor. The rotor of the induction motor is not energized separately. A multiphase alternating current is passed through the stator windings, producing a rotating field which cuts the conductors of the rotor. The winding of the rotor forms a closed circuit, and in the winding is induced an electromotive force causing a flow of current. The reaction between

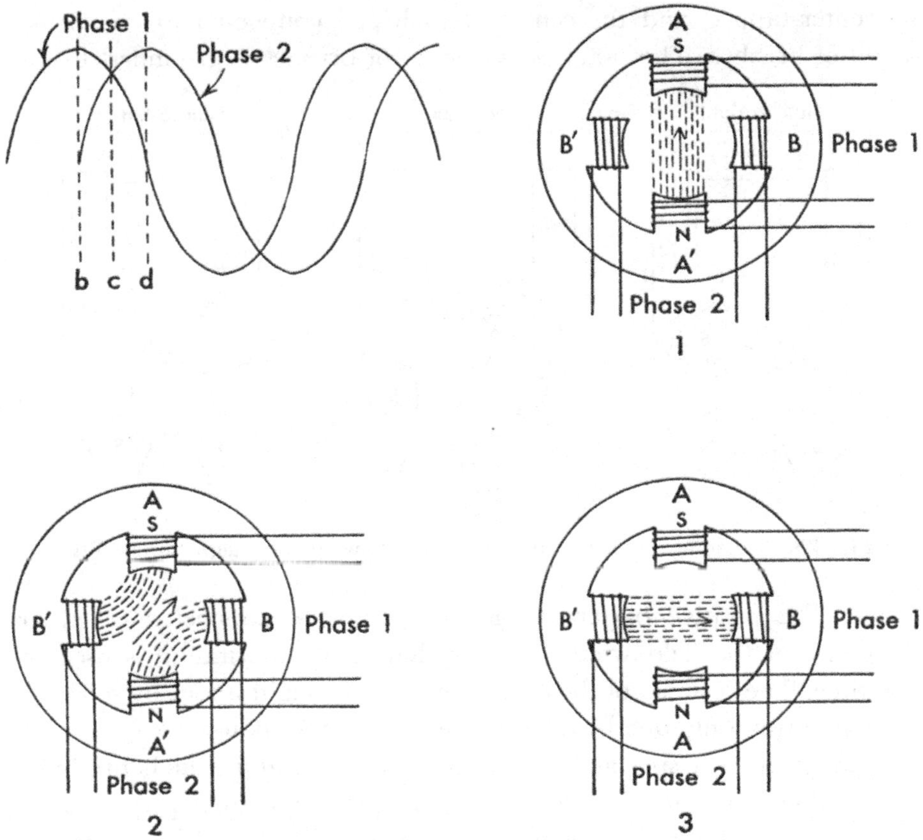

Fig. 72. Drawings to show how a rotating field is produced.

the fields set up in the rotor and the field of the stator causes the motor to turn. The induction type of motor serves a greater variety of uses than any other type of a-c motor.

In aircraft electrical systems, there are a number of places in which reversible motors are used, such as in the operation of controlled surfaces and the electric controllable-pitch propeller. The winding on the armature of these motors is such that they may be caused to rotate in either direction. Such motors are always series-wound, and the reversal of the armature may be brought about either by the reversal of the armature field or by the use of a split field. In reversing the armature field, the direction of the current flowing through either the field winding or the armature is reversed. When this is done, the armature reverses its direction of rotation.

A double-throw switch is usually used to bring about the reversal of current direction. When the split field is used, each field winding may be center-tapped and the center of each coil connected to one brush or set of brushes. The other brush or set of brushes is grounded to the

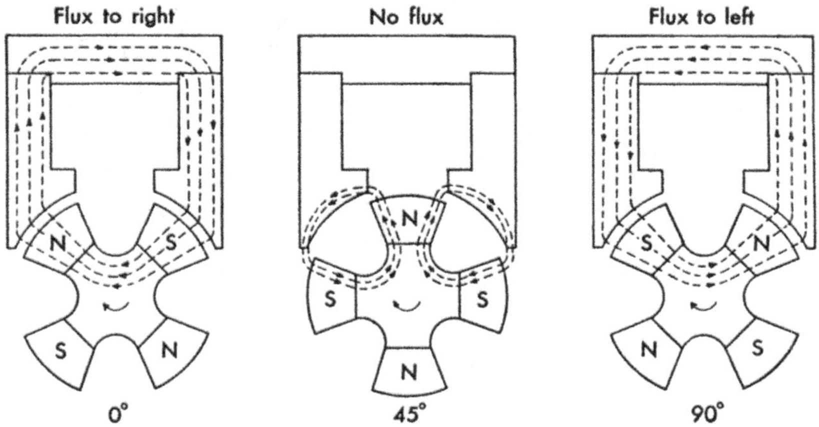

Fig. 73. A diagram of a four-pole magneto showing the magnetic circuits.

frame. The positive battery or generator pole is connected to either extremity of the field windings. Only half of each winding is used at one time. The reversal of the current may be brought about by solenoid switches using only one heavy positive wire to the motor.

A magneto is a special form of electrical generator which produces high voltage current. The electric field in the magneto is produced by means of magnets rather than by means of field coils. In a magneto, the magnets themselves usually rotate, and the coils are stationary. A four-pole permanent magnet is arranged for rotation between the pole pieces of unmagnetized soft iron. The yoke of soft iron is so placed that the pole pieces are opposite magnetically opposed poles of the rotating magnet.

An induction coil formed of a primary and a secondary winding is formed about the soft iron yoke. One end of the primary coil is permanently grounded, and the other end is connected to the breaker system by means of the breaker points. The primary system is shorted when the breaker points are closed. A primary condenser is connected across the breaker points and parallel with them. When the breaker points are open, the secondary current flows to the ground through the primary coil, which has a very low resistance. The other end of the secondary coil is connected to the spark plugs through a distributor.

When the poles of the magnet are opposite the yoke, a current is induced in the primary winding. As the magnets continue to rotate, this current decreases to zero, then builds up in the opposite direction to a maximum, and again falls to zero. There are four positions during each rotation of the magnet in which the field builds up to a maximum and dies down to zero. This varying field induces an alternating current in both the primary and secondary coils. The alternating current induced in the secondary coil is too weak to jump the gaps in the spark plugs. The secondary circuit remains open, and no large amount of current leaves the magneto as a result of the rotation of the magnet.

The breaker points are opened and closed by a cam which is located on the shaft carrying the magnets. The cam is adjusted to open and close the primary circuit at certain positions of the rotating magnets. The breaker points and the primary circuit are closed shortly after each maximum field has been built up. While the circuit is closed, a current is built up in the primary. This surge of current alternates in direction and occurs four times per rotation. If the breaker points are opened, the primary current falls rapidly to zero and, due to the action of the condenser, momentarily flows in the opposite direction. This sudden dying-down of the current in the primary induces a current in the secondary circuit. Due to the sudden dying-down of the magnetic lines of force, a current of high voltage is built up in the secondary circuit, causing a spark either in the spark plug or across a safety gap within the magneto itself. A safety gap is provided within the magneto to prevent shorting within the coils.

A rectifier is an electrical device for producing a direct current from an alternating current supply. The principle of the rectifier is that it permits the flow of electric current through it in one direction only. A rectifier limits or prevents the flow of electric current in the opposite direction.

The copper oxide rectifier makes use of the principle that, when two unlike substances are in contact, an electric current will flow more easily in one direction through them than it will in the other. When an alternating current is applied, more of the current will flow in one direction than it will in the other. Figure 74 shows diagrammatically the construction of a copper oxide rectifier. This device consists of two pressure plates with a copper disc between them on one side of which is a layer of copper oxide. A combination of copper and copper oxide offers a low resistance when the current is flowing from the copper oxide to the copper. This combination, however, offers high resistance to current flowing from the copper to the copper oxide. This copper and copper oxide plate then becomes a "valve" which allows electric current to flow readily in one direction but not in the other. A lead plate is placed next to the copper oxide to maintain good electrical contact. The copper and copper oxide plate and the lead plate are held in close contact between pressure plates. The pressure on the plates should be about 1500 lb. per sq. in.

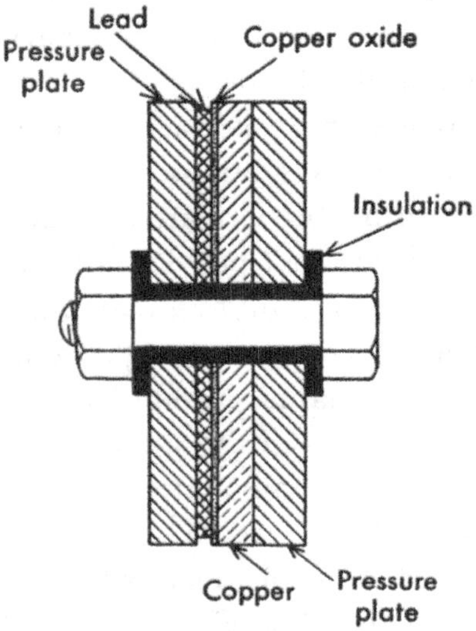

Fig. 74. A drawing to show the construction of a copper oxide rectifier.

After this combination has been used for two or three months, it becomes more efficient. The resistance to the flow of electric current from the copper to the copper oxide is about 25 per cent greater than

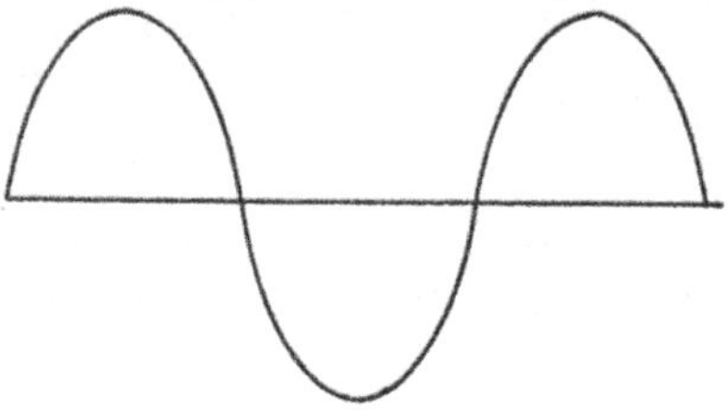

Fig. 75. A diagram to show the current input to a rectifier.

Fig. 76. A diagram to show the current output of a rectifier, half wave.

on a new unit. This increase in resistance is called "aging." The copper oxide type of rectifier is not perfect, but the resistance to the flow of electricity in one direction is approximately 50 times as great as in the other. Figures 75 and 76 show diagrammatically the current put into the rectifier and the current output. One half of the complete wave passes through the rectifier and is in the form of a series of pulses. Several units of this kind can be connected "back to back" to give full-wave rectification. Several units, when connected in parallel, will increase the capacity of the group.

The vacuum-tube rectifier depends upon the action or flow of electrons in a vacuum tube. The simplest form of vacuum tube consists of a filament and a plate. This is called a two-element vacuum tube.

Figure 77 shows a simple form of vacuum-tube rectifier. When heated, the filament in this type of tube gives off electrons in much the same

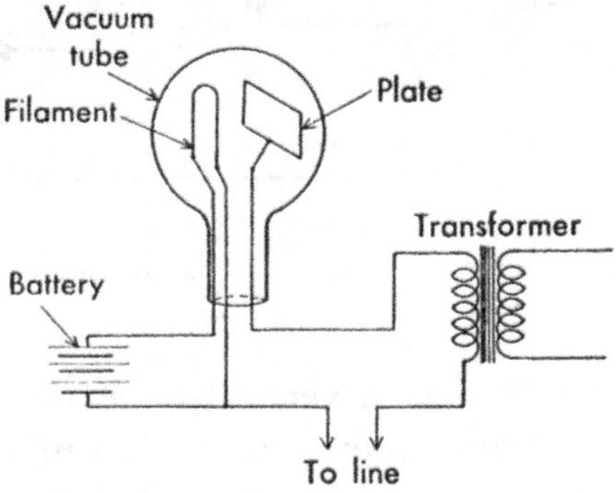

Fig. 77. A diagram to show the construction of a vacuum tube rectifier.

way that steam is given off from boiling water. Electrons are negative particles of electricity. If the rectifier is connected with an electrical system, such as a battery, in such a manner that the plate is positively charged, a stream of electrons will flow from the heated element through the semivacuum in the tube to the plate. If a supply of alternating current is applied to this arrangement, as shown in Figure 75, part of the current will flow from the filament to the plate when the current is flowing in that direction. When the current alternates to the opposite direction, no current will flow because the plate is positively charged and the negatively charged filament has no attraction for electrons. A battery is used to heat the filament and place a positive charge on the plate. A transformer is usually used in this device to regulate the voltage of the current flowing through the device. Since this device permits the flow of current in one direction only, it is used as a rectifier for alternating current. This device is a half-wave vacuum-tube rectifier. The heating of the filament has no part in the rectification of the current, other than furnishing a supply of "free" electrons.

A full-wave vacuum-tube rectifier circuit is shown in Figure 78. The full-wave rectifier makes use of two vacuum tubes. Each tube operates on every other alternation of the a-c supply. The current through the external d-c circuit is, however, always in the same direction. Two plates and two filaments may be placed in a single tube to produce a full-wave rectifier.

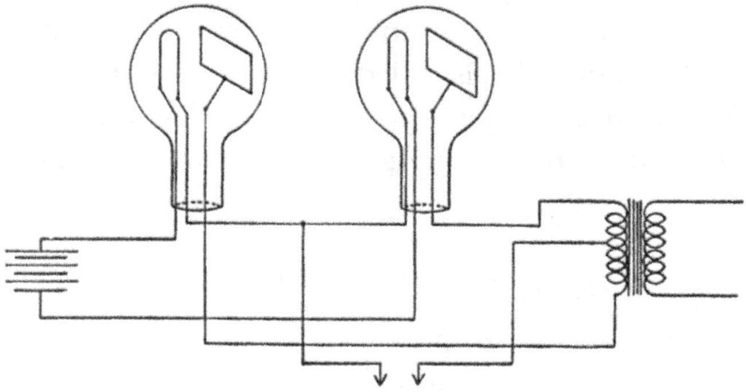

Fig. 78. A drawing to show the construction of a full-wave rectifier.

The mercury-vapor rectifier is very effective and is made by placing a small amount of mercury-vapor in a vacuum tube. This gas becomes ionized, and the movement of the additional electrons from the ionized gas particles produces a greater current flow than that produced by a simple vacuum tube arrangement. The Tungar rectifier is similar to the mercury-vapor rectifier except that, instead of mercury-vapor, argon gas is used in the tube.

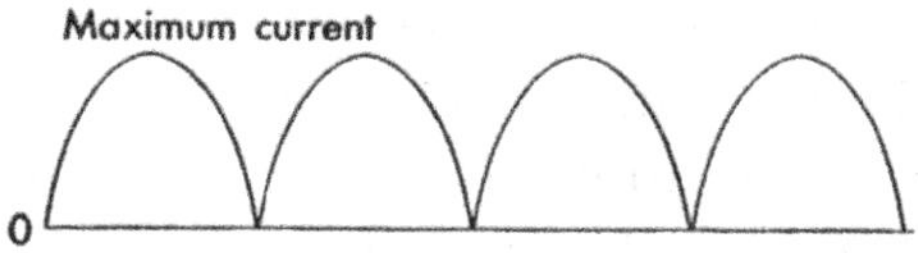

Fig. 79. A diagram to show output of a full-wave rectifier.

The mercury-vapor arc rectifier has a pool of liquid mercury within the vacuum tube. There are three anodes in the tube, one of which is the starting anode. The action of the pool of mercury is similar to the action of a filament in a vacuum. The three anodes act as plates. This rectifier must be started by tipping it so that a stream of liquid mercury contacts the starting anode. The mercury acts as a conductor from the starting anode to the mercury pool. When this conductor of liquid mercury breaks, an arc is formed which vaporizes some of the mercury. From then on, a constant stream of gas is given off by the mercury as long as the rectifier is supplied with current. This stream of gas appears as a hot spot moving around on the surface of the mercury pool. When an alternating current is applied, first one anode and then the other anode is positive and then negative with respect to the pool of mercury. One of the anodes will always be positive while the other is negative. During half the cycle when anode A is positive, electrons will flow from the mercury pool to A. During this time there will be no flow to anode B since it is negatively charged. During the next half cycle when A is negative and B is positive, the stream of electronic flow is from the pool to B. There is, therefore, an alternating flow of electrons from the pool to A and B. This produces a flow of current through the d-c circuit, but always in the same direction. A converter is an electrical machine which changes electrical energy to mechanical energy and then reconverts

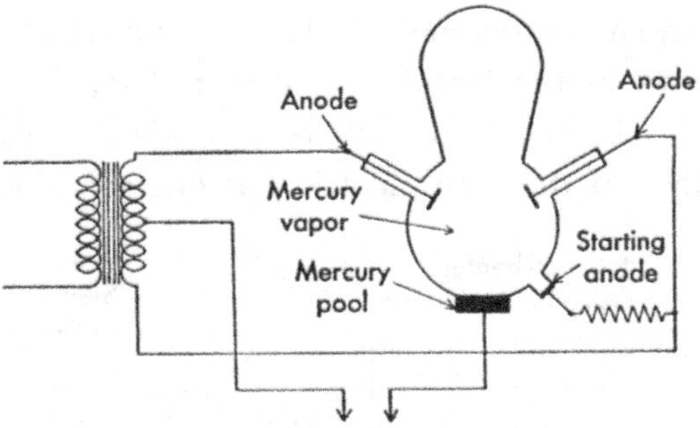

Fig. 80. A drawing to show the construction of a mercury vapor arc rectifier.

the mechanical energy to electrical energy. A machine of this type may change alternating current to mechanical energy and then to direct current. However, some converters change direct current to alternating current. This device is usually made up of a combination of an electric motor and an electric generator.

A motor generator consists of a motor and a generator which are connected by means of a flexible coupling. The motor may be run by either direct current or alternating current. The device may be used to change a low-voltage direct current to a high-voltage alternating current or any combination of currents.

A dynamotor is a d-c machine which has two windings on the armature and two opposite commutators. One winding is for high voltage, and one winding is for low voltage. Usually the low-voltage winding is connected to a storage-battery current which causes the armature to turn as a motor. The high-voltage winding produces a higher voltage than that obtained from the battery. The dynamotor may, of course, be used to change high-voltage current to low-voltage current.

VII ELECTRIC CIRCUITS

An electric circuit is a closed path along which an electric current can flow. The circuit includes, not only the wires, but the source of the electromotive force and any other electrical device through which the current flows. If any point in the circuit is not complete, the electric current cannot flow and the circuit is said to be open or broken. When the circuit is complete so that the electric current can flow, the circuit is

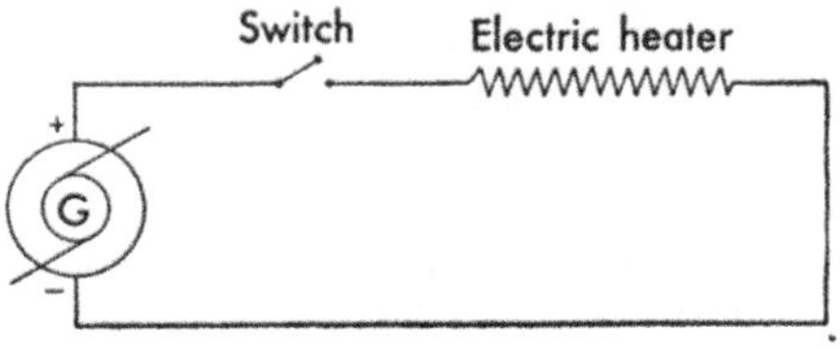

Fig. 81. A resistance load in the form of an electric heater.

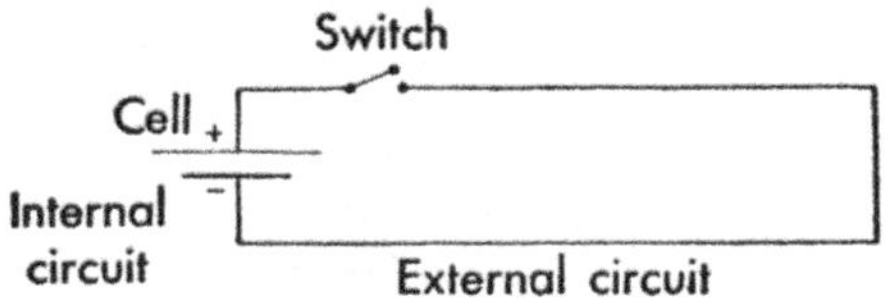

Fig. 82. A simple electric circuit consisting of a cell, a switch, and two pieces of wire.

said to be closed. The part of the circuit connected to the source of current, such as a battery or generator, is called the load circuit.

A resistance load is a circuit, not including the electrical source, in which most of the electric current is changed into heat. An electric heating device connected to a source of current would be a resistance load.

A switch is a device used to open or close a circuit.

An accidental path of low resistance which allows large amounts of electric current to shunt across from one part of the circuit to another, thus overloading it, is called a short circuit. Most circuits are protected from overloading by a fuse. A fuse is usually made of metal which melts at a comparatively low temperature. When a circuit becomes overloaded, the increased flow of current generates heat in the fuse and causes it to melt thus opening the circuit. When a fuse melts or burns out, it is said to "blow." Any short circuit in an electrical system should result in a blown fuse if the circuit is properly protected.

Figure 82 shows a very simple electric circuit. This circuit consists of a single piece of wire connecting the positive and negative terminals of a cell. The current flows from one terminal to the other, and the circuit is completed through the inside of the cell.

Most aircraft electrical systems are designed for 12-v. or 24-v. direct current. This current is usually furnished from a storage battery, and an engine-driven generator is usually installed to supply current to the electric system and to maintain a state of charge in the battery.

It should be noted that the direction of flow in all circuits is conventionally shown to be from positive to negative. It is well to remember, however, that the flow is actually from negative to positive. In all circuits, the instruments or electrical devices are connected as though the current flows from positive to negative.

Every electrical device requires a closed circuit through which the electric current can flow. This closed circuit may consist of two wires or one wire and a ground. For example, as shown in Figure 83, the

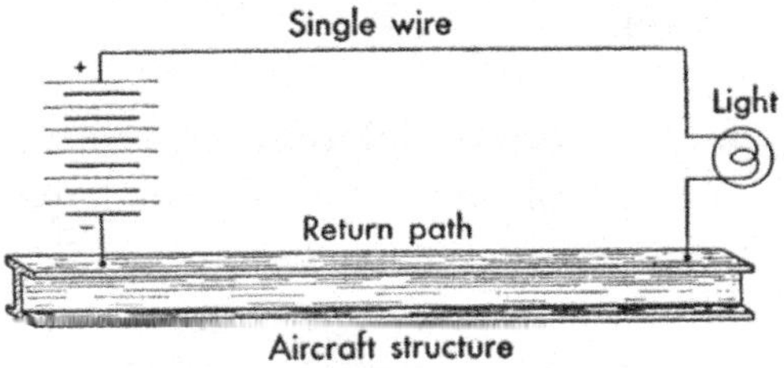

Fig. 83. Part of an airplane structure used as a ground, or return path.

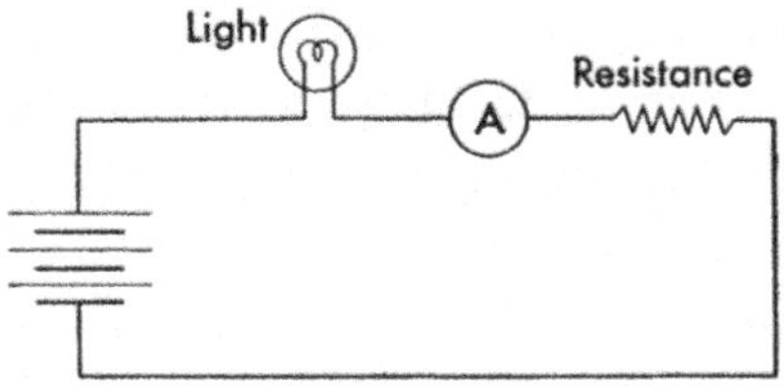

Fig. 84. A typical series circuit.

electric current is flowing from the battery through an electric circuit which is attached to a ground. A ground is a return path to the battery by means of the airplane frame or engine frame or some other conductor. In some electric circuits, the earth itself is used as a return path. The negative terminal of the battery is grounded to the structure of the airplane or engine, thus completing the circuit. When the airplane electrical system uses a ground as a return path, the system is called a "single-wire system."

Circuits are often classified as series circuits and parallel circuits. A series circuit is one in which the total current flowing in the circuit passes through each part of the circuit. Figure 84 shows a typical series circuit. The current flowing from the positive terminal of a battery passes through a lamp, an ammeter, a resistance coil, and back to the negative terminal of the battery. All of the current flowing through the circuit passes through each part of the circuit in series. A parallel circuit, as shown in Figure 85, is a circuit in which the electric current may flow over two or more paths between the terminals of the source of current. In this circuit, the current coming from the positive terminal of the battery is led by separate paths through a voltmeter, an ammeter, a lamp, and a resistance coil. Every circuit through which an electric current flows offers resistance to the flow of the current. The higher the resistance, the less current will flow with any given voltage. It is important in all circuits to know the total resistance of the circuit in order to determine the amount of electricity which will flow over the circuit.

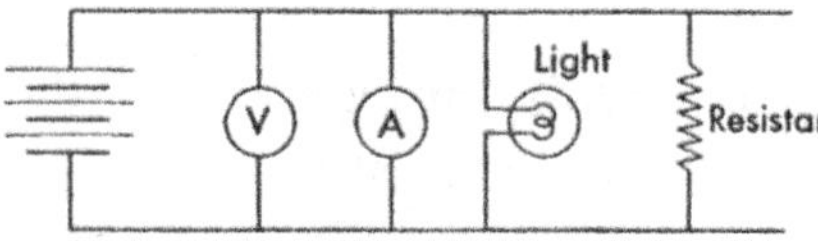

Fig. 85. A parallel circuit.

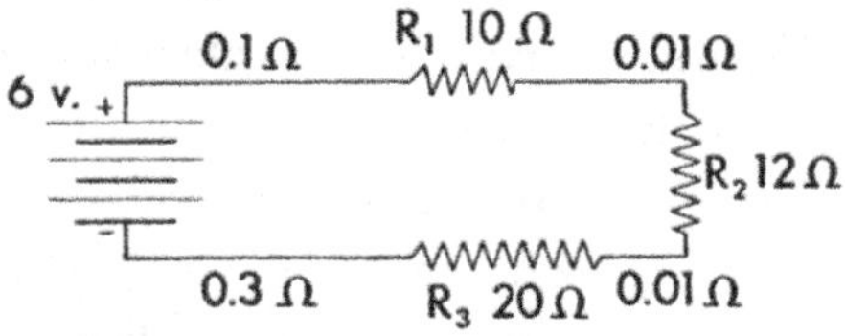

Fig. 86. Resistances in series in an electric circuit.

Figure 86 shows a circuit made up of a series of different resistances. Recall that resistance is measured in terms of ohms. Let R_1 equal the first resistance, R_2 equal the second resistance, and R_3 the third resistance. Then the amount of current which flows through each of these resistances will be I_1, I_2, and I_3. I = current in amperes. Assume that the part of the circuit between the positive terminal of the battery and R_1 is equal to 1/10 ohm. Each of the two segments of the wire between the three resistances has the resistance of 1/100 ohm, and that part of the circuit from R_3 back to the battery has a resistance of 3/10 ohm. If the battery has a voltage of 12, and if R_1 has a resistance of 10 ohms, R_2 has a resistance of 12 ohms, and R_3 has a resistance of 20 ohms, it is possible to figure the amount of current flowing in each part of the circuit as well as the total amount flowing in the circuit. The total resistance for any series circuit is the sum of all the resistances of the circuit. The total resistance for this circuit, then, is 42.42 ohms. The total current flowing through this circuit is in accordance with Ohm's Law which states that $I = \frac{E}{R}$. In this case where the voltage is 6 and the total resistance 42.42, the equation becomes $I = \frac{6}{42.42}$ (or 6 divided by 42.42). Therefore, it is found that approximately 0.14 amp.

of current will flow in this circuit. The current flowing in any part of the circuit can be found in the same manner.

It will be found that there is a drop in voltage across each part of the circuit or across any piece of electrical equipment using current in the circuit. Since the 0.14 amp. which flows in the circuit must pass through each part of the circuit, the current resistance drop may be found for each part of the circuit. Current resistance drop is represented by the symbol IR. This symbol is also used to indicate voltage drops. The freedom with which an electric current flows through a circuit depends upon the resistance. In connecting the various parts of the circuit, each connection adds resistance to the circuit.

If the contacts are loose, dirty, or are separated by thin layers of the oxides of the metals, considerable resistance is placed in the circuit. In making a connection, the surfaces which are to be in contact should be scraped or polished until the pure metal itself is exposed. Wherever possible, joints in an electric circuit should be soldered with a low-resistance solder.

In parallel circuits, such as shown in Figure 87, the three resistances are in parallel instead of in series. The current has three paths over

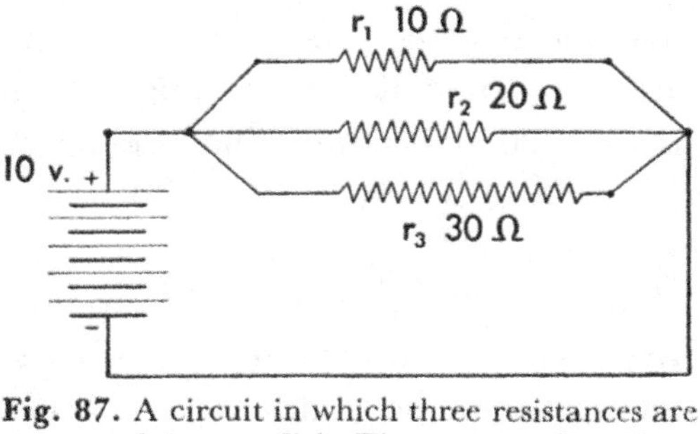

Fig. 87. A circuit in which three resistances are connected in parallel. The current has three paths over which it may flow.

which it may flow from one terminal of the battery to the other. Resistances in parallel, instead of increasing the resistance in the whole circuit as do resistances in series, decrease the resistance in the circuit. The total current in the circuit is equal to the sum of the currents flowing through each of the individual branches of the circuit. The resistance of the circuit is equal to the voltage divided by the total current.

If the first resistance, shown in Figure 87, which is equal to 10 ohms is called r_1 and the second resistance of 20 ohms is called r_2, and the third resistance of 30 ohms is called r_3, the total resistance to be found is called R. A formula may be written to find the resistance of this kind of circuit.

$$\frac{1}{R} = \frac{1}{r_1} + \frac{1}{r_2} + \frac{1}{r_3}$$

If we substitute the known resistances in this formula, it becomes

$$\frac{1}{R} = \frac{1}{10} + \frac{1}{20} + \frac{1}{30}$$

By adding the fractions together, it is found that

$$\frac{1}{R} = \frac{11}{60}$$

By cross multiplication, it is found that

$$11R = 60 \quad \text{or} \quad R = \tfrac{60}{11} \quad \text{or} \quad 5\tfrac{5}{11}$$

The total resistance of this circuit is $5\frac{5}{11}$ ohms.

There are a number of laws governing parallel circuits. First, the voltage is the same in all parts of a parallel circuit. Second, the sum of the currents through the various parts of the circuit is equal to the total current. Third, the total resistance of a parallel combination is equal to the voltage divided by the total current.

If, in the problem illustrated in Figure 87, the voltage is 10, the current which flows through the first resistance would be equal to 10 ÷ 10 or 1 amp. The amount of current flowing through the second resistance would be equal to 10 ÷ 20 or ½ amp. The amount of current flowing through the third resistance would be 10 ÷ 30 or ⅓ amp. The total current would be $1 + \frac{1}{2} + \frac{1}{3} = \frac{11}{6}$. The voltage, 10, divided by the total current $\frac{11}{6}$ shows a total resistance of $5\frac{5}{11}$ ohms.

Cells and batteries may be connected in various ways to obtain almost any required voltage or amperage, although the current and voltage from a cell or battery cannot be varied. The maximum voltage of a cell depends upon the materials from which the cell is made. While the maximum voltage for any type of cell is fixed, the current may be less than the maximum rate because of internal resistance or polarization. The capacity of a cell is the amount of current which can be obtained from the cell for a given length of time. This depends upon the size and other features of the battery.

The rating of a battery is usually given in ampere-hours. Two or more cells connected together form a battery. Common usage, however, gives the name, battery, to any electric cell or combination of cells. When cells are connected in series, the total voltage of the series is equal to the sum of the voltages of the individual cells. If 6 cells, each having a voltage of 1½ v., are connected in series, the total voltage is 9 v. This total voltage, however, is obtained only when no current is flowing or if the current flowing is very small. If the voltage of a series is tested when no current is flowing, the open-circuit voltage is found. As soon as any considerable amount of current starts to flow, there will be a voltage drop due to the internal resistance of the cells themselves.

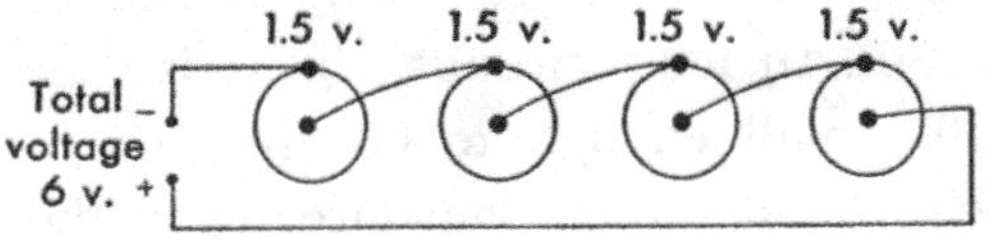

Fig. 88. Cells connected in series. Voltage equals sum of the cells.

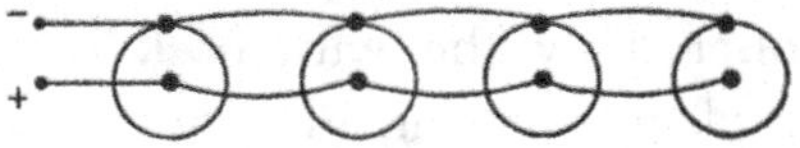

Fig. 89. Cells connected in parallel. Amperage equals sum of amperages of the cells.

The ampere-hour capacity of a battery made up of cells in series is the same as that of a single cell. This is true because the same amount of current flows through all of the cells. The whole battery wears out at the same rate as a single cell. All of the cells in the series are supplying the same amount of current. The advantage of connecting cells in series is to increase the voltage.

When cells are connected in parallel, the voltage of the battery is equal to the voltage of any individual cell. The ampere-hour capacity, however, is increased. The amount of current from the cells connected in parallel is equal to the sum of the currents from the individual cells. The advantage of this type of connection is to increase the amount of current, the voltage remaining the same. The ampere-hour capacity of cells connected in parallel, Figure 89, is equal to the sum of the ampere-hours of the individual cells.

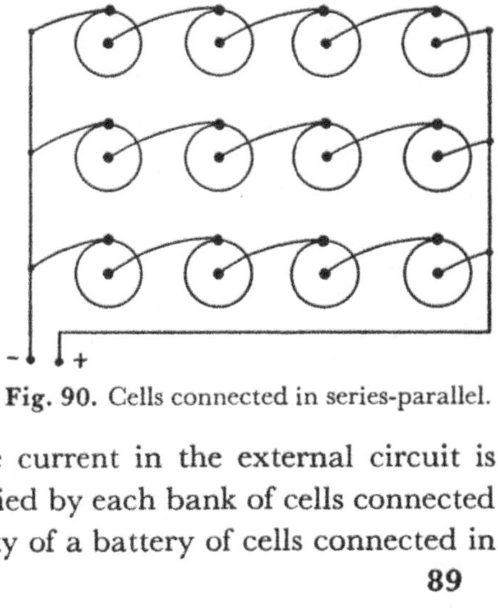

Fig. 90. Cells connected in series-parallel.

Cells may be connected in series-parallel as shown in Figure 90. The voltage obtained from a battery made up of cells connected in this manner is equal to the sum of the voltages of the cells connected in series in any one bank of cells. The current in the external circuit is equal to the sum of the current supplied by each bank of cells connected in parallel. The ampere-hour capacity of a battery of cells connected in series-parallel is equal to the capacity of one cell multiplied by the number of banks of cells connected in parallel.

In any electric system, the wires carrying the current from its source to the various electrical devices should be of such size that they offer very little resistance to the flow of current. When a wire is small enough to offer any considerable resistance, it tends to become heated. The amount of current a conductor will carry usually varies inversely with its temperature. The hotter the wire becomes, the greater the resistance it offers to the flow of electric current. In most circuits, the resistance offered by the wire itself is usually small enough to be ignored. In replacing wiring in a system, a wire smaller than that already in the system should never be used.

VIII ELECTRICAL DEVICES

Most electrical instruments and devices are operated by the effects of the electric current rather than by the current itself. An electric light gives off light because of the heating effect of the current. An induction coil produces a spark in the cylinder of an internal-combustion engine because an induced current has been set up by a magnetic field. Motors are turned by the effect of a magnetic field produced by an electric current. Voltmeters do not actually measure the pressure of the current flowing, but measure the magnetic effect caused by the current. This is true of most electrical instruments.

If a wire carrying a direct current is held in a north and south direction over a compass needle, the north end of the needle is deflected toward the west when the current is flowing from south to north. If the wire is placed under the compass needle from north to south and the current flows from south to north, the north end of the needle will now deflect toward the east.

D'Arsonval Galvanometer. An electric current is measured by measuring the effects of the current. There are a number of different effects which may be measured to indicate the amount of current flowing through a conductor. The heat generated by the current flowing may be measured. The electromagnetic effects of an electric current may be measured. Exact measurements of electric current may be made by the chemical action brought about when the current is passed through an electrolyte.

The D'Arsonval galvanometer is a device used to detect electric currents. The simplest form of this galvanometer consists of a coil of wire suspended between the two poles of a permanent magnet. The coil is free to turn about an axis which is perpendicular to the lines of force between the poles as shown in Figure 91. When there is no current flowing, the coil is held with its plane parallel to the lines of force in the magnetic field. The coil of wire is suspended under spring tension which returns it to its neutral position when no current is flowing in it. When a current of electricity is flowing through the coil, it attempts to turn until its plane is at right angles to the lines of force in the magnetic field, but the spring mounting prevents the free turning of the coil. The amount of current flowing determines the amount of turning against the spring pressure that the coil is able to accomplish. The greater the

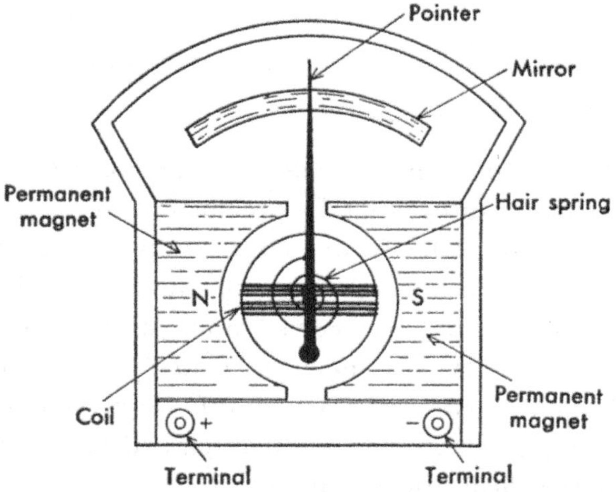

Fig. 91. Diagrammatic drawing of a galvanometer.

amount of current, the farther the coil will turn. When the current passes through the coil in one direction, the direction of turning is opposite to that in which it will turn when the current passes through the coil in the other direction. Some of these instruments are so arranged that, not only the amount of current is determined, but also the direction of the current. The needle is deflected to the right for a current in one direction and to the left when the current is reversed. Most galvanometers do not have graduations to indicate the amount of current flowing. They are used only to detect the flow of electric current. When the needle is arranged to move over a calibrated scale, this instrument may be used to measure volts or amperes.

Voltmeter and Ammeter. If a coil of wire is suspended between the poles of a magnet and a steady current of electricity is passed through the coil, the coil tends to arrange itself in such a manner that the largest number of magnetic lines of force can pass through the center of the coil, rotating itself to such a position that the plane of the coil is perpendicular to the lines of force. This movement is known as the Weston movement. The force with which the coil tends to rotate itself depends upon the strength of the field and the amount of current flowing through the wire in the coil.

Voltmeters and ammeters make use of this principle. A coil made up of a large number of turns of fine wire is suspended between the poles of a permanent magnet in such a manner that, when no current is flowing in the coil, the plane of the coil is parallel to the lines of force between the two poles of the magnet. The coil is so arranged that it is free to rotate but is held in position by means of a spring which resists the rotation of the coil. The rotating coil carries the indicating needle by which the voltage or amperage is indicated, and the movement of the coil is in direct proportion to the amount of current flowing. The force necessary to stretch the spring is in proportion to the amount of current flowing in the coil. As the amount of current increases, the indicator moves over the graduated scale which may be calibrated in

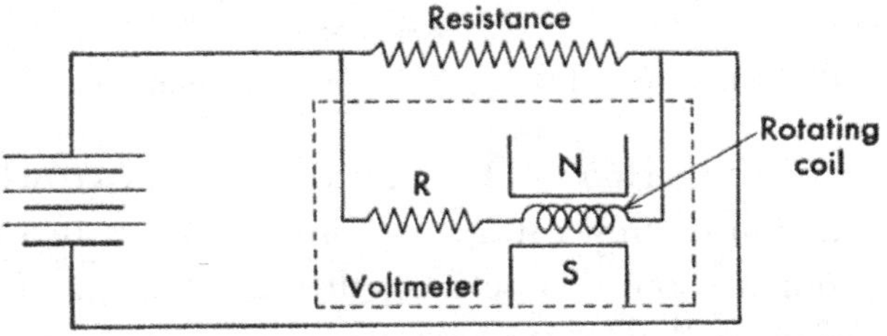

Fig. 92. A typical wiring diagram of a voltmeter.

either volts or amperes. In a voltmeter, it is not necessary that all of the current be led through the rotating coil, because the voltage of any part of the current is always equal to that in the main wire.

Voltage corresponds to pressure. A small pipe led off a water main carrying a pressure of 100 lb. per sq. in. will also show a pressure of 100 lb. per sq. in. The friction in the pipes, of course, must be ignored. In testing the voltage of a line, it is necessary to shunt only a small quantity of the current through the coil in the voltmeter. The shunt is connected in series with a high resistance to prevent injury to the instrument.

In measuring amperes, a different arrangement is used. Since amperes indicate the rate of flow, it is usually necessary that all of the current pass through the instrument. Many instruments, however, have an arrangement whereby only a small part of the total current flow passes through the coil. For example, the main current may be carried by a strip of metal, while a small part of the current, perhaps 0.001 of the current, flows through the coil. In other words, a known fraction of the current passes through the coil, and the instrument is calibrated so that the correct current flow is indicated. When the voltmeter or ammeter is used to measure direct current, care should be taken to connect the proper terminals to the plus and minus poles. If the opposite terminals are connected, the coil will rotate in the opposite direction.

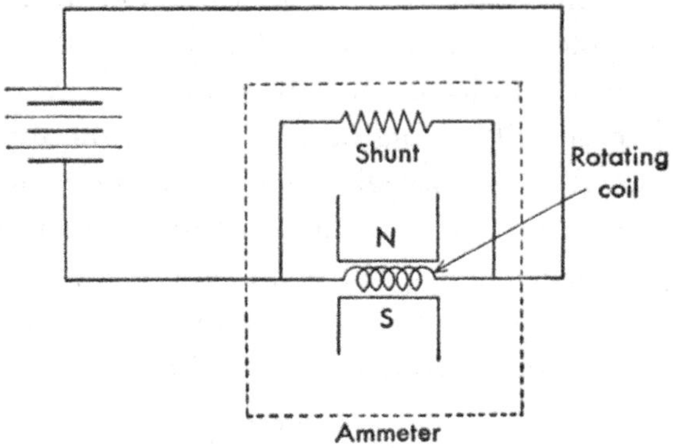

Fig. 93. A typical wiring diagram of an ammeter.

Thermocouple. The thermocouple is an electric device which is generally used to measure temperature. When the ends of two wires of different composition are joined closely together and this junction is heated, an electric current will flow through the wires if they are part of an electric circuit. The amount of voltage developed, as well as the amount of current, depends upon the two metals chosen and the temperature applied to the junction. The point where the heat is applied is called the "hot junction," and where the other ends of the wires complete the electric circuit is called the "cold junction." Both the voltage and amount of current flowing are very small, but they are great enough to obtain accurate indications upon an instrument such as a galvanometer. Almost any combination of dissimilar metals or alloys will give the thermocouple effect. One wire or lead of the thermocouple is usually of

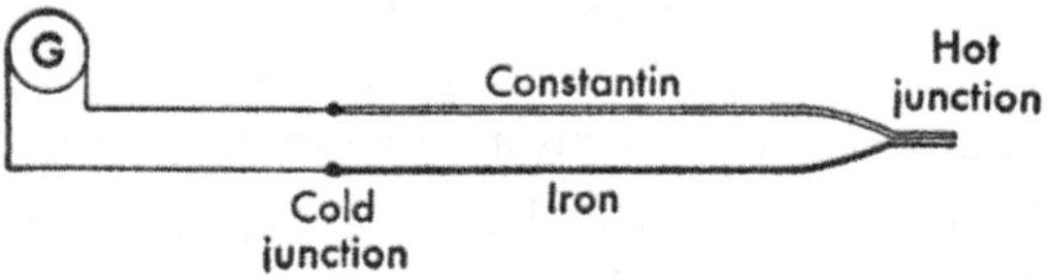

Fig. 94. A diagram to show the construction of a thermocouple.

iron, and the other an alloy of copper and nickel called "constantin." These two metals are selected because they develop a definite electromotive force per degree change in temperature. The cold junction must have a compensating device to cancel out the effect of temperature changes of the instrument itself, and changes in the temperature of the air surrounding the instrument must also be compensated for.

Measurement of Resistance. It is necessary, at times, to measure the resistance of a piece of electrical equipment or of an electric circuit. The resistance of the circuit influences the amount of current which will flow in the circuit. Faults in a circuit may often be discovered by checking the resistance in various parts of the circuit. There is no simple way of measuring resistance as there is of measuring voltage and amperage.

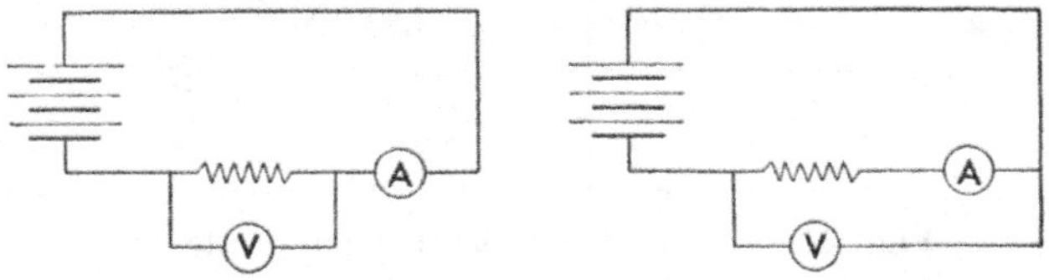

Fig. 95. A diagram to show how a voltmeter and an ammeter may be used to measure resistance.

Although there are methods which may be used to indicate resistance directly on the scale of the measuring instrument, most methods make use of the relationship between volts and amperes expressed by Ohm's Law. A simple and practical method of measuring resistance is by the use of an ammeter and a voltmeter. The ammeter and voltmeter are inserted in the circuit, and the resistance of the circuit without the resistance which is to be measured is found by applying Ohm's Law: Resistance equals amperes times volts. By inserting the unknown resistance into the circuit and again checking the readings of the voltmeter and ammeter, the resistance for the entire circuit may be determined. The difference between the two results gives a close approximation of the resistance of the unknown part.

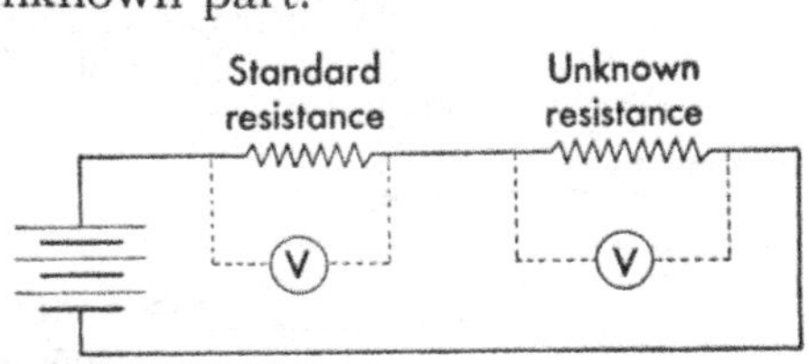

Fig. 96. A circuit for voltmeter and standard resistance method for measuring resistance.

A more common method of measuring resistance is by making use of a known or standard resistance. See Figure 96. In this method two readings are taken with the voltmeter: one with the unknown resistance out of the circuit, and the other with the unknown resistance in series with the voltmeter. No matter how a voltmeter is connected in a circuit, its reading is always the indication of the voltage across its terminals. The voltage across the terminals is the voltage across the internal resistance of the voltmeter. This reading is affected by any resistance which may be added in series with it. The current in the circuit is determined by dividing the total resistance of the circuit into the voltage of the circuit. The two voltage drops are proportional to the unknown resistance and the resistance of the voltmeter, since they are in series.

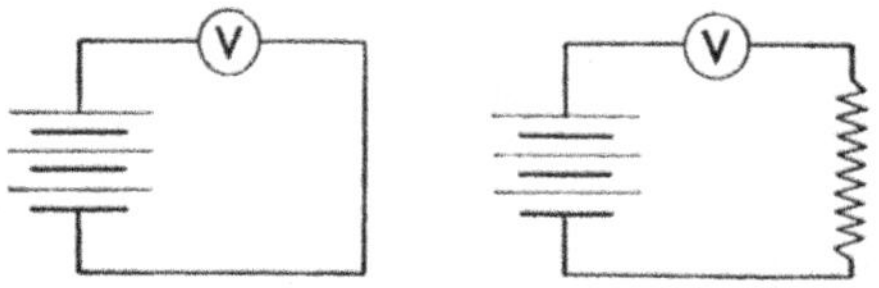

Fig. 97. A circuit for the known resistance voltmeter method for measuring resistance.

The common ohmmeter that is used in the field is of the known-resistance, voltmeter type. The battery and the meter are contained in one case. The resistance does not have to be figured as it is already set up on the calibrated scale in ohms. The instrument is usually made up of a volt-ohm meter and is called a voltohmmeter. This means that it may be used for the measurement of either resistance or voltage, and the scale is often calibrated for both ohms and volts. It is necessary, however, to have a switch to shift from one to the other.

Wattmeter. The wattmeter is an instrument used to measure the power being used in a circuit. The amount of power depends, not only upon

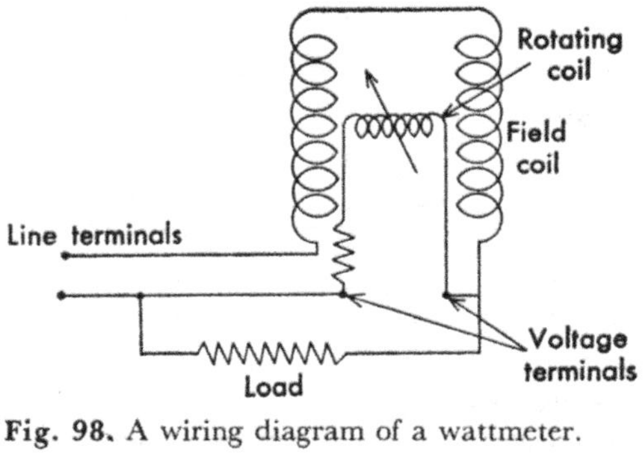

Fig. 98. A wiring diagram of a wattmeter.

the volts, but also upon the number of amperes flowing. The wattmeter is therefore a combined voltmeter and ammeter. It has two fixed coils in series which provide a magnetic field, and there is a movable coil which turns in this field. All of the current flowing passes through the fixed coils. These coils are the current-measuring element of the wattmeter. The movable coil is connected in parallel with the load in the circuit. This movable coil is the voltmeter element of the instrument. The amount of deflection of the indicator needle attached to the movable coil varies with the amount of current flowing. The force acting on the coil is the result of the magnetic fields of both the fixed and movable coils. The amount the coil is deflected is proportional to the voltage times the current. This instrument should not be confused with the electric meter, such as is used in a house circuit, which measures the watt-hours and kilowatt-hours which have been used. The wattmeter simply shows the rate at which the watts are being used.

Wheatstone Bridge. The Wheatstone bridge is a device used for remote indications of temperature on aircraft instruments such as the exhaust-gas analyzer. For this purpose, the Wheatstone bridge and a sensitive

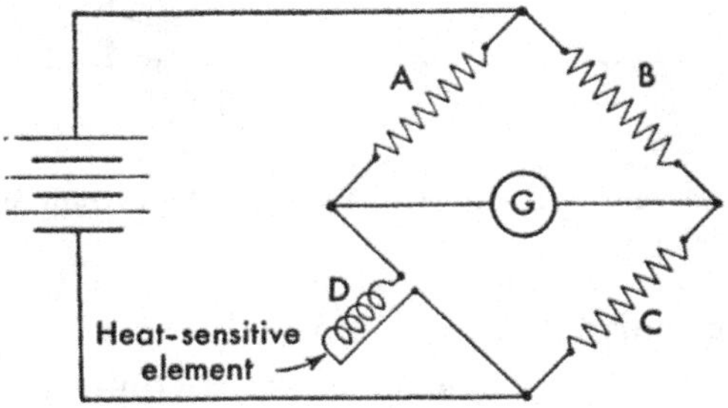

Fig. 99. A wiring diagram for a Wheatstone bridge.

galvanometer are used. The galvanometer is a simple voltmeter used to indicate variations in current or voltage. The Wheatstone bridge shown in Figure 99 consists of four resistances, A, B, C, and D. The resistances are connected, as shown in Figure 99. Three of the resistances, A, B, and C, are standard resistances having, for example, an exact resistance of 150 ohms, and the fourth resistance is the sensitive element, in this case, a coil having a resistance of exactly 150 ohms at 0° C. The resistance of this coil changes with variations in temperature. When the three coils and the sensitive element are at a temperature of 0° C., no current flows through the galvanometer connected across the bridge. If the temperature of the sensitive element rises, the resistance increases, and current will flow through the galvanometer, which is calibrated in terms of degrees, to show an increase in temperature. If the temperature of the sensitive element decreases, the resistance decreases and current will flow through the galvanometer in the opposite direction, causing the indicating needle to show a decrease in temperature.

Potentiometer (Voltage Divider). The potentiometer or voltage divider is used principally in the aircraft electrical system to indicate the fuel level. The principle upon which it operates is shown in Figure 100. This instrument consists of a resistance unit over which a sliding contact moves. The sliding contact and one end of the resistance unit are connected to a system of current supply. A switch is inserted in this circuit. When the sliding contact moves back and forth over the resistance unit,

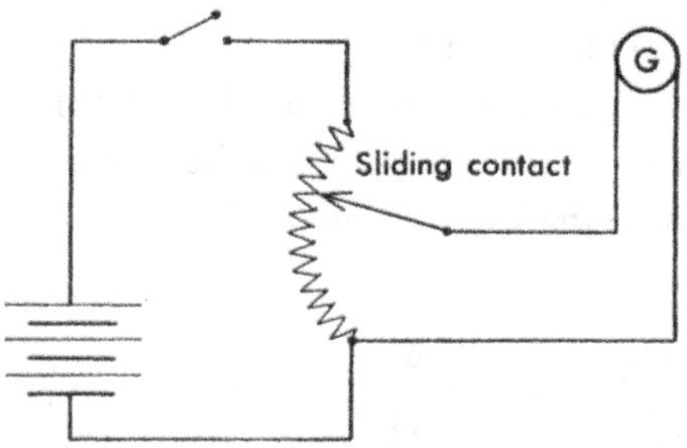

Fig. 100. A wiring diagram of a potentiometer or voltage divider.

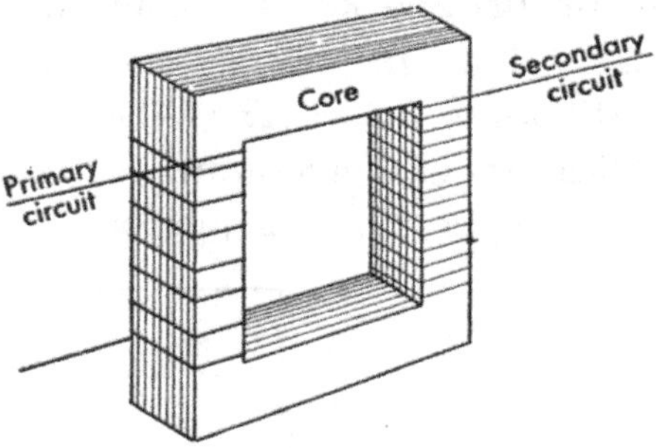

Fig. 101. A diagram of a transformer

the voltage of the current passing through the voltmeter varies. The sliding contact is moved by a lever element connected with a float in the fuel tank. The needle in the voltmeter varies with the movement of the sliding contact and indicates the amount of fuel in the tank.

Transformer. Because of the fluctuating character of an alternating current, it is possible to transfer the current or power from one alternating current circuit to another. This is done by means of mutual induction. A direct current cannot be transferred to another circuit by induction because the value of the current is constant. It is, however, possible to get a similar effect from a pulsating direct current. The transformer is an electrical device for transferring electric power from one circuit to another by means of mutual induction. A transformer is made up of a core of steel alloy or iron which provides a path of low reluctance for magnetic lines of force. A transformer also has two windings, one of which is called the primary winding and the other the secondary winding.

The primary winding consists of a number of turns of wire wound about the core or a part of the core. These turns must be insulated from each other. The primary winding carries the original alternating current.

The secondary winding is similar to the primary winding, but receives its current by means of induction. Figure 101 shows the arrangement of the windings about the core. When an alternating current passes through the primary winding, a fluctuating magnetic field is set up about the coil. This field expands and contracts with each fluctuation of the alternating current. As the expanding lines of force in this field cut across the secondary winding, a current is induced in that winding which is opposite in direction to the current in the primary winding. As the field collapses, it again cuts the secondary winding and again induces current in this winding. Because of this action a fluctuating current is set up in the secondary winding which corresponds to the current in the primary winding. There is some loss of current between that applied to the primary winding and the current produced in the secondary winding. However, this loss is small enough not to interfere with the practical use of the transformer.

A transformer may not only be used to transfer electric power from one circuit to another, but also to change the voltage of the current. The voltage in the two coils is proportional to the number of turns in each coil in the transformer. For example, if there are 100 turns in the primary coil and 1000 turns in the secondary coil, the voltage in the secondary coil will be 10 times that of the voltage in the primary coil; and if a 110-v. alternating current is applied to the primary coil, a current of 1100 v. is produced in the secondary coil. If this action is reversed and the 110-v. current is applied to the coil having the most turns, this coil becomes the primary coil and a current of 11 v. will be produced in the other coil, which now becomes the secondary coil.

If a transformer is used to increase the voltage produced, it is called a "step-up" transformer. When it is used to reduce the voltage, it is called a "step-down" transformer. The transformer does not generate power. It simply transfers power from one circuit to another. When the voltage is stepped up in the secondary coil, the number of amperes flowing in the secondary coil, compared with the number of amperes flowing in the primary coil, is reduced. For example, in a one-to-two step-up transformer, a current of 10 amp. at 110 v. applied to the primary winding will produce in the secondary winding a current of 220 v., but the amperes flowing will be reduced to approximately 5. As the voltage is raised or lowered by means of a transformer, the amperage is raised or lowered in inverse proportion. The volts in one winding multiplied by the amperes flowing in that winding must always equal the volts in the other coil times the amperes flowing in that coil. Power losses are ignored in this statement. Most transformers have a power loss of considerably less than 10 per cent.

Induction Coil. An induction coil is very similar to a transformer. It is made up of two coils of wire: The first one, carrying the primary current, is usually of heavy wire which carries a current of low voltage. The secondary coil is made up of many turns of fine wire and carries the secondary current which is of high voltage.

If an alternating or pulsating current is passed through the primary circuit, a secondary alternating current is set up in the secondary circuit. These secondary alternations flow in the opposite direction to the alternations of the current of the primary circuit.

The voltage in the secondary circuit is in direct proportion to the ratio of the number of turns of wire in the primary and secondary coils. If compared to that in the primary circuit, it is in the same proportion as the number of turns of wire in the two circuits. For example, if there are 10 turns of wire in the primary circuit and 1000 turns of wire in the secondary circuit, the voltage in the secondary circuit will be 100 times as great as the voltage in the primary circuit.

An iron core is usually placed inside the two coils to increase the strength of the field. This type of coil is actually a transformer and may be used to transform current from a low voltage to a high voltage or from high voltage to a lower voltage.

The induction coil used in ignition systems is similar to the induction coil described above. In the ignition system, a current from either a magneto or a storage battery flows through the primary circuit. In order to give the same effect as an alternating current, a direct current is interrupted many times a second by an arrangement called a vibrator. This cutting-off and building-up of the current causes the alternate build-up and collapse of the electric field surrounding the primary coil. This build-up and collapse cause the lines of force to cut the wires in the secondary coil, inducing in it an alternating high-voltage current. The breaker points act similarly to a vibrator.

If an ordinary 6-v. storage battery is used, such as that in an automobile, the voltage of the current in the primary circuit is 6 v., while the voltage of the current usually produced by the secondary circuit of an ignition system is approximately 20,000 v. The amperage of the primary current may be 5 or 6 amp., and that in the secondary coil will be only a few thousandths of one ampere or a few milliamperes. A milliampere is a thousandth of an ampere.

Capacitance. Every electric circuit has capacitance in addition to resistance and inductance. Capacitance means the ability to store a quantity of electricity. A capacitor is a device designed to store a quantity of electricity and therefore it has capacitance. The unit of capacitance or storage ability is called a farad. The farad is the capacity to store a coulomb of electricity. A coulomb is the amount of electricity which passes a point in a conductor in one second when a current of one ampere is flowing. A capacitor which would store one farad of electricity would be much too large for any practical use. The microfarad, which is equal to one-millionth of a farad, and micromicrofarad, which is equal to one-millionth of a microfarad, are the units more commonly used to measure capacitance in practical electricity.

When two metal plates or other conductors are placed close together but separated by a nonconductor, a capacitor is formed. The insulation material is called the dielectric. This dielectric may be air, glass, mica, porcelain, paper or other similar material. Condensers used in ignition systems and radios are capacitors. The variable condenser used to tune a radio is a condenser or capacitor made up of a number of metal plates separated from each other by a thin layer of air. The alternate plates are connected to separate conductors. These plates are so arranged that one set may be turned out from between the other set, reducing the capacitance or the ability of the capacitor to hold electricity. Thus the capacity of the condenser may be varied. The condenser used in ignition systems or telephones is usually made up of two long narrow strips of tinfoil separated by a thin layer of waxed paper. The strips of tinfoil form the plates and each strip is connected to a separate conductor.

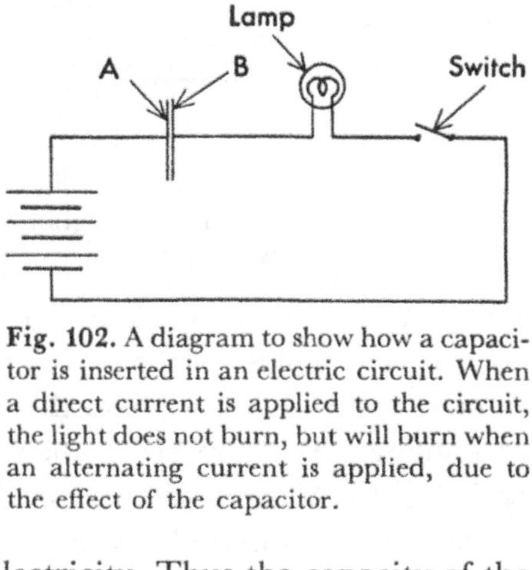

Fig. 102. A diagram to show how a capacitor is inserted in an electric circuit. When a direct current is applied to the circuit, the light does not burn, but will burn when an alternating current is applied, due to the effect of the capacitor.

Figure 102 shows a simple capacitor, consisting of two plates, inserted in an electric circuit. In this circuit a light and a switch are also inserted. A battery furnishes the electric current. When the switch is closed the circuit is closed. Although no current can flow through the capacitor, the capacitor accumulates a considerable quantity of electricity. The capacitor seems to act somewhat like two small rubber balloons partly filled with air and pressed together. If air is blown into tubes connecting the two balloons, air may be forced into the balloons but cannot flow through them from one to the other. As the air pressure is released, part of the air flows from the balloons back into the tubes. A capacitor seems to work in somewhat the same manner.

In Figure 102, the plates A and B will be at zero potential if the switch has been open for some time. When the switch is closed plate A takes on a considerable charge of positive particles of electricity, while plate B takes on a corresponding charge of negative electricity. This is due to the difference in potential or electric pressure between the terminals of the battery. If the switch is opened the capacitor retains the quantity of

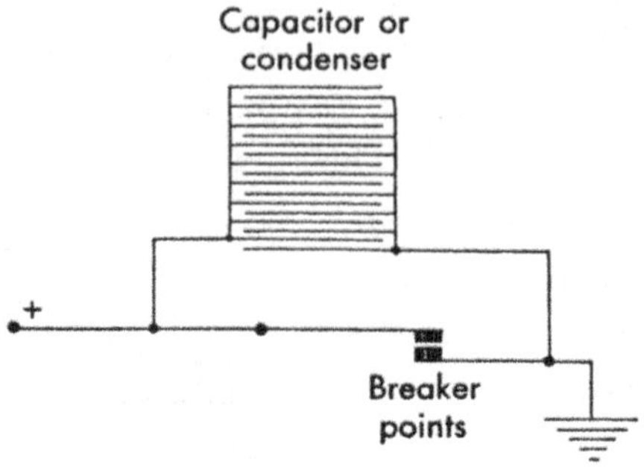

Fig. 103. A diagram showing how a condenser is connected in parallel with the breaker points.

electricity on the plates. The charge on each plate attracts the charge on the other plate. This electricity, however, gradually leaks away through the dielectric or into the air until the plates are again at zero potential as they were before the switch was closed. When the switch is first closed a maximum charge of electricity is forced into the capacitor by the sudden surge of current. If the switch remains closed, part of this charge flows back into the circuit from the capacitor.

If instead of being connected in series with the switch, the capacitor is connected in parallel with it, a considerable charge of electricity is forced into the capacitor when the switch is opened if a current is flowing in the circuit. If the switch remains open, part of this charge flows back into the circuit almost instantly. This is the function of a condenser in the primary circuit of an ignition system. The condenser in an ignition system is connected in parallel across the breaker points and receives surges of electric current each time the points are opened. This relief of electrical pressure in the circuit prevents arcing across the points which would cause burning of the points and weakening of the spark delivered to the spark plugs.

When an alternating current is applied to a circuit such as that shown in Figure 102, the light may be made to burn even though no current flows through the capacitor or condenser. The alternate surges of current force a quantity of electricity into the condenser, and then withdraw it. These alternate surges passing through the lamp cause it to light in the same manner as an ordinary alternating current. The lamp will not light if a direct current is applied to the circuit.

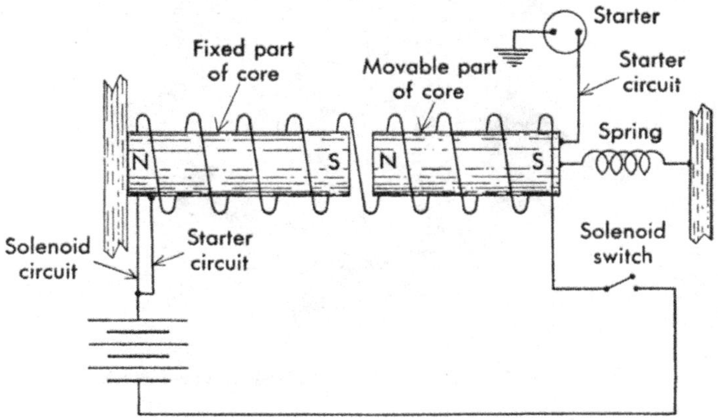

Fig. 104. A diagram of a solenoid switch and starter circuit.

Solenoid. A solenoid switch has many uses in aircraft electrical systems because of its ability to operate as a remote-control device. A solenoid switch consists of a solenoid coil in which have been inserted two soft iron cores, as shown in Figure 104. One core is fastened firmly in place in the coil, while the other is free to move. When the coil is energized by the passing of an electric current through the coil, the two cores of iron become strong electromagnets. No matter which direction the current flows through the coil, the iron cores become electromagnets having unlike poles toward each other. For example the inside end of one core becomes a strong north pole, and the inside end of the other pole becomes a strong south pole. This attraction pulls the cores together. A spring is inserted in such a manner that it pulls one core away from the other when no current is flowing through the coil. This sliding action of the one core may be used to close a switch, engage starters, or operate signals or warning devices. Only a small amount of current is necessary to energize the solenoid.

Relay. A relay is another device used for remote control. The relay consists of an electromagnet made up of a soft iron core about which is a coil made of many turns of wire. The core is fastened firmly in the coil. Placed close to one end of the core is a piece of soft iron plate. This plate is mounted on a pivot in such a manner that, when the coil is energized and the core becomes an electromagnet, the plate is attracted to it. The soft iron plate is attached to a device which closes an electric

Fig. 105. A relay switch. (Courtesy Leece-Neville Company)

circuit, as shown in Figure 106. As long as the coil is energized, the circuit will remain closed. The circuit may carry a heavy current of electricity, such as that used by a starter, while the coil may be energized by a very small current carried by small wires from a remote point, usually from the aircraft battery by way of a push button or switch on the instrument board. As soon as the current stops flowing in the coil, the spring attached to the lever arrangement on the relay pulls the iron plate away from the electromagnet and opens the circuit.

Vibrator. A vibrator is similar to a relay. The only difference is that as soon as the contact is made which closes the circuit, the circuit about the coil is broken. Figure 107 shows a vibrator diagrammatically. This device will open and close a circuit many times a second. It is commonly used in an ignition system in connection with an induction coil to produce a rapidly fluctuating direct current. As the iron plate is attracted to the core of the coil, the points completing the external circuit close but, at the same instant, the other pair of points break, opening the coil circuit. This releases the iron plate from the pull of the electromagnet.

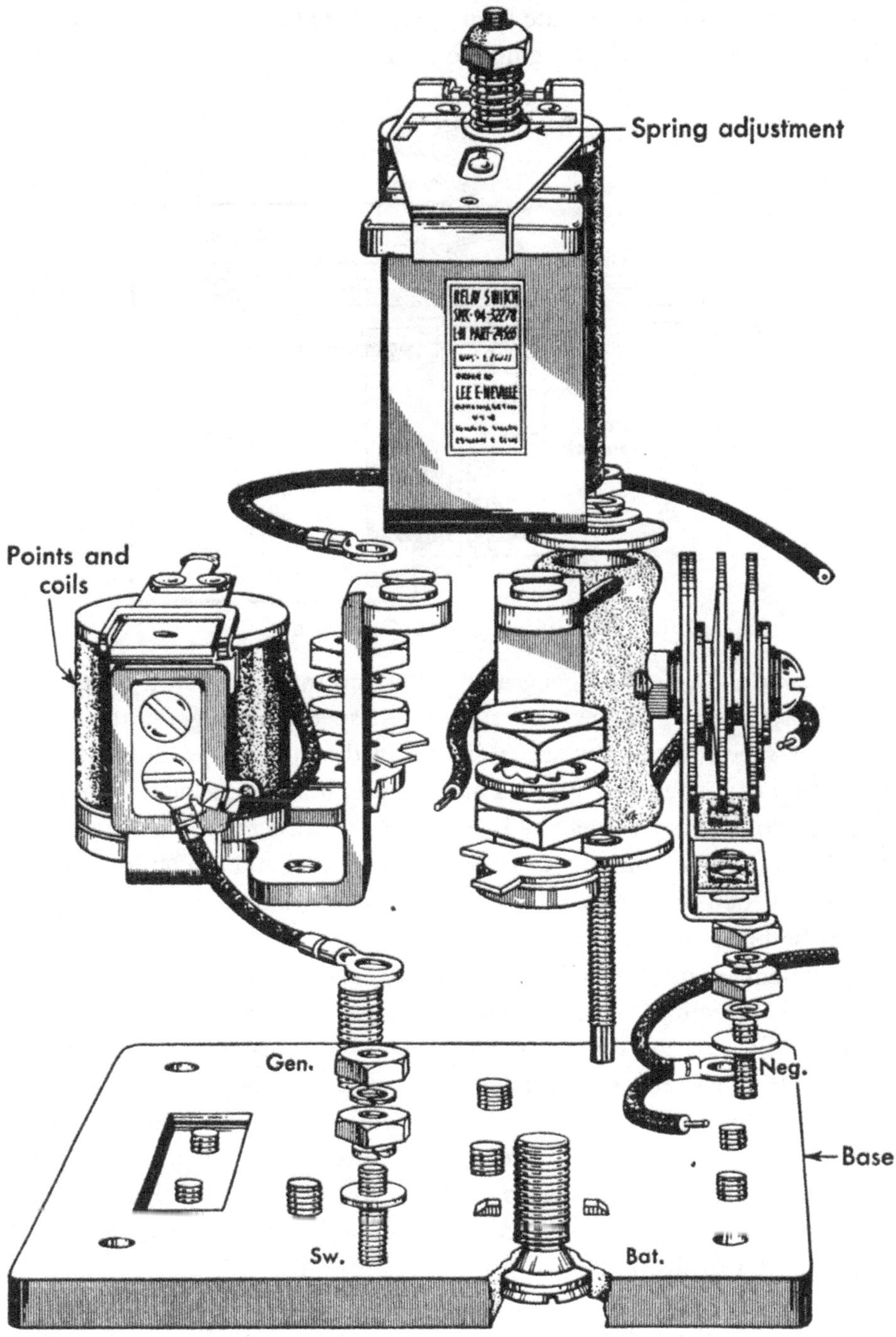

Fig. 106. Exploded view of a switch relay. (Courtesy Leece-Neville Company)

A spring attached to the plate then opens the external circuit and brings the points together, which again closes the coil circuit.

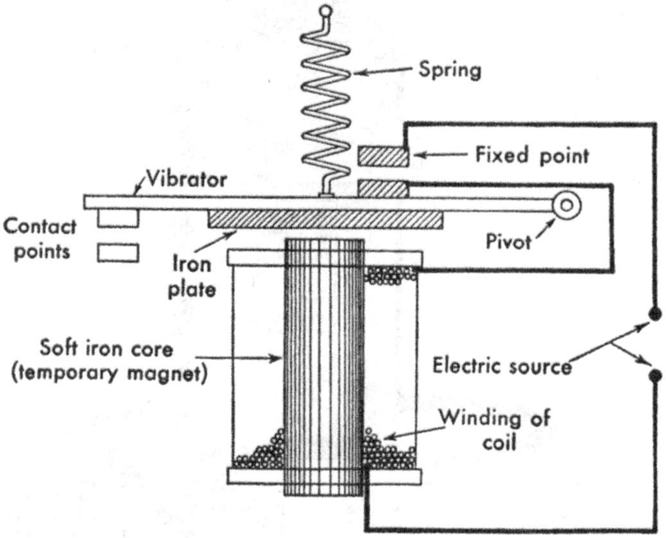

Fig. 107. Diagrammatical drawing of a vibrator.

Voltage Regulators. The generator in the aircraft electrical system does not always operate at the same speed. The speed of the generator varies with the speed of the aircraft engine; and the current produced by the generator varies with its speed. For example, if a generator will produce a current of normal voltage at 1600 r.p.m., and the engine is speeded up to such an extent that the generator rotates at 2300 or 2400 r.p.m., a current of such high voltage is produced that various units in the circuit might be burned out.

A voltage regulator is installed in the generator circuit to prevent variations in voltage. A generator charging a 12-v. battery must supply current at a voltage of approximately 14.25 v. This voltage is just high enough to overcome the voltage of the battery and keep the battery charged. The function of the voltage regulator is to maintain the generator-current voltage at approximately this value. As the voltage increases, a resistor is introduced between the plus side of the armature and the plus side of the field, reducing the strength of the field so that the voltage will drop.

The simple type of voltage regulator cuts in a resistance between the armature and the field by means of a vibrating point. When the voltage is above that required, this point vibrates so rapidly that the voltage is held at approximately a constant value. Normally, the contact points are held in a closed position by a spring. A coil having an iron core is so arranged that, as the voltage increases, the iron core, acting as a magnet, opens the points. One end of the core winding is connected to the positive side of the armature, and the other end is grounded. This places the coil across the generator. Any change in generator voltage will cause a change

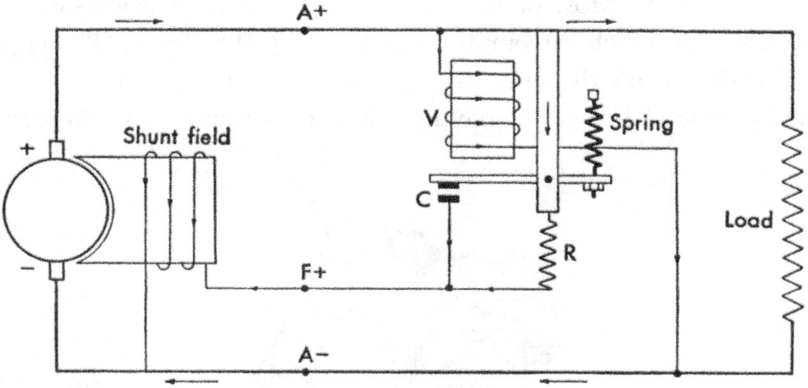

Fig. 108. A diagram to show construction of a voltage regulator of the vibrator type.

in the strength of the electromagnet formed by the coil. As the voltage builds up, the magnetic strength of the core increases, pulls the points apart, and opens the circuit. This causes the current to flow through the resistor placed in the circuit. The current flowing through the resistor reduces the output of the generator, and the voltage falls. As soon as the voltage falls, the magnetic strength of the iron core decreases and allows the points to close. The voltage of the current can be regulated by adjusting the spring tension on the points. If higher voltage is required, the tension on the spring is increased, and the tension is decreased to lower the voltage.

Another type of voltage regulator makes use of a pile of carbon disks. The resistance of the carbon pile is decreased by squeezing the carbon disks together and increased by releasing the pressure. This action varies the amount of current flowing through the carbon pile which, in turn, regulates the voltage of the current produced by the generator.

Current Regulator or Limiter. In order to control the current to protect the generator and the rest of the circuit, a current limiter is installed on the generator-control panel. The current limiter is similar to the voltage regulator and operates by regulating the amount of current flowing through the field. The generator's current should never be allowed to rise much above the rated output. The current regulator controls the current by controlling the generator field current. Figure 110 shows the schematic drawing of a complete control panel including the current limiter, reverse-current cutout, and voltage regulator.

When the contact points are closed, part of the current flows from the plus terminal of the generator back into the field through the plus stud on the generator. Most of the current flows through a low-resistance heavy iron coil which surrounds an iron core in the circuit. The current limiter differs from the current regulator in that this core is inserted in the main lead instead of having one end grounded. As the current

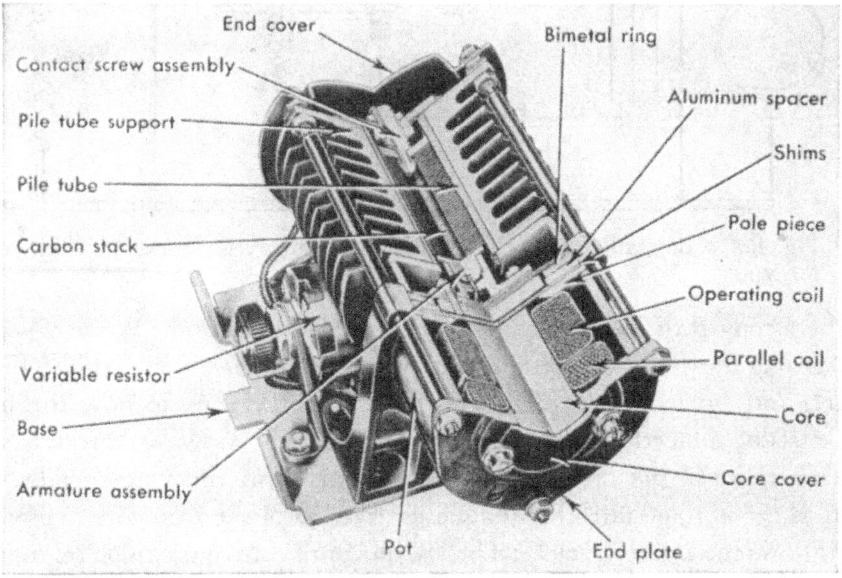

Fig. 109. A cutaway view of a carbon pile voltage regulator. (Courtesy Leece-Neville Company)

builds up above the desired point, the iron core, acting as an electromagnet, opens the points and forces the current to the field to flow through the resistor. Reducing the field current lowers the output of the generator. As the output of the generator is decreased, the strength of the magnet in the coil is lessened and the points close. This action allows the current to build up again. The coil in this circuit is in series, while the similar coil in the voltage regulator is in parallel across the generator. The current limiter is operated by a change in the line current, while the voltage regulator is affected by variations in generator voltage. A spring attached to one of the contact points regulates the amount of current put out by the generator.

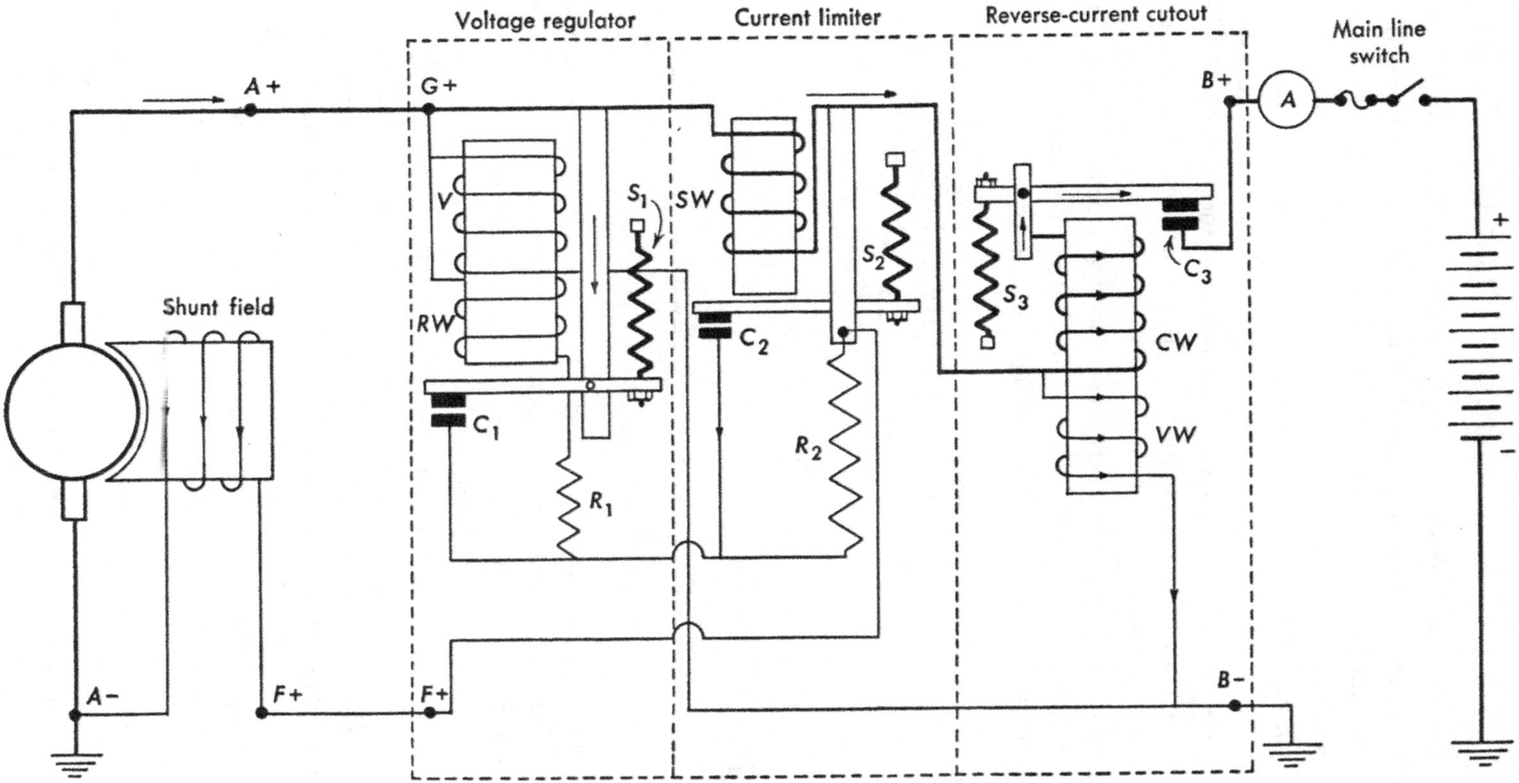

Fig. 110. A generator control panel system, consisting of a voltage regulator, current limiter, and a reverse-current cutout.

Reverse-Current Cutout. With the generator circuit closed and the generator not operating, current would flow from the battery through the generator and might burn out the generator armature. Until the generator reaches its operating speed, the current produced is at a lower voltage than the voltage of the battery. To prevent feedback from the battery through the generator, it is necessary to install in the circuit a reverse-current cutout. A reverse-current cutout is one of the units making up the generator control panel. Figure 111 is a schematic drawing of a reverse-current cutout.

When the aircraft engine is operating at normal speeds, the generator produces a current of higher voltage than that of the battery and the current from the generator flows through the battery. As the aircraft

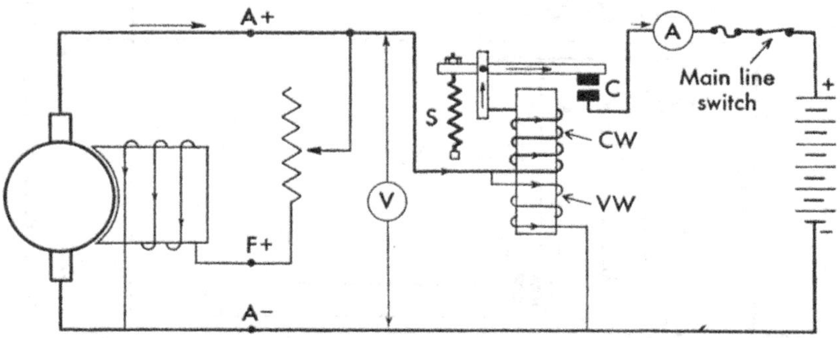

Fig. 111. A schematic drawing of a reverse current cutout.

engine is slowed down and the voltage falls below that of the battery, the reverse-current cutout points open, thus preventing the battery current from flowing to the generator. The contact points are normally held open by a spring on the pivoted point arm. When the points are open, the circuit is broken between the generator and the battery. An iron core is inserted in the circuit around which two windings are placed. One winding consists of the main lead between the generator and the battery; the other winding is a leadoff from the main lead, the other end of which is grounded. As the generator increases its speed, the voltage is also increased. The core with its take-off winding is connected across the generator, one end to the main lead from the plus side of the armature and the other end to a ground. As the voltage builds up, the iron core acts as an electromagnet and closes the contact points, allowing the current to flow through the main lead to the battery. The two windings pass around the iron core in the same direction. When current is flowing only from the generator to the battery and to the ground, the core acts as an electromagnet. If, however, the current in the main lead becomes less than the battery voltage, current flows from the battery to the generator. This current, passing around the core in the opposite direction from the grounded lead, neutralizes the effect of the ground lead, demagnetizes the core, and allows the points to open. This current would be shown on the ammeter for discharge. Whenever the current is flowing from the magneto to the battery, the ammeter in the circuit shows CHARGE. When the points open, no current can flow in the main lead. The reverse-current cutout automatically protects the generator from the battery current.

The generator control panel may have on it only two units consisting of the voltage regulator and the reverse-current cutout, or it may consist of three units: (1) the voltage regulator, (2) the current limiter, and (3) the reverse-current cutout. The generator control panel is usually mounted to be as free as possible from the effects of vibration and is usually attached to the engine side of the fire wall. It should be located so as to be easily accessible through an inspection opening.

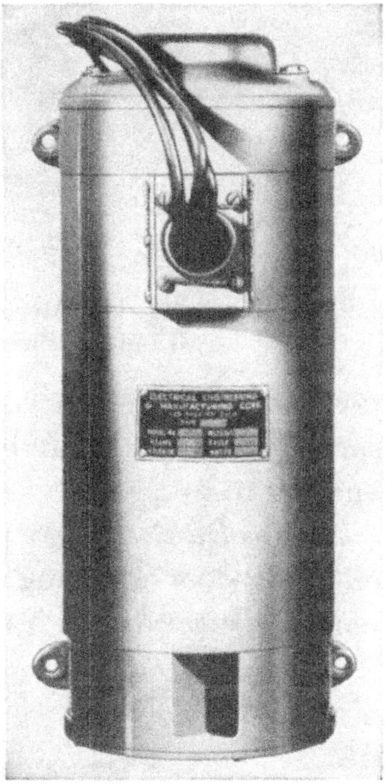

Fig. 112. A splashproof inverter. (Courtesy Electrical Engineering & Manufacturing Corporation)

Splashproof Inverter. It is often necessary to change one voltage to another. The inverter shown in Figure 112 may be used to change 32-v. direct current into 110-v., 60-cycle alternating current. The inverter is fan-cooled and rated at a capacity of 500 v.a. for continuous duty. It may be used for intermittent duty up to a peak capacity of 900 v.a. The inverter can also be adapted for use with a power supply of 110-v. direct current.

Electrically Actuated Aircraft Stabilizer Unit. The horizontal stabilizer actuating unit, shown in Figure 113, is used to balance an airplane about its longitudinal axis by moving the horizontal stabilizer electrically. With this screw-jack actuator, the stabilizer surfaces are now raised or lowered, eliminating the need for trim tabs on the elevator surfaces. One actuator is required, and a single switch located in the pilot's compartment operates the actuator which raises or lowers the stabilizer at the rate of ⅝ in. per sec. A total of 11 in. of movement is provided. Loads up to 16,500 lb. in one direction and 8000 lb. in the other direction are permissible under maximum load conditions. This unit, including a 1¼-hp. motor which operates on a 28-v. direct current, a gear

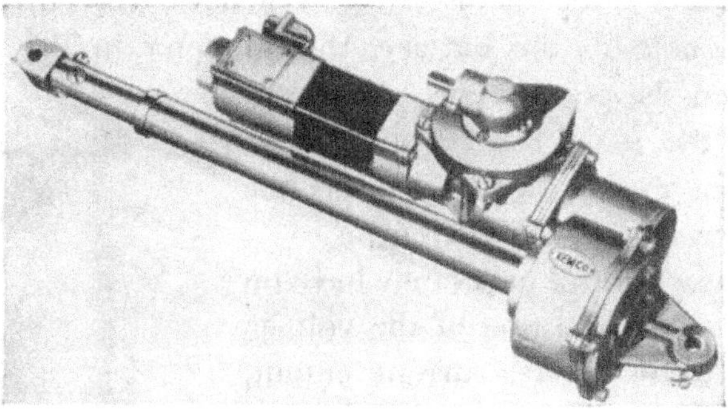

Fig. 113. An electrically actuated aircraft stabilizer unit. (Courtesy Electrical Engineering & Mfg. Corporation)

box, and a drive screw, weighs about 22 lb. A magnetic clutch and spring-loaded brake are built into the unit, which is entirely dust, sand, and salt-spray proof.

Electric Oil-Cooler-Door Actuating Unit. Figure 114 shows an electric oil-cooler-door actuating unit. It consists of a ⅛-hp. thermally protected motor, equipped with a magnetic clutch and brake. The unit contains

Fig. 114. An electric oil-cooler-door actuating unit. (Courtesy Electrical Engineering & Mfg. Corporation)

a motor-driven gear-reduction unit having a compound planetary reduction of 450 to 1, with an additional worm and sector reduction of 30 to 1, giving a total reduction of 15,000 to 1. The maximum torque on the drive shaft is 6000 in.-lb. This unit operates on a 28-v. direct current.

Electric Cowl-Flap Actuating Unit. Figure 115 shows an electric cowl-flap actuating unit. The unit consists of a 1/6-hp. motor equipped with a magnetic clutch and brake. The shaft output is 8 in.-lb. at 1200 r.p.m. A straight 8-to-1 gear reduction is provided for opening and closing the flaps. This unit operates on a 28-v. direct current and weighs 4 3/4 lb.

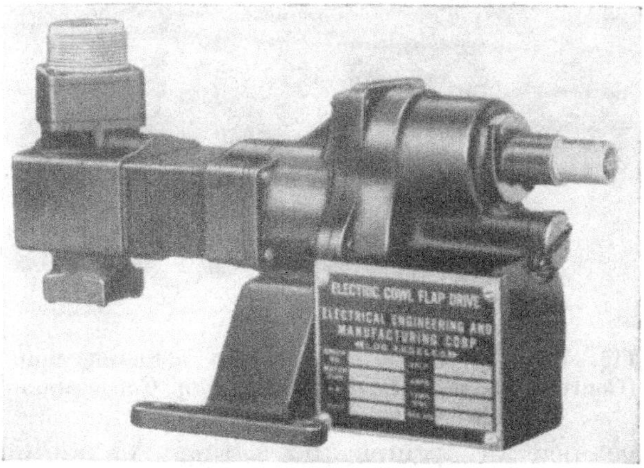

Fig. 115. An electric cowl-flap actuating unit. (Courtesy Electrical Engineering & Mfg. Corporation)

Electric Wing-Flap Drive Unit. Figure 116 shows an electric wing-flap drive unit designed to operate the flaps on heavy, fast aircraft. The unit consists of a 2-hp., 7500-r.p.m. motor, which is equipped with a magnetic clutch and brake.

Fig. 116. An electric wing-flap drive unit. (Courtesy Electrical Engineering & Mfg. Corporation)

Electric Dive-Brake Actuating Unit. Figure 117 shows an electric motor-driven actuating unit to operate the dive brakes on extremely fast airplanes. These flaps are used to overcome the phenomenon of compressibility and assist in controllability at extremely high speeds. The unit consists of a 1 1/4-hp. electric motor which operates at 10,000 r.p.m. It is equipped with a magnetic clutch and brake to prevent overtravel.

A screw actuator is operated by a double gear-reduction unit, which exerts a maximum force of 3000 lb. with 2½ in. of travel on the jack in less than 2 sec. A control gear-reduction drive turns limit-travel switches

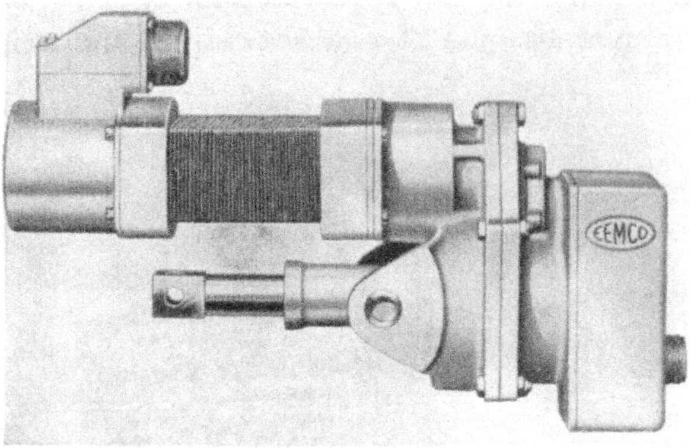

Fig. 117. An electric motor-driven actuating unit. (Courtesy Electrical Engineering & Mfg. Corporation)

especially developed for opening and closing. An additional take-off shaft is provided for a positioning indicator. The unit is operated by a push button in the pilot's compartment.

Electric Angle-Blower Drive Unit. Figure 118 shows an electric angle-blower drive unit. This unit is designed for cabin-heater drives and consists of a ½-hp., fan-cooled motor, which operates at 7000 r.p.m. It is equipped with a right-angle gear box and drive shaft. The unit operates on a 28-v. direct current.

Fig. 118. An electric angle-blower drive unit. (Courtesy Electrical Engineering & Mfg. Corporation)

IX A TYPICAL AIRCRAFT ELECTRICAL SYSTEM

The complete electrical system of a two-engine aircraft is described in this chapter. The parts which go to make up the system have been described in detail in other parts of this text.

The current supply is a 24-v. d-c, and the wiring system is a single-wire, ground-return type.

Ignition System. The ignition system consists of a switch unit mounted on the engine-control stands. This unit controls the ignition system and consists of a master switch and two ignition switches. Each ignition switch is connected by a ground wire to one of the magnetos on each

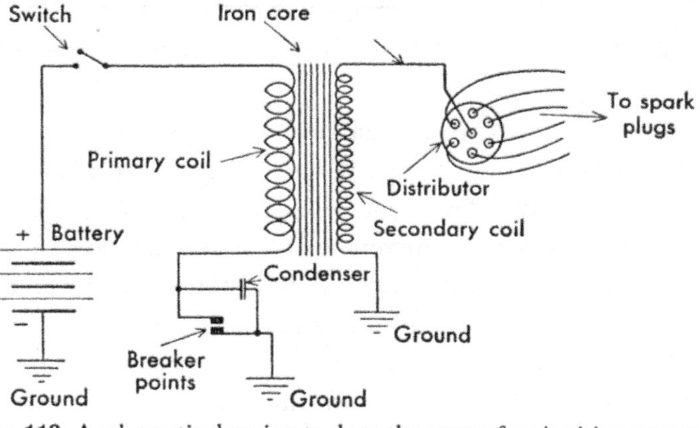

Fig. 119. A schematic drawing to show the parts of an ignition system.

of the engines. The airplane structure itself acts as a ground or return line. A cannon plug is mounted on the fire wall to assist in easy removal of the conduit. When this plug is disconnected, the magnetos are automatically grounded. All engine ignition wiring is rubber insulated and coated with a protective covering to prevent failure due to oil or wear. Figure 119 is a schematic drawing showing the ignition system and its parts.

The ignition system is completely shielded to prevent interference with the radio system.

There are two magnetos mounted on each engine, generally at the rear of the engine. The righthand magneto fires one set of plugs, usually the front set, while the lefthand magneto fires the other set. The spark advance to be used is given on the engine-data plate. The magnetos are compensated for equal spark advance on all cylinders.

A booster coil to provide a hot spark during starting is connected to the righthand magneto of each engine.

Generator System. A generator is mounted on each engine. It is driven by a gear train from the crankshaft. A generator control box is included in each generator circuit. A flexible conduit carries wires from the generator to the engine junction boxes and to the connector plugs on the

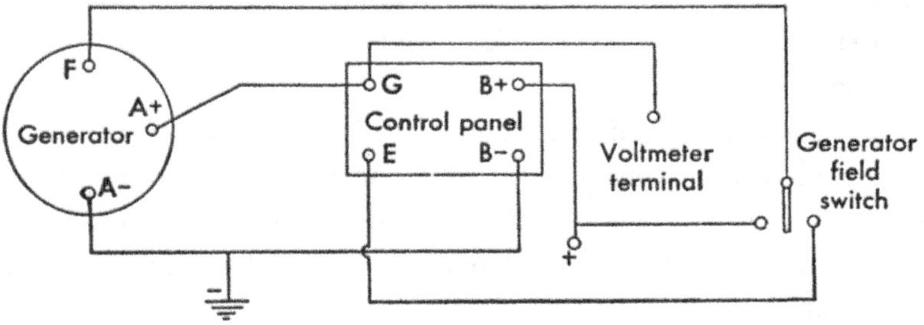

Fig. 120. A schematic drawing to show the generator and control panel circuit.

fire wall. The wires or leads are carried from the fire wall through the center section of the leading edge of the wing. These leads are carried to the main switch box. The generators have an output of 1500 w. at 50 amp., and furnish all the current used for the aircraft, in addition to maintaining the storage batteries in a full state of charge. Figure 120 is a schematic drawing showing the generator circuit.

The generator control box, which is included in each generator circuit, consists of a voltage regulator, a current limiter, and a reverse-current cutout.

The voltage regulator is a relay which governs the voltage output of the generator. In this system, the relay is adjusted by spring tension to open at 28.5 v. (see Figure 108).

The current limiter is a relay which limits the amount of current put out by the generator. The control spring on the relay is set to open when the current from the generator reaches 50 amp. When a current output of 50 amp. is reached, the relay opens and the current flows through a resistor in series with the generator field. This causes a decrease in the field current of the generator which causes a corresponding drop in the current output of the generator (see Figure 110).

The reverse-current cutout is a relay which allows current to flow from the generator to the battery only when the generator-current voltage is greater than that of the battery. When the circuit is closed and the generator voltage is less than the voltage of the battery, the current would flow back through the generator. Since the resistance of the generator is small, the heavy load from the battery could burn

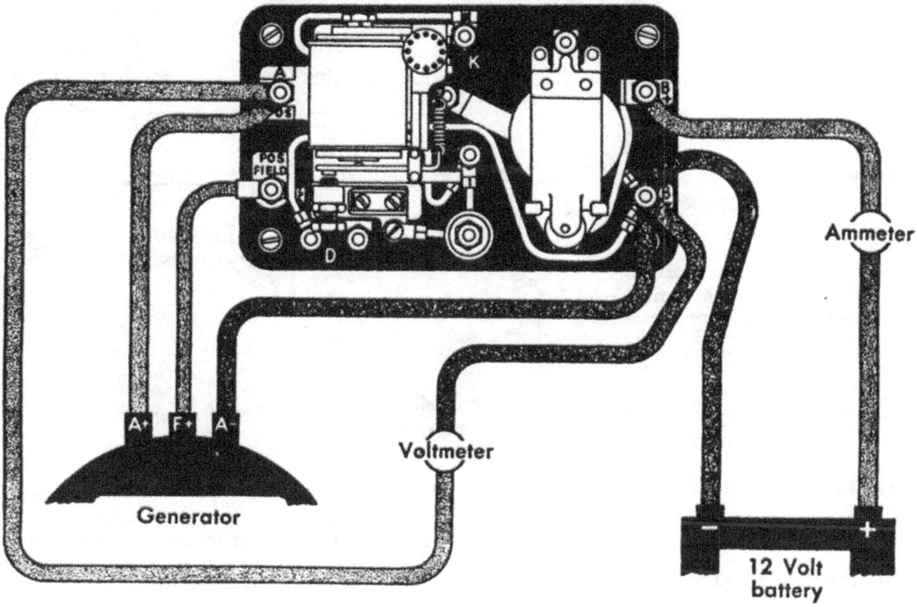

Fig. 121. A generator control-box circuit. (Courtesy Leece-Neville Company)

out the generator armature. As soon as the voltage of the generator current becomes less than the voltage of the battery, the cutout automatically opens the circuit and prevents the current from flowing back through the generator (see Figure 111). The contact points close when the generator voltage reaches 27.2, and the relay points open when the voltage falls to 26.8. When the current from the battery flows to the generator, the field of the relay coil is reversed, which neutralizes the field strength of the first coil, allowing the relay to open. Figure 121 shows a generator control-box circuit.

Starting and Propeller Feathering. A combination starter and propeller-feathering pump motor is installed on each engine. When the starter switch is in the ON position, either engine may be rotated by the use of the selector switch. When contact to either engine is made, the starter solenoid in that circuit completes the starter-motor circuit. The starter motor rotates the engine in the proper direction. Figure 122 is a schematic drawing showing the starting- and feathering-motor circuit. When it is desired to feather either propeller, the propeller-feathering switch for that propeller is placed in the proper position. The operation of this switch causes the feathering solenoid to close the motor circuit. The combination starting and feathering motor then rotates in the opposite direction and drives the feathering pump. A switch solenoid holds the

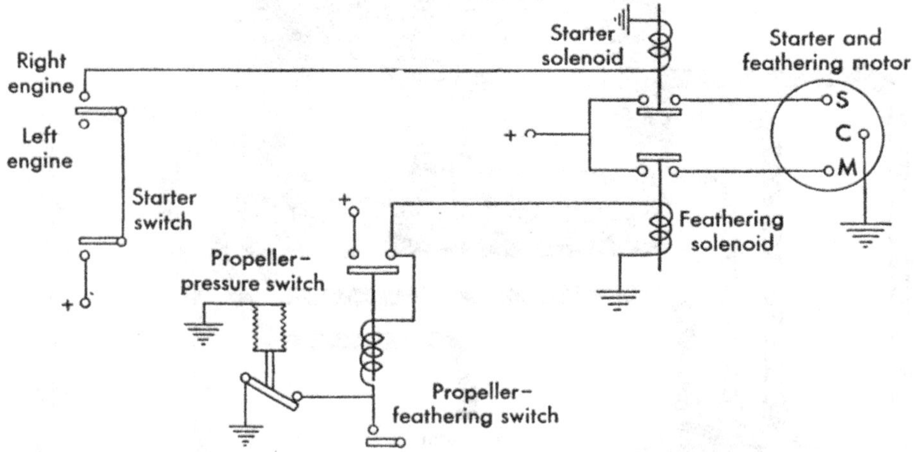

Fig. 122. A schematic drawing to show the starting- and feathering-motor circuit.

propeller switch in the contact position. When the propeller has reached the feathered position, the hydraulic pressure in the feathering system builds up until it forces the propeller pressure switch to open. This breaks the feathering-switch solenoid circuit and stops the rotation of the feathering motor.

Radiator-Shutter Circuit. Each engine is equipped with an individual radiator for cooling the engine oil. The air flow through each radiator is controlled by a separate electric circuit for operating the radiator shutters, which are controlled by switches located on the instrument panel. Figure 123 shows a schematic drawing of the oil radiator-shutter circuit. The shutters are operated by an electric motor. When a shutter switch is placed in the CLOSED position, a solenoid closes the motor circuit, and the motor rotates in the proper direction to close the radiator shutter. As the solenoid closing the circuit is energized, another solenoid in the circuit opens its motor-to-ground contact, and as the shutters reach the CLOSED position, a limit switch is opened which breaks the

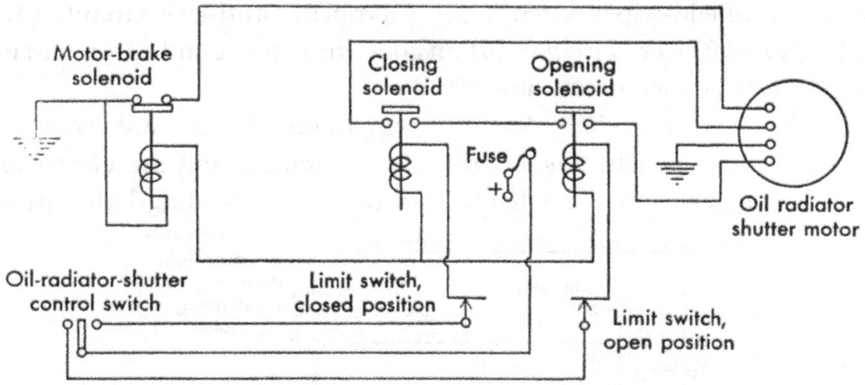

Fig. 123. A schematic drawing of the oil radiator-shutter circuit.

circuit. The breaking of the circuit causes the contacts on the second solenoid to close and the first solenoid to open. The shutters may be stopped at any desired position by operating the shutter switch. Throwing the switch to the OPEN position causes a third solenoid to close the circuit, allowing the motor to rotate in the opposite direction to the circuit closed by the first solenoid. Figure 124 is a schematic drawing of

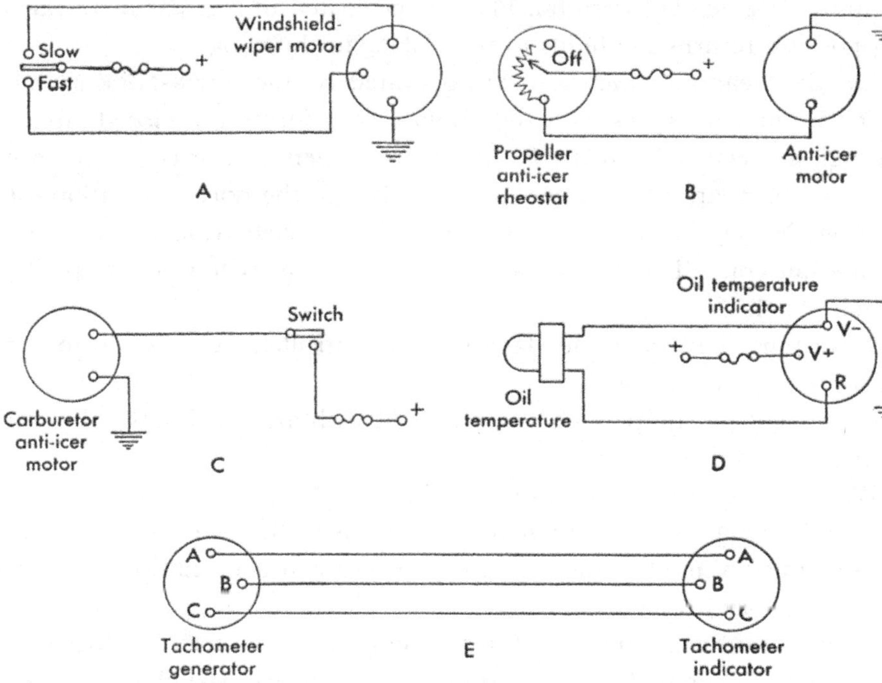

Fig. 124. A schematic drawing of (a) windshield-wiper circuit, (b) propeller-anti-icer circuit, (c) carburetor-anti-icer circuit, (d) oil-temperature-indicator circuit, (e) tachometer-indicator circuit.

(1) a windshield-wiper circuit, (2) a propeller-anti-icer circuit, (3) a carburetor-anti-icer circuit, (4) an oil-temperature-indicator circuit, and (5) a tachometer-indicator circuit.

Landing Lights and Other Aircraft Lights. Landing lights are located in the outer wing panels. These lights are retractable and are electrically operated. Figure 125 is a schematic drawing of a landing-light circuit.

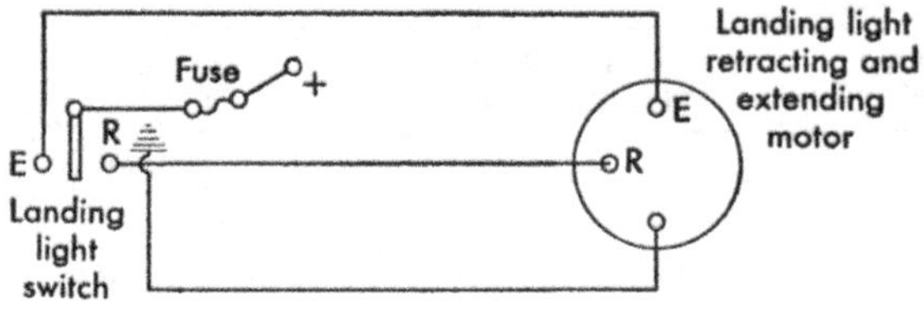

Fig. 125. A schematic drawing of a landing-light circuit.

The switches controlling the operation of the landing lights are of the single-pole, double-throw type. The vertical position of the switch is OFF. Moving the switch forward extends the lights and turns them on. Returning the switch to the OFF position turns the lights off, leaving them in the extended position. Placing the switch in the rear or retracting position returns the lights to their place in the wing.

Landing-gear warning lights are operated by the DOWN-LOCK switch which is wired in series with the DOWN switch forming a closed circuit through a green light in the flight station. When both green lights are on, both gears are fully extended and locked in the normal position for landing. Navigation lights are located, one on each wing tip and one on the tail cone. These lights are controlled by a switch on the pilot's switch panel.

A warning light in the fuselage nose is controlled by a switch in the pilot's switch panel.

Each baggage compartment is equipped with an individual light and switch.

Wheel lights are mounted on each side of the fuselage and directed at each landing-gear wheel when the wheel is in the extended position. These lights are used at night to assist the crew in determining whether or not the gear is extended.

In addition to the individual instrument lights, the instrument panel is directly illuminated by two lamps on the electric switch panel. An indirect lighting system for the instrument panel is controlled through a rheostat.

A light for the emergency panel located at the right of the pilot is provided for proper illumination.

At night the cabin is illuminated by three lights. These lights may be operated from either the pilot's compartment or from the cabin switch box.

The lavatory compartment is provided with a light and light switch. Figure 126 is a schematic drawing of the indirect instrument panel and tab control lights.

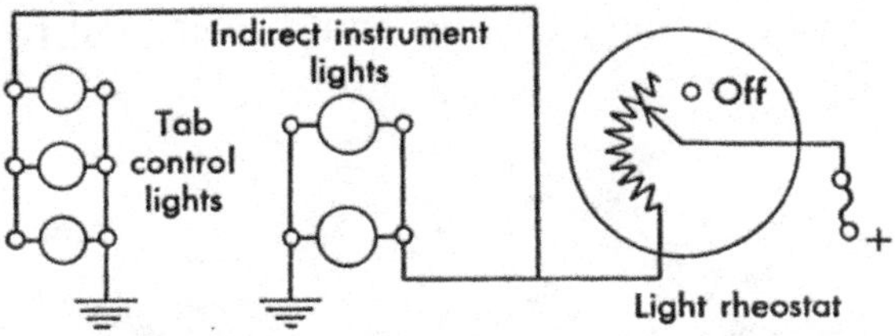

Fig. 126. A diagram of the indirect instrument lights and tab control light circuit.

A light mounted on a flexible lamp holder is provided in the radio compartment for the radio operator when radio equipment is installed.

Two flares are carried within the bottom of the fuselage. The flare-release switches are located in the pilot's compartment. To release a flare, it is first necessary to break the safety wire on the switch guard. After breaking the safety wire, the switch guard is released and allows access to the flare master switch for contact. It is possible to release either or both flares by the use of their respective switches. The flare is ejected by the propellant charge which is fired by the burning of the small resistance wire which passes through the electric ignitor. Firing

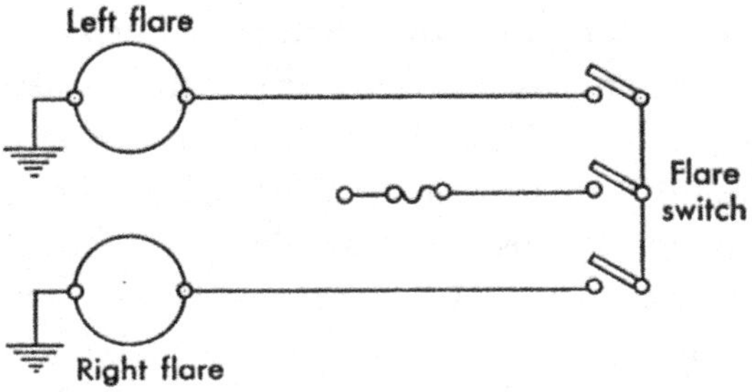

Fig. 127. A schematic drawing of the flare circuit.

this cartridge furnishes the pressure necessary to expel the flare candle and parachute away from the airplane. Figure 127 is a schematic drawing of the flare circuit. The flare candle is not lighted until the flare has completely cleared the tube. The lighting of the candle is controlled by a pull wire. One end of the pull wire is fastened to the tube, and the other end is attached to the ignitor on the candle. The wire must be extended to its full length before the candle is lighted by the delayed fuse. A delay of two seconds is provided before the candle is lighted in order that the flare may be some distance from the airplane. The flare floats to the earth by means of its parachute.

The external lighting system of an aircraft usually consists of the navigation lights, which include a green light at the tip of the right wing, a red light at the tip of the left wing and a white tail light. Each unit

Fig. 128. A landing gear retraction motor unit. (Courtesy Jack & Heintz, Inc.)

usually consists of two lamps, in case of the failure of either one. The lights are streamlined into the aircraft surface to which they are attached. The lamp units are connected in parallel and operated by a single switch. These lights are sometimes operated by means of a two-way switch having two ON positions. In one ON position, the switch remains in position. In the other, it operates against a spring which opens the switch as soon as it is released. The spring switch is used for signaling. The landing lights are powerful lights and are directed at an angle to illuminate the landing area when the aircraft is in the landing position. The lights are usually located midway along the leading edge of each wing and streamlined into the wing surface. The landing lights draw a large amount of current from the system and may be operated by means of a relay. The lights are usually wired so that one or both may be used. The lamp circuits are generally fused for safety. Some systems have the retractable lamps located on the undersurfaces of the wing. These lights may be stopped in any desired position. Retractable lights are usually operated by means of small reversible electric motors. Formation lights or passing lights may be included in the external lighting.

The internal lighting of an aircraft consists of the instrument-board lights and lights in the cabin or other parts of the aircraft. Most instruments have built into the instrument case an individual light consisting of a small 3-v. bulb. Resistors are inserted into the light circuit to decrease the battery current voltage to prevent burning out these lights. Fluorescent lighting may be used to decrease glare, and a device may be installed to produce ultraviolet light which causes the markings on the instrument to glow without visible light. The system may be arranged so that no light may be seen from the outside.

Electrical Instruments. The temperature and fuel gauges are electrical and are operated by battery current. A cylinder head or engine temperature indicator operates by means of a thermocouple mounted beneath a spark plug of each engine. Pressure-type switches operate the fuel and oil warning lights. The fuel-pressure warning switch is set to give warning when the fuel pressure falls below 14 pounds per sq. in. The oil pressure warning switch is set to give warning when the oil pressure falls below 65 p.s.i.

A 12-v. battery is carried in a battery compartment on each side of the fuselage. The battery compartments are located just below the leading edge of the center wing section. The two batteries are connected in series to produce a 24-v. system. The battery compartments are lined with acid-resistant paint and are ventilated. Provision is also made for draining.

A master switch disconnects the battery from all electrical equipment except the landing gear warning-horn circuit and the main-cabin warning-light circuit. When the airplane is resting on the ground, the warning-horn circuit is open, and no current will flow.

All conduit, plumbing lines, control surfaces, and all separate metal parts are bonded. This bonding is done with flexible tinned copper braid, cadmium-plated copper sheet, integrally bonded fair leads, or integrally bonded mounting clamps.

The proper fuses are located in a fuse box.

When the landing gear is retracted by means of an electric motor, a train of gears is used to reduce the speed and increase the force applied by the motor. Limiting switches which open and close solenoid switches to prevent overtravel are usually installed. A magnetic and gauging clutch may be installed in the system to engage the motor with the retracting mechanism when the motor is energized.

Warning lights are usually connected in such a manner that they will show when the landing gears are fully down or when the landing gears should be lowered. As the throttle is closed beyond a certain point, a switch mounted on the throttle control will cause red lights to show on the instrument board or may sound a Klaxon horn or operate both light and horn warning the pilot that the landing gear should be lowered. In case the pilot wishes to close the throttle when he does not intend to land, a special switch may be released which silences the horn or turns off the lights. As the throttle is opened and again closed, the mechanism opens as before.

Aircraft Wiring System. Wiring systems are not exactly the same on all aircraft, but certain fundamental principles are followed in each system. Feeder lines from the battery and generator are carried to the main junction box and may be attached to a bus bar. From this point, the electric current is carried to the various electrical devices. Main switches are generally provided in both the battery and generator lines, and there is usually a master switch which enables the operator to cut off all current flow in one operation.

The switches may be controlled by means of relays that may be remotely controlled. The main junction box is made of metal and is so located that it is easily accessible for service. All circuits should be protected by fuses to prevent overloading, and these are usually placed in the main junction box. Electric lines lead from the fuses to the control switches in the cockpit and then to various locations throughout the aircraft. There may be several bus bars which are fed from the main bus bar. When the battery master switch is closed, any electric device may be operated by its individual switch located in the cockpit. Sometimes various devices are connected by means of a safety switch to the auxiliary bus bar. This bus bar is sometimes referred to as a "safety bus bar."

Magnetos are independent of the generator battery system but often have a safety switch included in the same assembly operated by the master switch. A single-engine safety ignition switch may have three terminals for the magneto system. Two of the terminals included in this group may be designated for battery or auxiliary.

Such devices as a Pitot heating element, an electric starter, control surface, or landing-gear motors should be connected to an auxiliary bus bar. This prevents accidental turning-on of these devices. Devices that should not be turned on before the engine is in operation should be so wired that the individual control switches are inoperative until the ignition switch has been turned on. The main ignition switch should turn off these circuits automatically.

The lead wire from the positive battery terminal leads directly to one terminal of the motor starter switch. This arrangement is necessary because the starter motor circuit draws the heaviest current and should

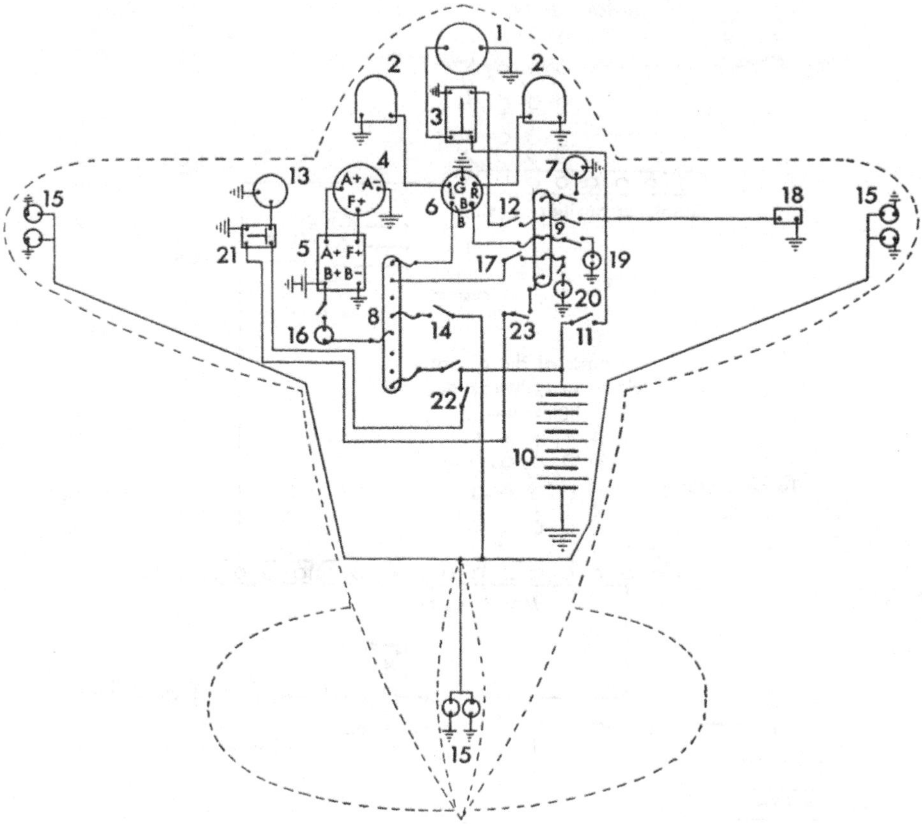

Fig. 129. A schematic drawing of a complete aircraft electrical installation: (1) electric starter; (2) magnetos; (3) starter solenoid; (4) generator; (5) control panel; (6) ignition switch; (7) starter, cartridge type; (8) main bus bar; (9) safety bus bar; (10) battery; (11) starter switch; (12) starter solenoid switch; (13) hydraulic pump motor; (14) navigation light switch; (15) navigation lights; (16) ammeter; (17) safety switch; (18) Pitot heater; (19) fuel indicator; (20) warning horn; (21) hydraulic pump motor solenoid; (22) hydraulic pump motor switch; (23) hydraulic pump motor solenoid switch.

be connected with the battery by means of a low-resistance circuit. The battery, starter motor, and starter switch are generally located close to each other. This arrangement is to allow the heavy leads to be as short as possible. Lines from the bus bar to the battery may be lighter as they do not usually carry heavy current. These lines are usually connected to the positive battery wire where it is attached to the starter-switch terminal. A two-position battery master switch is provided in some installations. This switch is arranged either to connect the starter directly with the battery or with the battery carried on a portable battery cart.

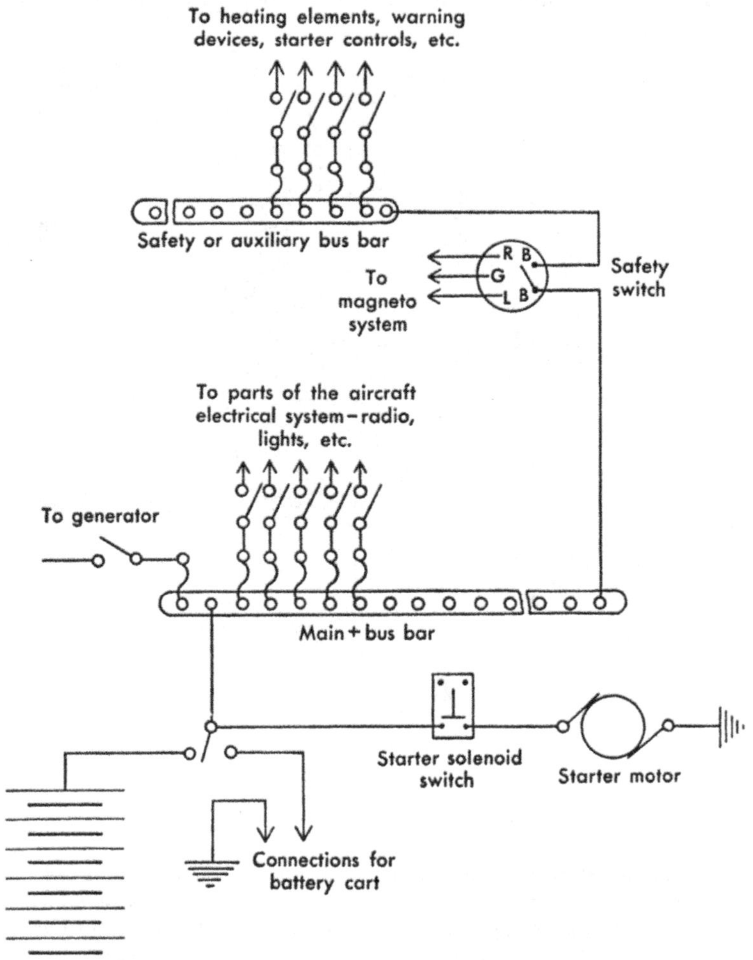

Fig. 130. A schematic drawing of the distribution system of an aircraft electrical installation.

All wires in an aircraft electrical system should be easily identified by means of an attached number or by means of differently colored wires. No splicing is permitted when replacing a defective wire. All replaced wires must be of the type, length, and size recommended in the specifications. Before a wire is installed, a band of tape should be wrapped around it near each of its terminals and the identifying number written on this tape. India ink should be used, and the number should be protected by a coat of shellac.

A connector panel is a panel of insulating material by means of which the terminals are insulated from each other and from the aircraft structure. Wires should be joined only by means of terminals. A connector plug is usually installed to simplify the removal and installation of an electric unit. When this system is used, any electric unit may be removed and reinstalled by simply removing the plug or by plugging in. This makes the breaking of connections unnecessary. The plug and receptacle should be fastened together with a threaded coupling. The contact pins and the plugs are usually staggered so that they will fit into a socket in one way only and may not be inserted in a wrong socket. The sockets and pins should be identified by the same number that is on the wire itself.

Each airplane should include as a part of its equipment the manufacturer's wiring diagram for the electrical system. This wiring diagram is usually in the form of a blueprint and should be kept in the aircraft at all times. Marked on this diagram should be the type and serial number of the aircraft. There should be noted on this diagram any information which the mechanic should have available for the maintenance and service of the electrical system. The blueprint and drawing should be clearly marked by means of standard symbols to designate the various parts of the system and the items of electrical equipment. The serial number of each separate item should be printed near its symbol, and a complete table of equipment should be placed on the blueprint.

X STARTING SYSTEMS

Most early aircraft engines, although of comparatively low horsepower and compression, were rather difficult to start. The magneto with which these engines were equipped delivered a weak spark because of

Fig. 131. A light aircraft engine equipped with two magnetos, a generator and a starter. (Courtesy, Continental Motors Corporation)

the low speed with which the propeller was "pulled through." Priming systems and accurately adjusted carburetors had not come into use. Some early aircraft were equipped with a booster magneto located in the cockpit. As the propeller was pulled through, the pilot turned this magneto with a hand crank. Later, magnetos were equipped with an impulse coupling. This consists of a spring arrangement which prevents the magnetos turning when the propeller first starts to rotate. At the time the spark is to be delivered to the cylinder which is under compression, the spring is suddenly released, allowing the armature to snap

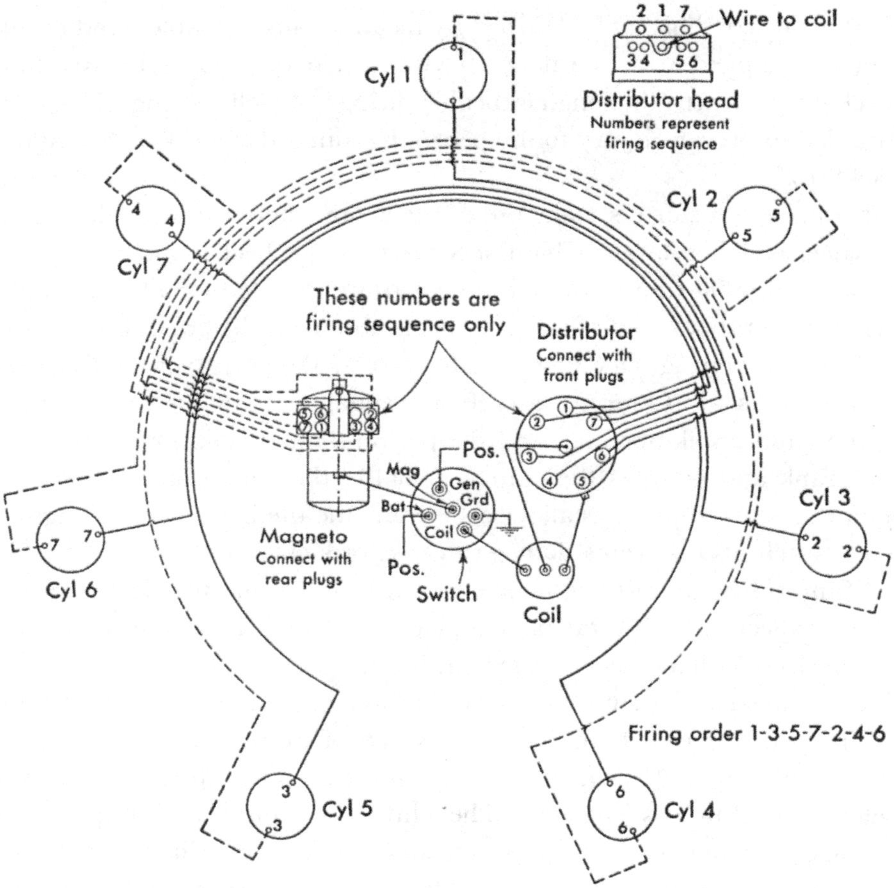

Fig. 132. A wiring diagram of a battery-magneto ignition system. (Courtesy Jacobs Aircraft Engine Company)

forward at high speed. This action delivers a hot spark to ignite the mixture in the cylinder. On many engines at the present time, one set of plugs is connected with a battery system which delivers a hot spark at low speeds. In one position the ignition switch connects the battery system with one set of spark plugs. As soon as the engine starts, the switch is turned to the regular magneto ignition system. Some engines continue to operate one set of plugs on the battery system.

The engine may be started in a number of different ways other than by pulling through. These methods include direct hand cranking, hand inertia starters, electric inertia starters, direct electric starters, air-injection starters, and the cartridge starter.

The direct hand-crank arrangement consists of a shaft equipped with a worm gear which engages a gear on the crankshaft when the starter shaft is turned. This shaft is turned by means of a detachable hand crank and is equipped with a ratchet device which disengages the starting mechanism when the engine begins firing. A self-locking device is installed to prevent injury to the person handling the crank if the engine backfires.

Hand inertia starters make use of the energy stored in a small, heavy flywheel which rotates at high speeds. This flywheel is speeded up to 10,000 to 12,000 r.p.m. through a gear train operated by a hand crank. When the proper speed of the flywheel is obtained, the crank is removed and the pilot engages a clutch which transfers the energy of the flywheel to the crankshaft. On some aircraft, the engaging clutch is located close to the hand-crank opening, and the person doing the cranking removes the crank and engages the clutch. Usually, the clutch arrangement is such that some slipping is allowed to overcome the inertia of the engine. This clutch also prevents damage in case of kickback.

Many other aircraft-engine starters operate on the inertia principle. The flywheel may be rotated by an electric motor. The rapidly moving flywheel, as in hand inertia starting, is connected to the engine crankshaft by means of a clutch and should have enough energy to turn the engine over three or four times at a speed of from 80 to 100 r.p.m. If the engine does not start, the flywheel must be built up to speed again before the clutch is engaged. The clutch is usually a multiple-disk-arrangement under spring pressure, and it allows a certain amount of slipping to take place to prevent damage due to sudden loading or engine kickback. If, for any reason, the engine cannot be rotated, the clutch will slip until the flywheel comes to rest. This clutch consists of several disks fastened to the cranking shaft and another set of disks fastened to the barrel. When the clutch is engaged, springs press the plates together with sufficient force to crank the engine. These springs may be adjustable. The two sets of clutch plates are usually made of different kinds of material to prevent excessive air. The starter is usually automatically disengaged when the engine starts.

Some electric inertia starters have a hand crank which may be used

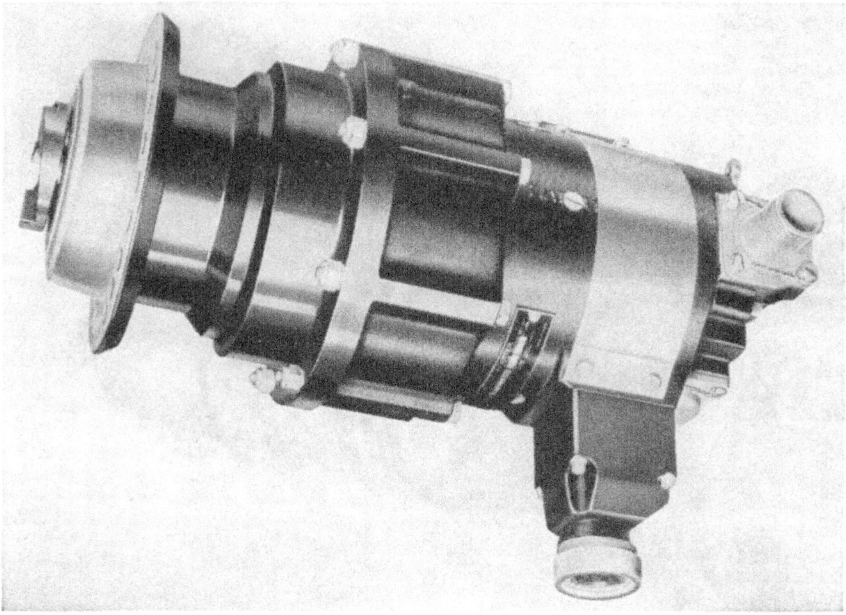

Fig. 133. An electric inertia or hand cranking starter. (Courtesy Eclipse-Pioneer Division, Bendix Aviation Corporation)

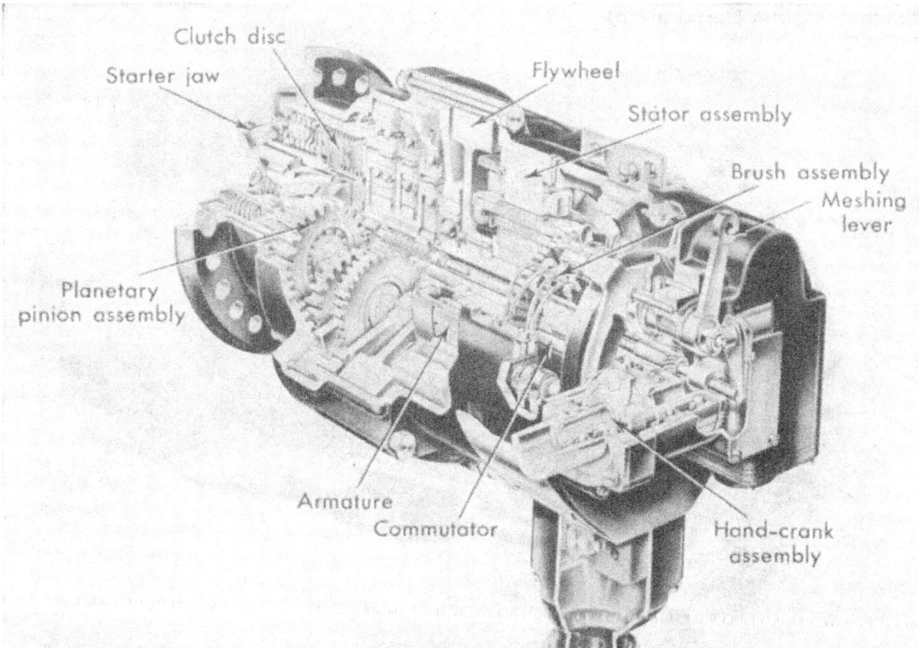

Fig. 134. A cutaway view of an electric inertia or hand cranking starter. (Courtesy Eclipse-Pioneer Division, Bendix Aviation Corporation)

Fig. 135. An electric direct cranking starter. (Courtesy Eclipse-Pioneer Division, Bendix Aviation Corporation)

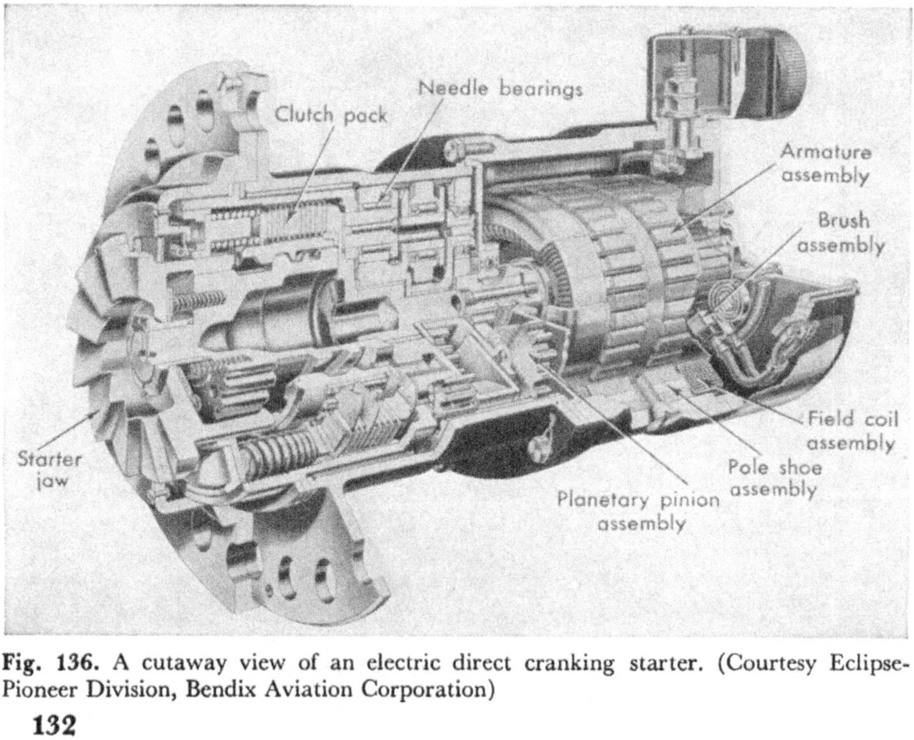

Fig. 136. A cutaway view of an electric direct cranking starter. (Courtesy Eclipse-Pioneer Division, Bendix Aviation Corporation)

in case the source of electric current fails. When the hand crank is used, the electric motor is disconnected and inoperative. When the starter is cranked by means of an electric motor, a movable jaw is engaged by means of a helically splined motor shaft similar to that used in an automobile starter. As the motor starts, the jaw is moved forward and engages the flywheel. This type of starter should not be re-engaged, in case the engine fails to start, until the flywheel of the starter comes to a complete stop.

The motor in the electric inertia starter is usually series-wound for either 12- or 24-v. current. These motors are similar to the d-c generator. The electric resistance of this type of motor is very small and allows a heavy flow of current, resulting in a powerful starting torque. As the motor gains speed, an induced electric effect causes less current to flow. One of these motors may draw as much as 350 amp. when starting and only about 75 amp. at high speed. Care should be taken not to allow this type of motor to race, and it should *not* be tested without a load on it. The motor brushes are usually remotely controlled by means of a solenoid arrangement. When the switch in the cockpit is closed, the solenoid is energized, bringing the brushes into contact with the armature. The switch in the cockpit is usually spring-operated and must be held in position until the starter has built up to its full speed. Releasing the switch breaks the connection with the solenoid and cuts off the current from the motor. The engaging clutch on the starter may be operated by means of a solenoid.

An arrangement is sometimes made whereby a booster coil is connected to the solenoid meshing switch. When the switch which operates the solenoid is closed, the booster is automatically energized. The booster coil supplies high voltage to the trailing distributor finger of the right-hand magneto until the cockpit control switch is released.

An electric motor may be used to crank the engine directly. This type of starter, however, requires a heavy current to develop sufficient torque. It consists of an electric motor, an automatic engaging and disengaging mechanism, and reduction gears, and it operates through an adjustable clutch. When this type of starter is used, the engine is cranked directly like an automobile. It is not necessary to store up energy, as it is when the inertia type of starter is used. When the starter switch is pressed, the cranking jaw automatically engages before the mechanism begins to rotate. When the engine starts, the starter is automatically disengaged.

The air-injection type of starter is not in common use. It consists of an air tank and a high-efficiency air compressor operated by the engine. The air pressure, which is built up to approximately 450 p.s.i. by the compressor, is regulated by automatic valves which prevent excessive pressure. A control located in the cockpit allows the air to be released to the cylinders in their regular firing order. The air lost from the supply tank is replaced by the compressor while the engine is in operation. This starting system may be used on light airplanes and weighs approximately 30 lb.

The cartridge starter is sometimes used to start an engine. This starter consists of a chamber in which a cartridge is placed. The cartridge is ignited by electric contact. The explosion of the cartridge in the chamber

Fig. 137. The jaws of an aircraft engine starter. (Courtesy Jack & Heintz, Inc.)

of the starter forces a cylinder to move along spiral grooves, rotating the crankshaft. These starters may be used in places where extremely cold weather makes starting difficult. A special cartridge is necessary, and ordinary shotgun shells should *not* be used. These starters are equipped with a safety diaphragm which blows out if the starter builds up excessive pressures.

At the present time, most aircraft, other than light aircraft, are equipped with some type of electric starter. The type of starter used depends upon the type of engine.

Before starting an aircraft engine by hand cranking, the pilot should be sure that the wheels are properly chocked or that the parking brakes are firmly set. The engine should never, except in an emergency, be started without a qualified person at the engine controls. The oil and fuel supply should also be inspected. With the throttle closed and the switch off, the proper number of priming strokes, if the engine is equipped with a primer, are given on the priming pump. The engine is then rotated by hand through four or five revolutions in the direction of normal rotation. The propeller is put in the proper position to pull through, and the ignition switch is turned on. The throttle is then cracked, which means that it is opened slightly, and the propeller is pulled through as rapidly as possible. If the engine fails to start, the

Fig. 138. An electric inertia and direct cranking starter. (Courtesy Eclipse-Pioneer Division, Bendix Aviation Corporation)

switch is turned off and the engine rotated as before, but without additional priming. The switch is then turned on and the engine again pulled through. If the cylinders of the engine become flooded with fuel, the engine is rotated backward through five or six complete revolutions with the switch off and the throttle fully opened. Then, with the throttle closed, the starting procedure is repeated without priming.

On large engines equipped with a starter, the engine is primed and, with the switch off, the engine is started to rotate by means of the starter. As soon as the engine begins to rotate, the ignition switch is turned on. This procedure assists in preventing kickback which might cause damage or place excessive strains on the structure.

Before starting a newly installed engine, a thorough inspection of all parts and their installations must be made. The ignition switch on aircraft works somewhat differently than does the ignition switch on automobiles. On the automobile, the ignition switch, when in the OFF position, simply opens the primary circuit. On the aircraft, the ignition switch or magneto switch grounds the magneto by means of a ground wire. It is important that the ground wire be securely in place for, if it becomes loose, the ignition system will operate and the engine will continue to run

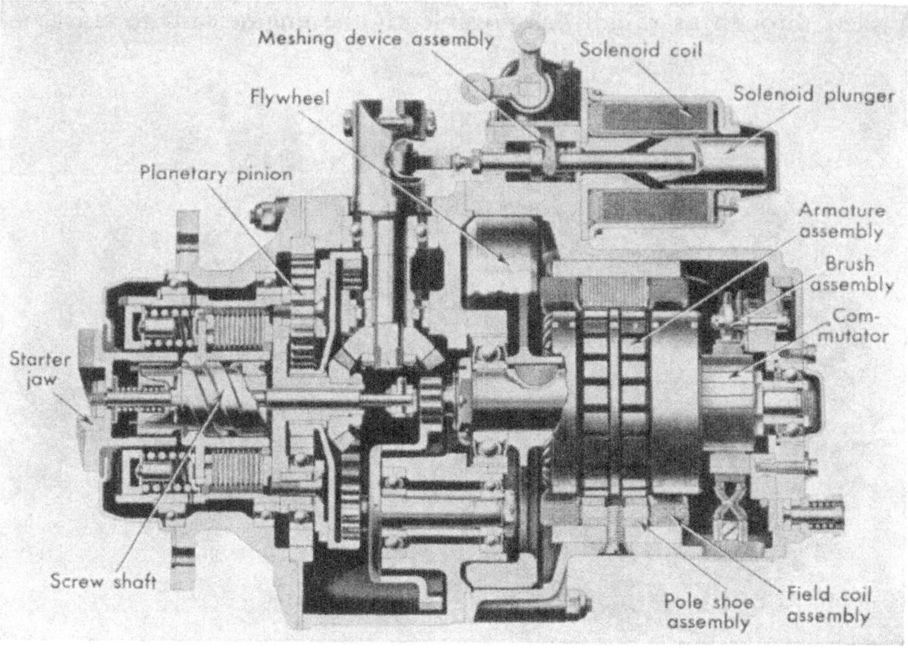

Fig. 139. A cutaway view of an electric inertia and direct cranking starter. (Courtesy Eclipse-Pioneer Division, Bendix Aviation Corporation)

although the ignition switch is in the OFF position. If the ground wire is loose or disconnected, the engine may start if, for any reason, it is rotated. In starting the engine after the proper priming charges have been drawn into the cylinder, unless the aircraft is equipped with battery ignition on one set of plugs, the ignition switch is placed in the BOTH ON position. As soon as the engine starts, the oil-pressure gauge should be inspected and, unless the oil pressure begins to register within a few seconds, the engine should be immediately stopped. The oil system should then be checked before any further attempt is made to start the engine. After the engine is started, the ignition switch should be turned to the RIGHT and the LEFT position, testing each magneto and its corresponding set of plugs. Operating on each magneto separately should not cause a loss of more than 50 to 150 r.p.m. The engine should be warmed up at a speed slightly in excess of idling speed for a sufficient length of time to allow the oil to become warm and all engine parts thoroughly lubricated. The engine should not be operated for periods in excess of approximately 30 sec. at full throttle while on the ground. This "revving up"

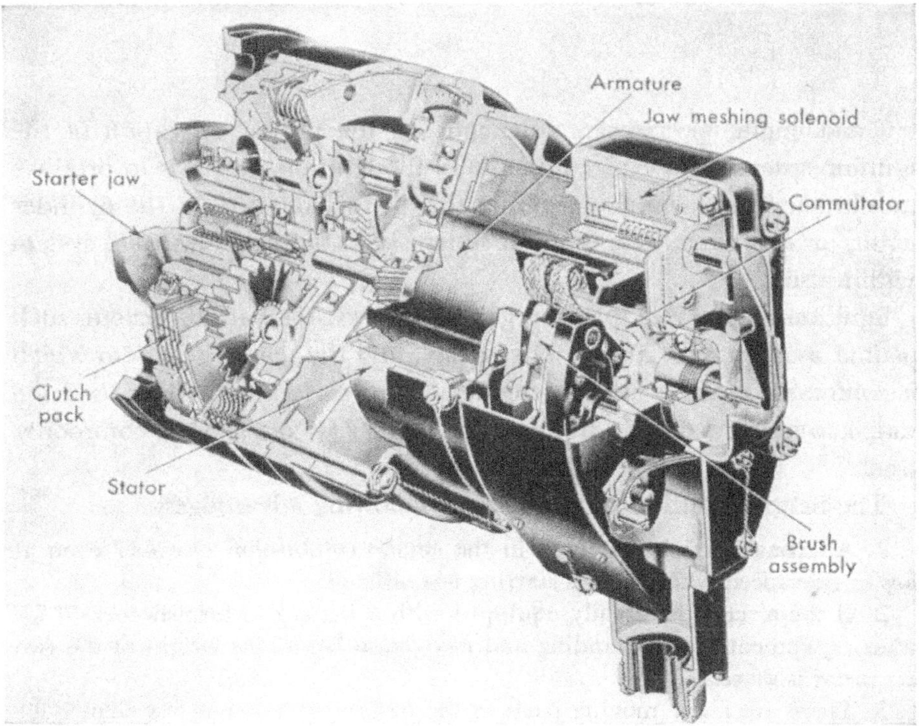

Fig. 140. A cutaway view of a direct cranking electric starter. (Courtesy Jack & Heintz, Inc.)

should be done, however, to ascertain whether or not the engine is reaching its maximum r.p.m. and to check the proper operation of all engine instruments. All movements of the throttle should be smoothly and gradually performed. Sudden opening of the throttle should be avoided as this may cause the engine to stop or, if the engine takes hold, will cause excessive strain on the structure. Engines should not run for extended periods on the ground unless provision for extra cooling are made because most engines will seriously overheat under these conditions.

XI ENGINE ELECTRICAL SYSTEMS

Good engine performance depends on the proper operation of the ignition system. The entire function of the ignition system is to produce an electric spark which will ignite the air-fuel mixture in the cylinder at the proper time. The ignition system is a complete electrical system within itself.

Ignition systems fall into two general classes: the battery system, such as that used on most automobiles today, and the magneto system which is commonly used on aircraft engines. Most early automobiles used the magneto ignition system, but now the battery system is more commonly used.

The battery ignition system has the following advantages:

1. A strong spark is obtained in the engine combustion chamber even at low engine speed which makes starting less difficult.
2. If the aircraft is already equipped with a battery to furnish current for other equipment such as landing and navigation lights, the weight of the two magnetos is saved.
3. There are fewer moving parts in the battery system and less chance for mechanical failure.
4. The battery generating equipment furnishes a convenient supply of electric current for instrument lights and other electrical equipment in the aircraft.

The magneto system offers the following advantages:

1. When magnetos are used, a saving of weight is accomplished because a storage battery is not needed.
2. There is less danger of fire when magnetos are used.
3. There are fewer external connections which might become loose.
4. The magneto system occupies less space and has less wiring than the battery system.
5. There is no danger of ignition failure due to the battery becoming discharged.

Battery Ignition Systems. The battery ignition system consists of (1) the battery which is the source of electric current, (2) an induction or high-tension coil, (3) the breaker mechanism, (4) a condenser, (5) a distributor, (6) the spark plugs, and (7) the necessary wiring and switches.

In the operation of the battery ignition system, the current from the battery is led through the primary circuit. The voltage of the current in the primary circuit corresponds to the voltage of the battery. The voltage of an aircraft battery may be 6, 12, or 24 v. The primary circuit consists of the battery lead to the induction coil; the primary winding in the induction coil; and the leads to the breaker mechanism, one side

Fig. 141. Three-fourths left rear view of an inverted six-cylinder aircooled aircraft engine. The two magnetos are mounted at the top rear of the engine. The hollow magneto drive shaft carries oil to the main bearings. (Courtesy Ranger Aircraft Engines)

of which is grounded to the engine as is one terminal of the battery. When the breaker points are closed, the current flows from one terminal of the battery through the primary windings of the coil, through the breaker mechanism, and back through the ground to the battery. With the exception of the main switch which opens and closes the circuit, this completes the primary circuit.

The secondary circuit of the battery ignition system consists of (1) the secondary windings in the induction coil, (2) the lead from the induction coil to the distributor and to the rotor within the distributor head, and (3) the wires leading from the distributor head to the spark plugs and the ground, back through the engine from the spark plugs, to one terminal of the secondary winding in the induction coil which is also grounded to the engine.

The battery furnishes a low-voltage direct current. As the current from the battery flows through the primary winding of the induction

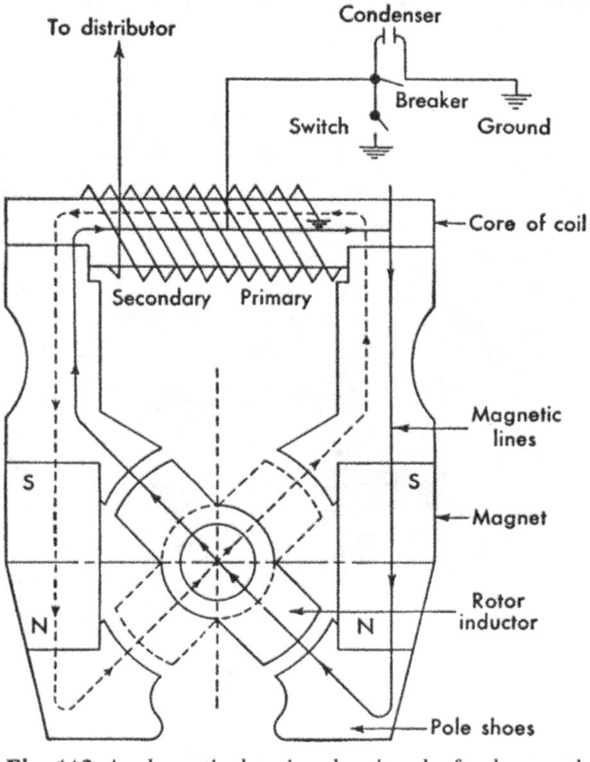

Fig. 142. A schematic drawing showing the fundamental operations of an aircraft magneto.

coil, a magnetic field is set up around the primary winding and expands outward into space, cutting the secondary winding. This field remains constant when there is no change in the amount of current flowing through the primary winding. As the points in the breaker mechanism close, the current starts to flow through the primary circuit, and the electric field is built up. This build-up is so gradual that the current generated in the secondary winding is not intense enough to jump the spark gap in the spark plugs. At the instant a spark is desired, the points are rapidly opened, causing the sudden collapse of the magnetic field. This, in turn, cuts the secondary winding and induces a high voltage, some 20,000 to 25,000 v., in the secondary circuit.

When an electric current is suddenly interrupted, self-induction is set up by the circuit, which tends to keep the current flowing. This self-induction would cause an arcing across the points were it not for the condenser which is connected in parallel across the points. The surge of electric current, due to self-induction, is absorbed in charging the condenser, and this prevents arcing across the points. Arcing would

Fig. 143. A modern aircraft engine magneto. (Courtesy Eisemann Corporation)

cause burning of the points which would decrease the intensity of the current generated in the secondary circuit. As the self-induced current in the primary circuit dies down, the condenser discharges back into the primary circuit. The points in the breaker mechanism are opened and closed by means of a cam arrangement on the rotor shaft. The rotor is attached to the top of the rotor shaft in the distributor head and is connected by means of a sliding contact to the lead from the high-tension or secondary winding of the coil. Each spark-plug lead is connected to a metal block in the distributor head. The rotor has a metal point which clears the metal blocks in the distributor head by a few thousandths of an inch. This point is opposite the proper block which connects with the lead to the spark plug at the time the spark is desired in that plug. The high-voltage current set up in the secondary circuit jumps across the gap between the rotor and the metal block and is carried to the spark plug where it jumps the spark-plug gap, producing the spark in the combustion chamber. The current then travels back through the ground

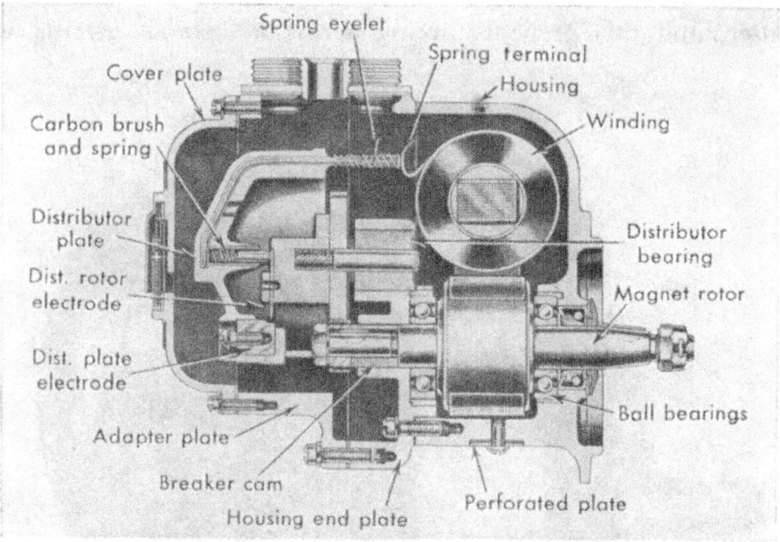

Fig. 144. A cutaway view showing construction of a magneto. (Courtesy Eisemann Corporation)

to the terminal of the high-tension coil which is grounded to the engine structure. There is usually a safety gap installed in the high-tension coil to protect the high-tension winding from overheating or from being damaged by a short circuit, or in case a wire becomes loose and the secondary circuit is opened.

Magneto Ignition System. The magneto ignition system operates in a manner similar to the battery system, except that the magneto is the source of the current instead of a battery. The magneto furnishes a low-tension alternating current which flows through the primary circuit in a manner similar to the primary circuit of the battery ignition system. The modern aircraft magneto has included in the magneto itself the high-tension coil, the primary circuit, the condenser, the safety gap, the distributor head, the rotor, and the breaker mechanism. All parts of this system operate in the same manner as in the battery system.

In the magneto system, the switch, when in the OFF position, closes

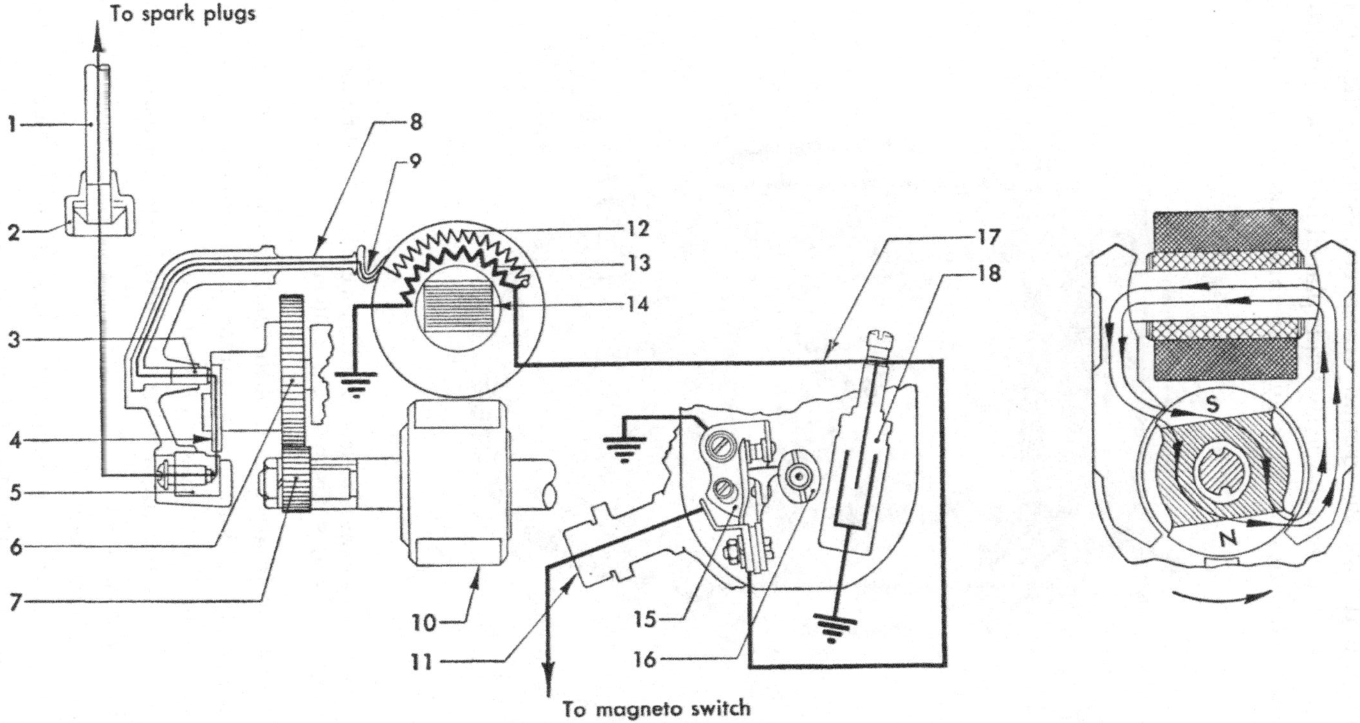

Fig. 145 Schematic diagram of electric and magnetic circuits of an aircraft magneto. (1) High-tension cable; (2) high-tension terminal; (3) carbon brush and spring; (4) distributor electrode; (5) distributor-plate electrode; (6) distributor gear; (7) pinion gear; (8) spring eyelet conductor; (9) spring clip; (10) magnet rotor shaft; (11) adapter; (12) secondary winding; (13) primary winding; (14) coil core; (15) breaker assembly; (16) breaker cam; (17) winding lead to condenser; (18) condenser. (Courtesy Eisemann Corporation)

the circuit grounding the magneto so that no current can flow through the primary circuit. In case the ground wire becomes loose, the engine will continue to run even though the switch is placed in the OFF position. When this occurs, the pilot stops the engine by cutting off the fuel supply.

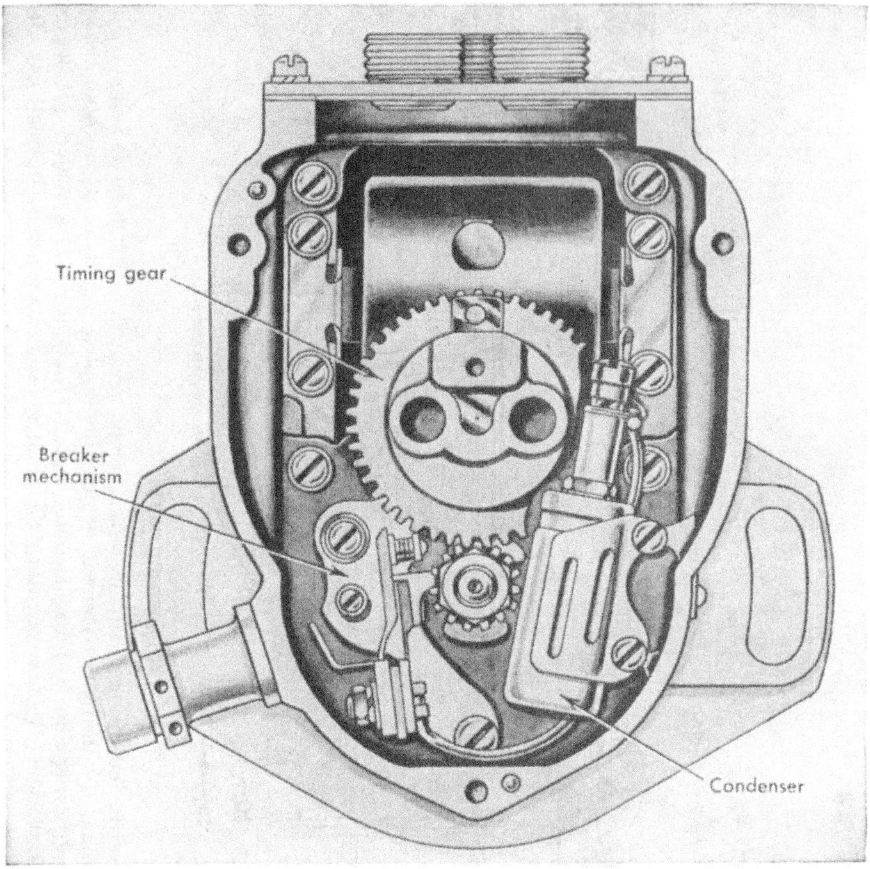

Fig. 146. A cutaway view showing condenser, breaker mechanism, and timing gears of an aircraft magneto. (Courtesy Eisemann Corporation)

The current in the primary circuit of the magneto system is produced by electromagnetic induction. Current may be produced by electromagnetic induction in three ways: (1) by moving a conductor through a magnetic field, (2) by moving a magnetic field in such a manner that it cuts a stationary conductor, and (3) by a fluctuating magnetic field cutting a stationary conductor.

In the ordinary electric generator, the magnetic field, which is produced by means of stationary magnets, is cut by coils of wire rotated in this field. The rotating coils are called the armature. In most aircraft magnetos, the coils are stationary and the magnetic field produced by permanent magnets is rotated past the coils. The rotating armature is made up of permanent magnets which rotate and carry the field past the coils. This armature may be made up of two, four, or eight poles. As the poles of a magnet pass a coil, a current is set up in one direction in the coil and, as the other pole of the magnet passes the coil, the current

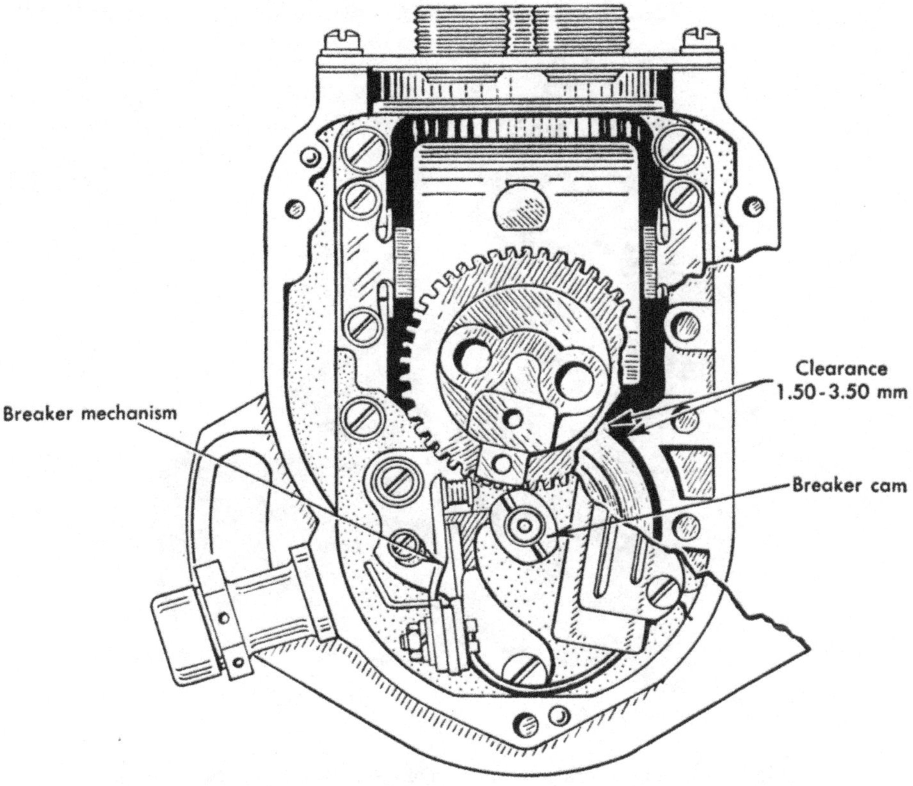

Fig. 147. Drawing of a magneto showing breaker cams, breaker mechanism, and clearance between armature and field coils. (Courtesy Eisemann Corporation)

dies down to zero and then is set up in the opposite direction in the coil. This action produces a low-voltage alternating current. This current is interrupted by the opening of the breaker points, causing a collapse of the field and setting up the high voltage required for the spark in the secondary windings of the coil.

Due to the fact that when the magneto is rotating at low speeds, a weak spark is generated, a booster arrangement is usually included in the magneto system. This system consists of an auxiliary high-tension coil and a vibrator. The booster switch may consist of a push button or may be connected to the starter switch and operate only while the engine is being started. The booster segments in the magneto are usually a short distance behind the main segment in order to retard the spark, thus preventing kickback while the engine is being started.

Spark plugs are of a number of different types, although they all function on the same principle. A spark plug consists of a threaded metal shell that screws into the spark-plug bushing in the cylinder head. The

Fig. 148. An aircraft engine magneto. (Courtesy Scintilla Magneto Division, Bendix Aviation Corporation)

opening in the bottom of the shell has from one to four small electrodes fastened around its circumference. These electrodes project inward toward the center of the opening. Inserted into this shell is the spark-plug core, which consists of insulating material surrounding the main electrode of the spark plug. This electrode projects into the opening between the electrodes fastened to the spark-plug shell. The space between the center electrode and the electrodes on the shell is called the spark gap. This gap is about 0.015 in. However, it should agree with the manufacturer's specification. The spark plug is assembled before it is placed in the engine. The high-tension wire carrying the current to the plug is connected to the main electrode which extends through the insulating core. The insulating material used in aircraft spark plugs is either mica or a special type of porcelain.

Spark plugs are classified as hot, medium, or cold, and these types are used in various kinds of engines. The hot plugs are usually longer than the cold plugs and protrude farther into the combustion chamber. These

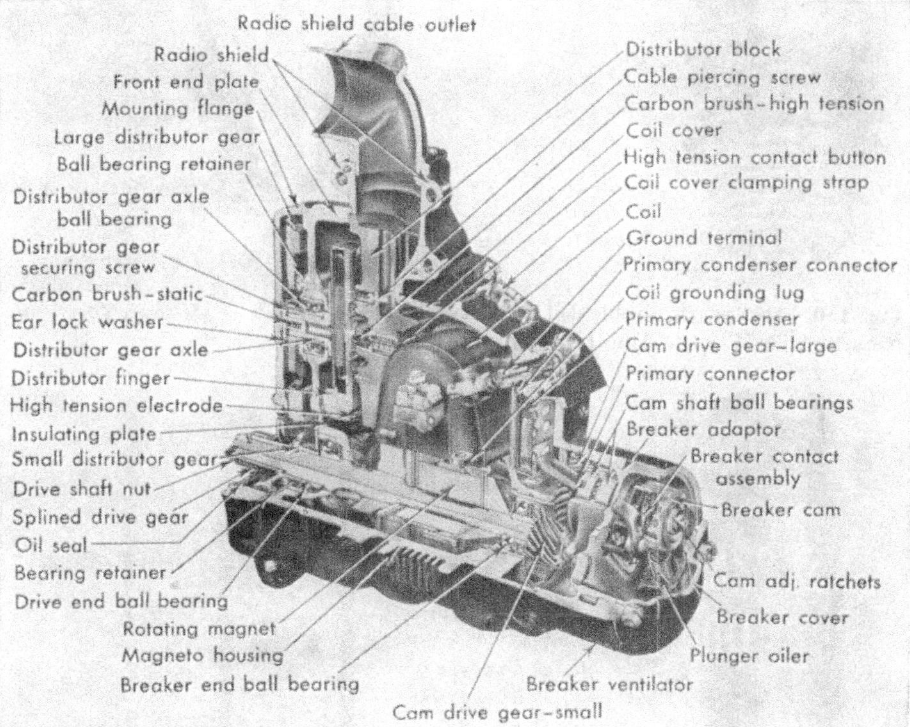

Fig. 149. A cutaway view of an aircraft engine magneto showing its many parts. (Courtesy Scintilla Magneto Division, Bendix Aviation Corporation)

plugs do not cool as rapidly as the cold plug. Hot plugs are more generally used on liquid-cooled engines. On air-cooled engines, arrangements are sometimes made to cool the plug itself. One plug has cooling fins as a part of the insulator. The contact between the center electrode and the high-tension wire leading to the plug must be such that it will remain secure at all times. The fastening may be made by means of a knurled nut, a spring pressure plate, or a spring clip. Whatever the type of fastening, it must be such that it will not work loose because of the vibration of the engine.

The Civil Aeronautics Administration requires that all aircraft engines

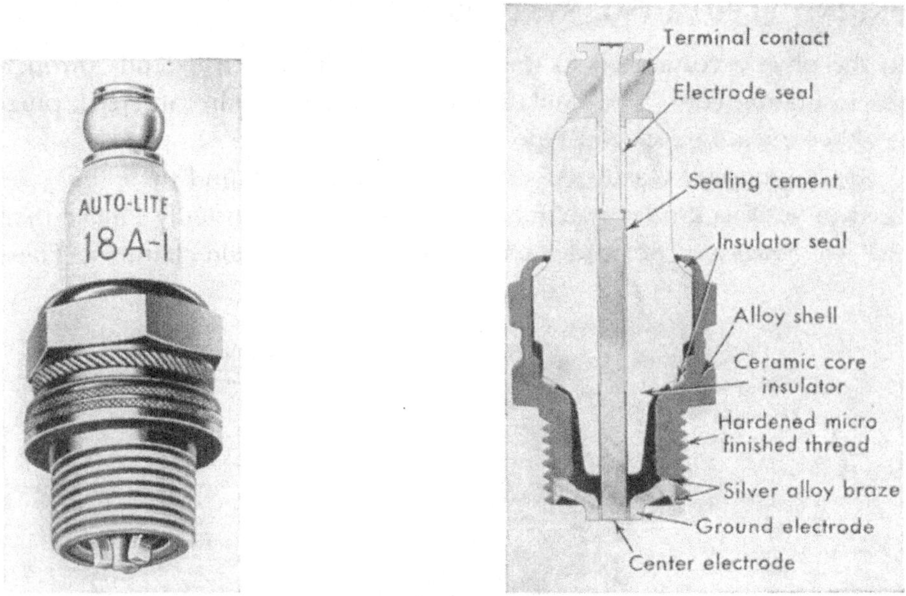

Fig. 150. Short reach, unshielded "cold type" aircraft engine spark plug. Two views. (Courtesy The Electric Auto-Lite Company)

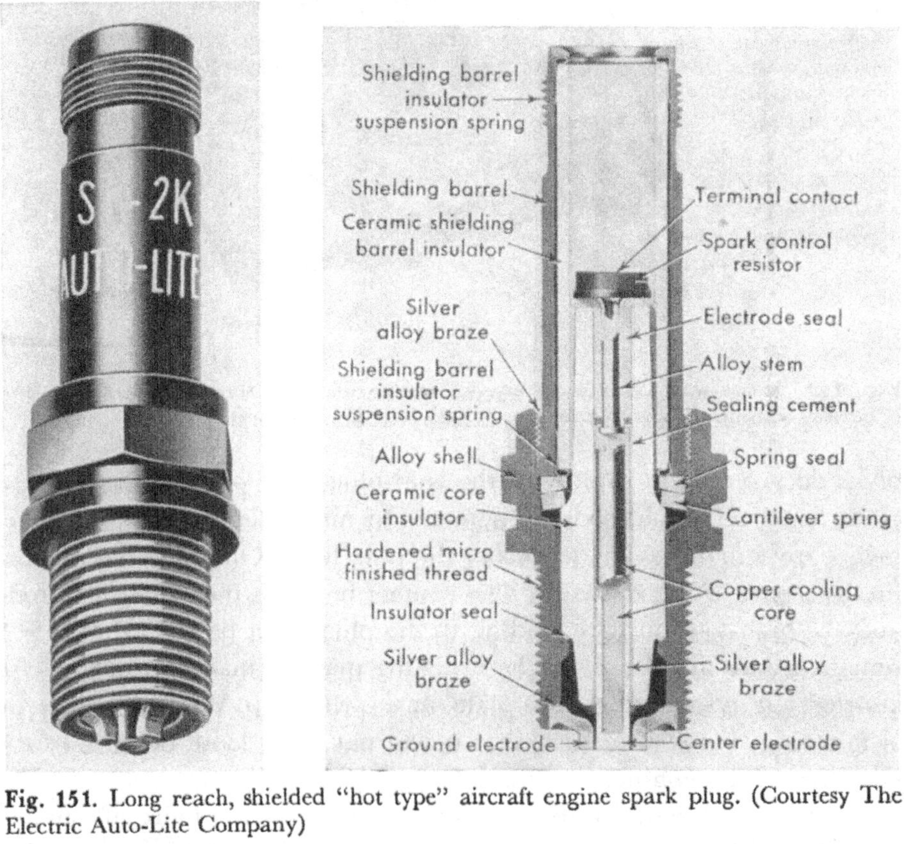

Fig. 151. Long reach, shielded "hot type" aircraft engine spark plug. (Courtesy The Electric Auto-Lite Company)

be equipped with dual ignition. This means that there are two spark plugs in each cylinder combustion chamber, each connected with a separate ignition system. In other words, these engines have two completely separate ignition systems. These separate sources of ignition may

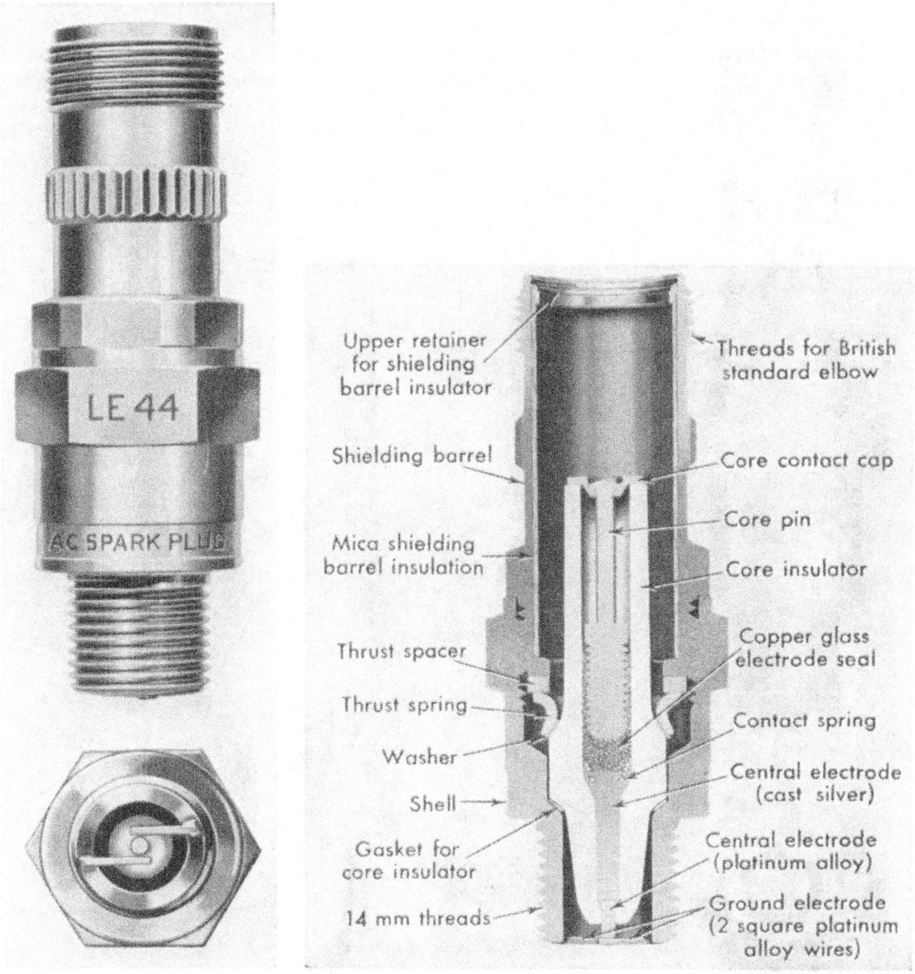

Fig. 152. Shielded aircraft engine spark plug having two "shell" electrodes. The "shell" electrodes are of platinum. Two views. (Courtesy AC Spark Plug Division, General Motors Corporation)

consist of two separate magnetos, two separate battery ignition systems, or a magneto for one set of plugs and a battery ignition for the other set. This dual-ignition arrangement is primarily a safety factor, but it is also a method by which the engine power is increased. An aircraft engine will operate on either set of plugs, but when either set of plugs is cut out, such as by turning off one magneto, the engine operates at a loss of from 50 to 100 r.p.m. Using only one set of plugs may also cause detonation to take place, particularly when the engine is operating under full power.

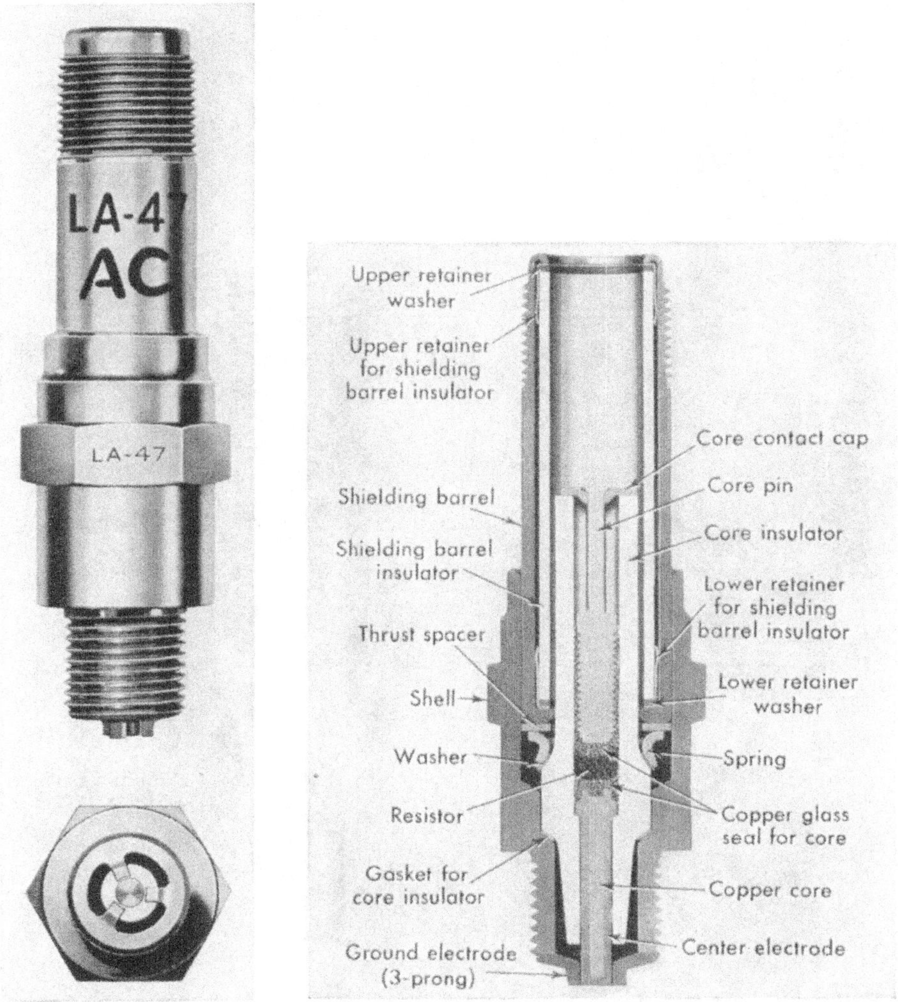

Fig. 153. Shielded aircraft engine spark plug having three "shell" electrodes. Two views. (Courtesy AC Spark Plug Division, General Motors Corporation)

Dual ignition ignites the fuel charge in two places, usually on opposite sides of the combustion chamber. This leads to more rapid burning of the charges and a more nearly equal distribution of the forces developed. Many engines have an arrangement whereby a battery system may be connected with one set of plugs during the time the engine is being started. This assists in starting the engine, as a hot spark is developed at low engine r.p.m.

When a battery is used in an aircraft, it is necessary to have some arrangement whereby the battery will charge while the aircraft is in operation. The battery receives a continuous charge whenever the engine is being operated at ordinary speeds. In some of the older aircraft, as well as in some of the present-day light aircraft, a battery is installed to operate the lights or radio equipment. When no provision is made for charging the battery while in flight, the battery is often found to be run down or inoperative at critical times.

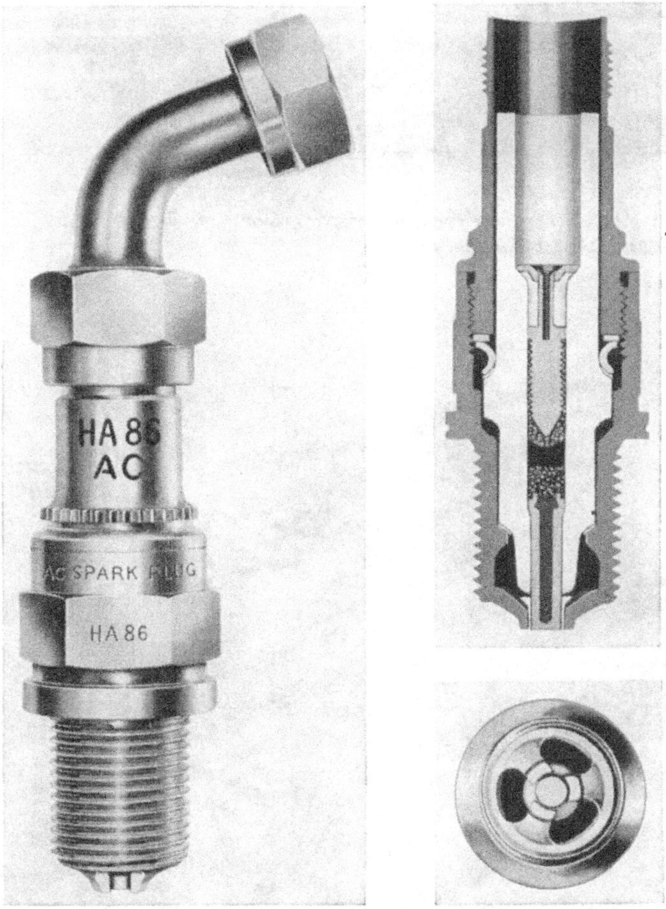

Fig. 154. High altitude aircraft engine spark plug. Three views. (Courtesy AC Spark Plug Division, General Motors Corporation)

The battery-charging system consists of a d-c generator, usually of the shunt-wound type. In this type of generator, the field circuit or

Fig. 155. An aircraft engine driven generator. (Courtesy Delco-Remy Division, General Motors Corporation)

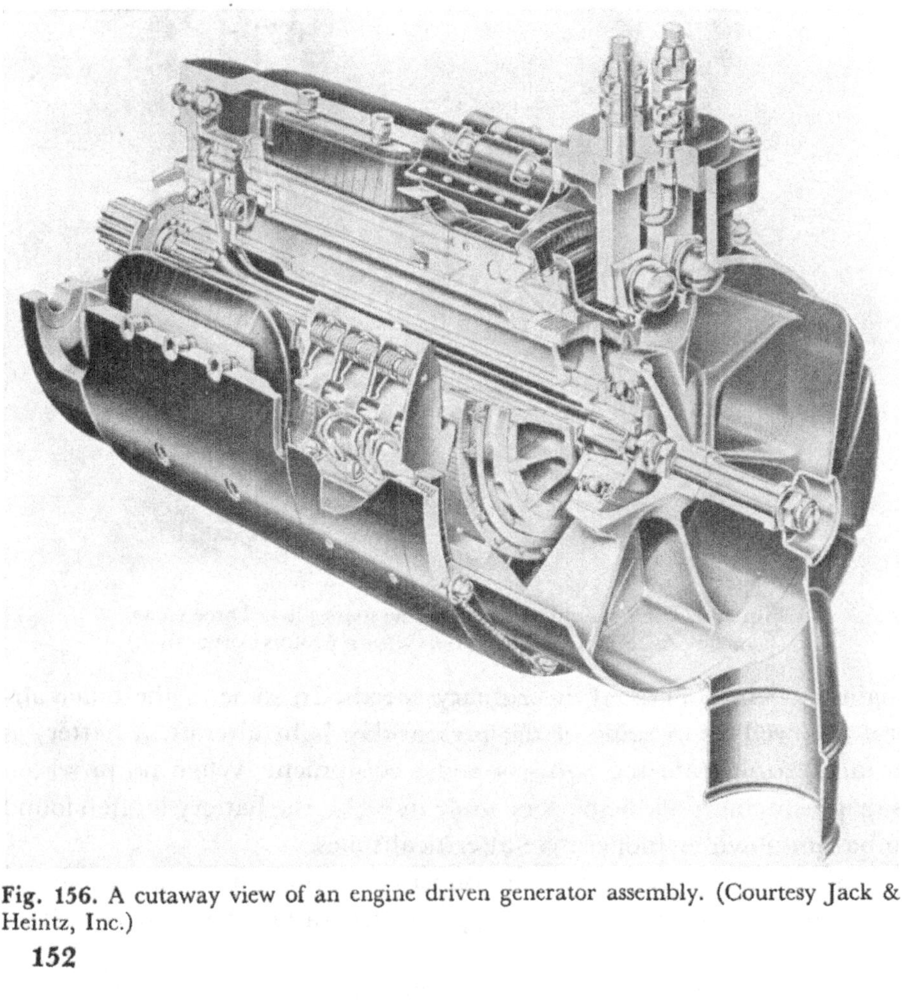

Fig. 156. A cutaway view of an engine driven generator assembly. (Courtesy Jack & Heintz, Inc.)

winding is connected across the armature circuit. This arrangement allows a variation in current output. The amount of current going into the battery depends upon the amount of current flowing through the field circuit. When the battery is low or when a considerable amount of current is used as in starting, the charging rate will be high. As the battery becomes nearly charged, the charging rate becomes approximately zero. The generator is operated by gears from the crankshaft of

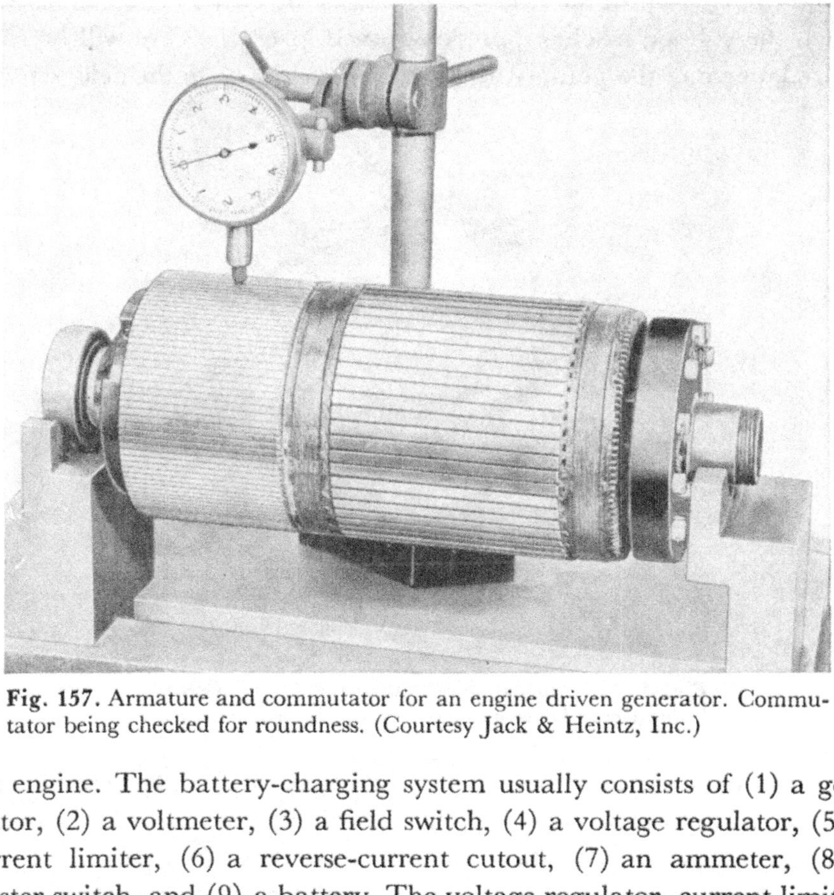

Fig. 157. Armature and commutator for an engine driven generator. Commutator being checked for roundness. (Courtesy Jack & Heintz, Inc.)

the engine. The battery-charging system usually consists of (1) a generator, (2) a voltmeter, (3) a field switch, (4) a voltage regulator, (5) a current limiter, (6) a reverse-current cutout, (7) an ammeter, (8) a master switch, and (9) a battery. The voltage regulator, current limiter, and reverse-current cutout are usually mounted together in a single unit called the *control panel*. The field switch may be mounted on the control panel if it is installed near the pilot in the cockpit, or on the instrument panel if the control panel is mounted elsewhere.

The voltage regulator consists of a core around which is wound a coil of fine insulated wire which is connected across the armature circuit. This means that the winding will be affected by the voltage in the circuit.

A flat soft-iron blade is mounted close to one end of the core. This blade is pivoted and has a pair of contact points attached to one end in such a manner that they may be opened or closed by changes in the voltage of the armature circuit. A suitable spring is attached to one end of this blade and also to an adjusting screw. A resistance is connected to the voltage regulator in such a manner that the current flowing to the generator field will pass through it when the points are opened. This reduces the flow of current to the field and regulates the output of the generator. When the voltage reaches a predetermined level, the core will be magnetized, opening the points which causes the current to the field winding

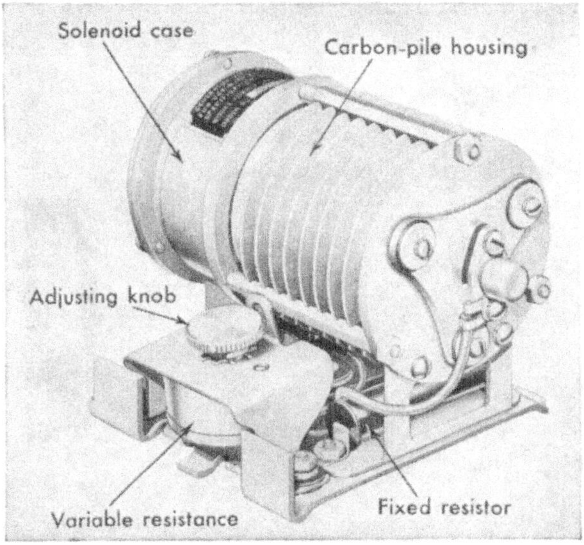

Fig. 158. A carbon pile voltage regulator. (Courtesy Delco-Remy Division, General Motors Corporation)

to be reduced. This holds the output of the generator to the desired limit. If the armature circuit is overloaded, the voltage will not be high enough to open the points and they will remain closed, thus permitting full current flow to the field and developing the maximum output of the generator. If the circuit is too highly loaded, the voltage regulator will not function and the generator may be burned out.

The current limiter is used to protect the generator in this case. The current limiter consists of a core with its winding, a pivoted blade, points and spring, and is quite similar to the voltage regulator. The coil, which is wound around the core, is connected in series with one side of the armature circuit. This coil is affected by the amperage or rate of flow of current in the armature circuit. The resistor and point circuits of the voltage regulator and the current limiter are connected to the field winding of the generator in such a manner that either one may control

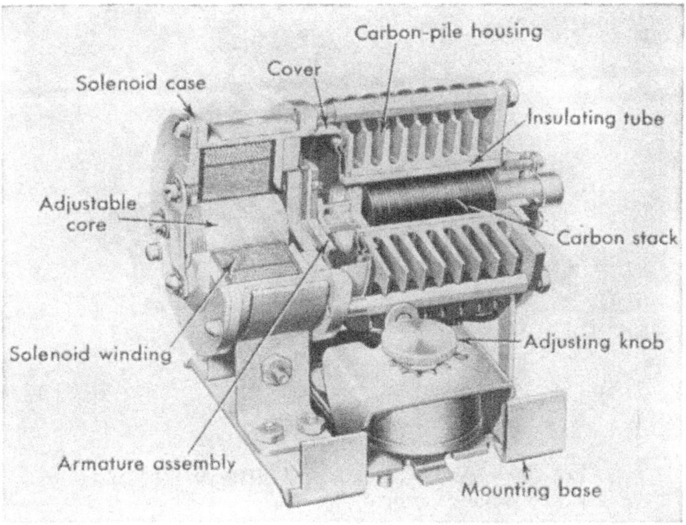

Fig. 159. A cutaway view of a carbon pile voltage regulator. (Courtesy Delco-Remy Division, General Motors Corporation)

the flow of field current and output of the generator. When the circuit load and the charge of the battery are normal, the voltage regulator will regulate the output of the generator by alternately opening and

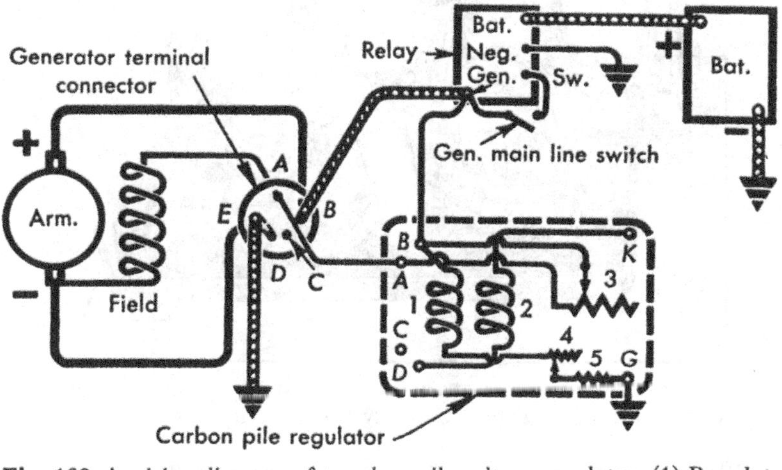

Fig. 160. A wiring diagram of a carbon-pile voltage regulator. (1) Regulating coil; (2) equalizing coil; (3) carbon pile; (4) variable resistance; (5) fixed resistance. (Courtesy Delco-Remy Division, General Motors Corporation)

closing the points at a rate of approximately 45 to 90 times per second. The resistor is cut in and out of the field circuit in order to maintain approximately 14.5 v. If the circuit is overloaded, there will be a voltage

Fig. 161. A relay current control. (Courtesy Leece-Neville Company)

drop, causing an increase in the flow of current to the field winding. This, in turn, increases the amperage or flow of current from the generator. The spring control is so adjusted that, when the output reaches the maximum for which the generator is designed, the current-limiter

Fig. 162. Reverse current cutout mounted on a control panel with a current regulator. (Courtesy Leece-Neville Company)

points will open, decreasing the flow of current from the generator. This device functions only when necessary to limit the output and protect the generator from being burned out.

The reverse-current cutout consists of a wire-wound iron core on one end of which is mounted a blade and contact point. A spring and adjusting-screw assembly is arranged in such a manner as to hold the point open when the system is not functioning. This device prevents the battery current from flowing back through the generator when the engine is stopped.

The field switch provides a means of preventing the flow of current to the generator field in case the electric load is very light and there is no need for current to be sent to the battery.

A voltmeter and ammeter are installed in the circuit to inform the pilot of the voltage of the current being generated and the amount of current flowing.

The master switch, which should be fused, furnishes a means of opening the circuit when the aircraft is stored or is not in use. This switch and fuse eliminate a possible fire hazard.

The firing order of an engine has considerable effect upon vibration. As a rule, the cylinders of an engine do not fire consecutively, but the firing is spaced to produce the least amount of vibration and torsional strain. Radial engines have always been built with an odd number of cylinders in each row, with the exception of some of the new two-cycle aircraft engines now being produced. The firing order for a nine-cylinder engine is usually 1–3–5–7–9–2–4–6–8. One cylinder is skipped each time. In in-line engines, opposed engines, and multiple in-line engines, the firing order is also arranged to produce a minimum amount of vibration. For example, one small, six-cylinder, opposed engine has a firing order of 1–5–3–6–2–4. In one of the large, two-row, radial engines of fourteen cylinders, the back row is numbered 1–3–5–7–9–11–13 and the front row is numbered 2–4–6–8–10–12–14. The firing order is 1–10–5–14–9–4–13–8–3–12–7–2–11–6. Note that the firing skips alternately from row to row.

The timing of the ignition of an engine consists of adjustments which determine the exact time that the spark occurs in the combustion chamber. The spark does not occur at the exact instant when the piston reaches top dead center but somewhat in advance of this point. This adjustment, which is known as spark advance, may be as great as 45° of crankshaft rotation ahead of top dead center. If the spark occurs after top dead center, which is usually the case while starting the engine to prevent kickback, the spark is said to be retarded.

From the time the spark ignites the mixture, a definite time interval

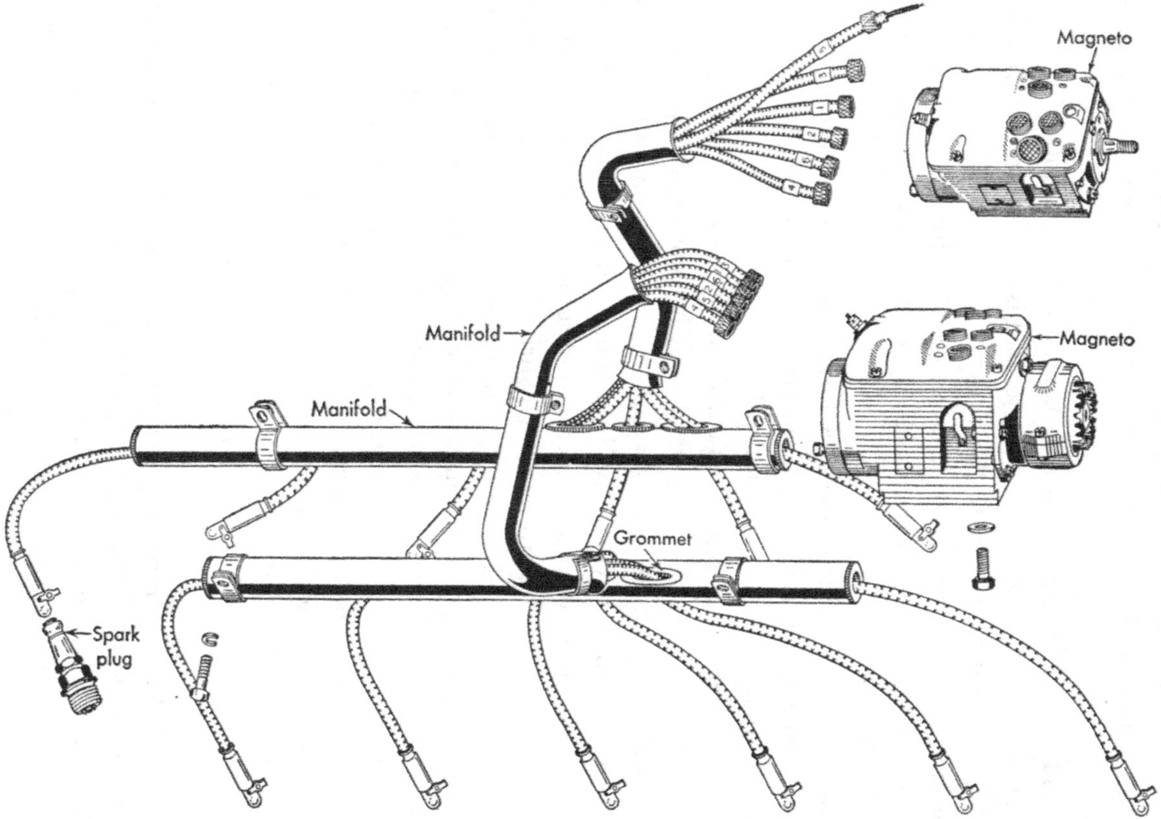

Fig. 163. Magnetos and standard (unshielded) wiring for an inverted six-cylinder aircraft engine. (Courtesy Ranger Aircraft Engines)

is required until the maximum temperature and pressure are reached in the combustion chamber. This period of time is affected by several factors such as (1) the speed of rotation of the engine, (2) the condition

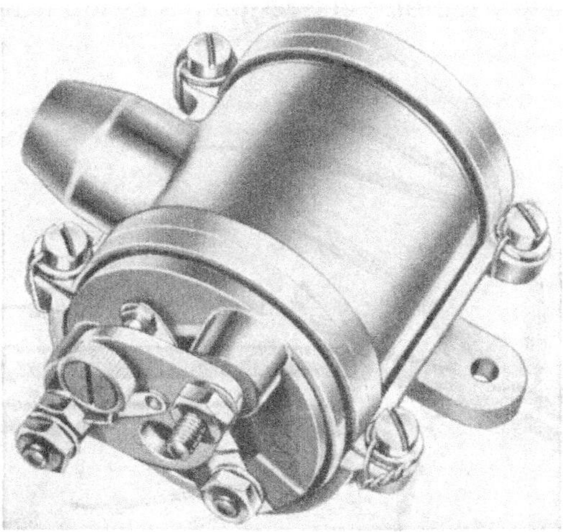

Fig. 164. An ignition booster coil. (Courtesy Ranger Aircraft Engines)

of the mixture in the combustion chamber, (3) the compression ratio, (4) the shape of the combustion chamber, (5) the spark-plug location, (6) the quality of the fuel, and (7) the fuel-air ratio. A period of time equal to about 0.0030 sec. is allowed for the burning of the fuel. An

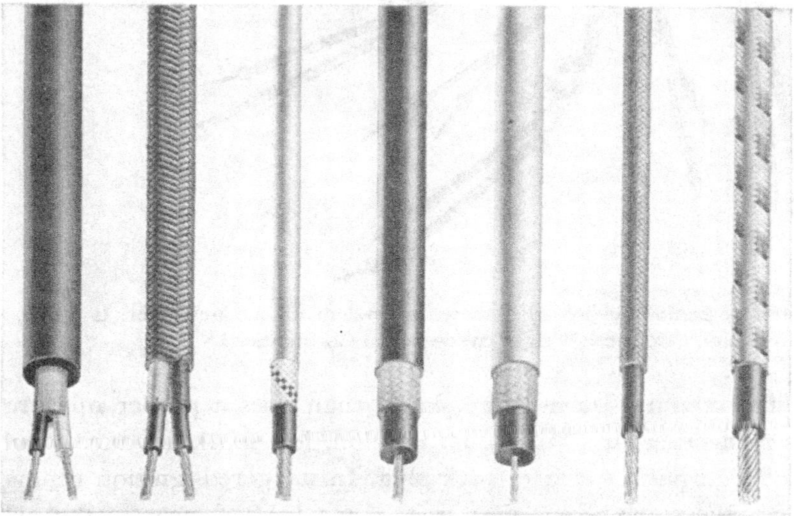

Fig. 165. Various types of aircraft engine ignition high tension wires. The third from the right is called "steelduct." (Courtesy Electric Auto-Lite Company)

engine operating at 2300 r.p.m. would have a spark advance of approximately 35° before top dead center. This advance would allow the maximum temperature and pressure to be developed in the combustion chamber at approximately the time the piston passes top dead center.

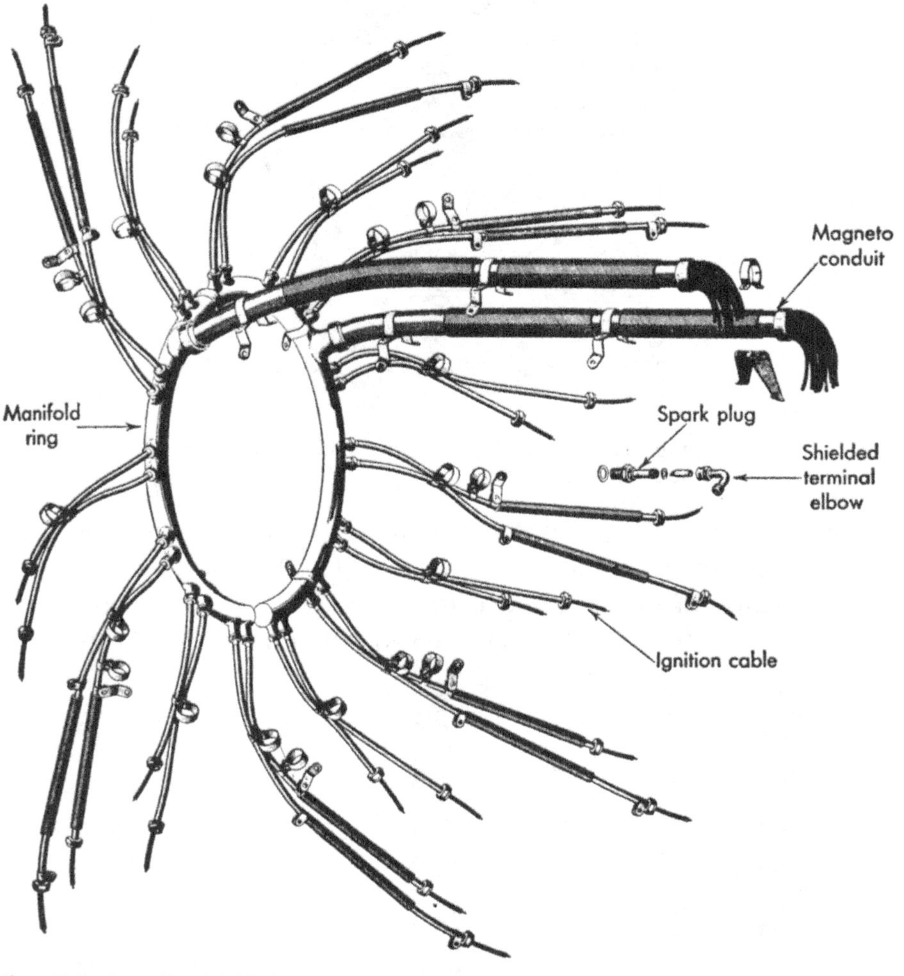

Fig. 166. A radio shielded ignition wiring system for a 14-cylinder, two-row, radial aircraft engine. (Courtesy Wright Aeronautical Corporation)

A lean mixture burns more slowly than does a proper mixture and leads to overheating. With dual ignition, less spark advance would be needed than with a single spark plug. In staggered-ignition timing, the exhaust-side spark plug, that is, the one located nearest the exhaust valve, will fire from 4 to 8° of crankshaft rotation before the plug located near the intake valve. If the spark plugs are located at nearly right angles to the valves, the firing may be synchronized, that is, two plugs fire at exactly the same time. In high-powered single and twin-row radial engines, because of the connecting-rod arrangement, it may be necessary to use compensated timed magnetos. The lobes on the breaker cam of these magnetos are spaced at unequal intervals to cause the magneto to fire unevenly but in proper relation to the variations in piston location caused by the elliptical travel of the knuckle pins. When timing this type of magneto to the engine, the lobe marked with a zero must be the one used in opening the breaker points and the magneto must be timed to the proper cylinder. When this is done, the unequally spaced lobes will be in the proper relation to the cylinders they are designed to fire.

The wires used to carry the high-tension current in the ignition system consist of several fine strands of wire enclosed in thick rubber or other proper insulation. One end of each spark-plug wire has a suitable terminal for attaching to the spark plug, while the other end is attached securely to the metal block in the distributor head. The other high-tension lead is from the coil to the distributor.

If the aircraft is equipped with radio, the entire ignition system must be shielded. This is done by enclosing all parts of the electrical system in the proper metal shields. The wires are usually covered with a flexible woven-metal conduit, and a shield is arranged to cover the spark plugs. Aircraft equipped with radio usually have the metal parts of the plane bonded, which means that all the metal parts of the aircraft are connected by means of suitable conductors that are in turn properly grounded.

PART II AIRCRAFT HYDRAULIC SYSTEMS

XII HYDRAULIC PRINCIPLES

For hundreds of years men have studied the flow of liquids, but it was not until recent years that the use of liquids under pressure came into common use. Hydraulics is the branch of science that deals with the study of liquids, particularly of liquids in motion. The word hydraulics is used when the liquids are in motion, and the word hydrostatic is applied when the liquids are in a state of equilibrium or at rest. The basic principle upon which aircraft and all other hydraulic systems operate is given in Pascal's principle which states, "Pressure exerted anywhere on a confined liquid is transmitted undiminished to every portion of the interior of the containing vessel."

Properties of Gases and Liquids. Gases are made up of particles which are free to move about and have no tendency to cling together. Any amount of gas confined in a container distributes itself evenly throughout the container. If unconfined, gases seem to expand indefinitely.

In liquids, the particles are free to move about but have a tendency to remain close to each other. A liquid in a vessel settles to the lowest part of the vessel. Some particles of the liquid escape from the surface. This process is called evaporation. Liquids and gases expand and contract due to changes in temperature. Gases may be compressed to almost any volume if enough pressure is applied. Liquids are practically incompressible. By decreasing pressure on a confined gas, a gas can be made to expand to fill any size of container. Liquids cannot be made to expand. Of course, a tremendous pressure or suction will cause a liquid to be compressed or slightly expanded. If a metal sphere is filled with a liquid, pressure applied to any point on this liquid would, as stated in Pascal's principle, be carried undiminished through the liquid to all the interior area of the sphere.

A pipe with a cross-section area of ½ sq. in. is inserted into a metal sphere having an inside area of 100 sq. in. This pipe is filled with liquid and, by means of a plunger in the pipe, a pressure of 10 lb. per sq. in.

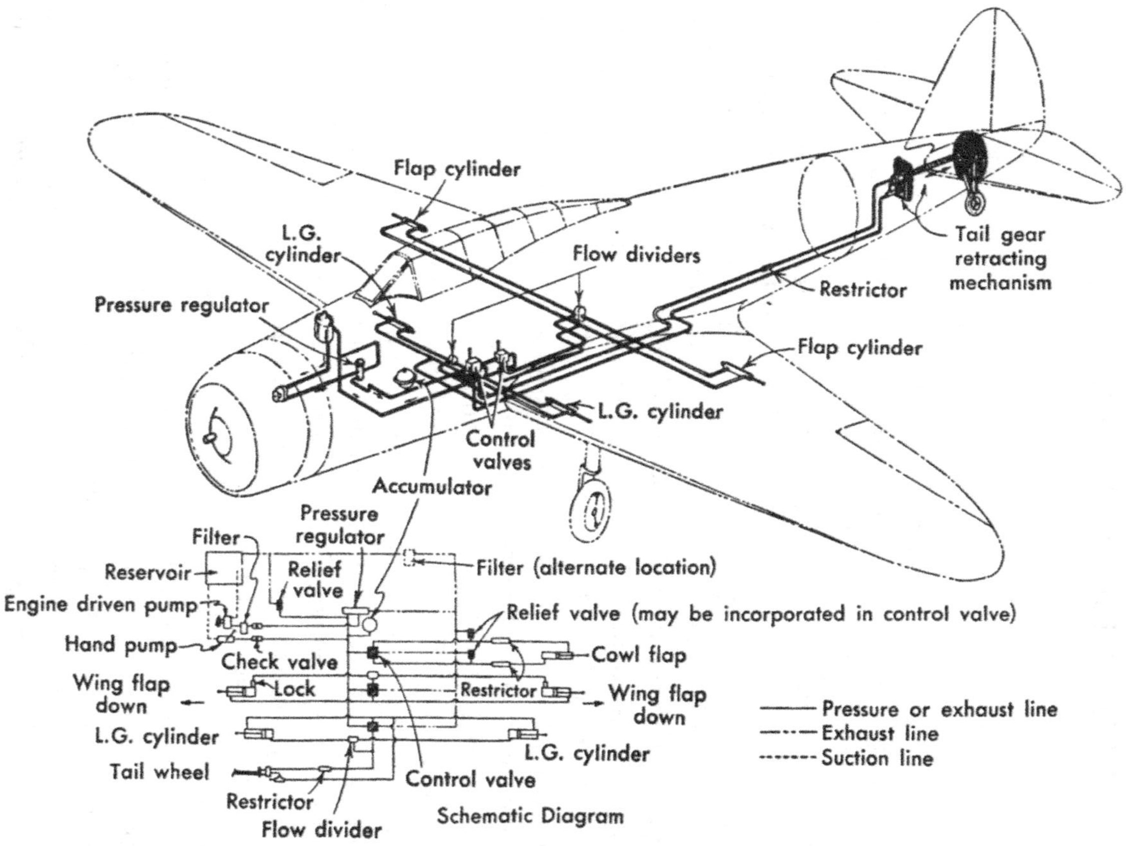

Fig. 167. A typical hydraulic installation in an airplane. (Courtesy Pesco Products Company)

is applied to the liquid. Since the area of the liquid in the pipe acted upon by the plunger is ½ sq. in., a pressure on the plunger of 5 lb. is all that is necessary to exert a pressure of 10 lb. per sq. in. on each square inch of the inside of the sphere. The liquid in the sphere will transmit to the walls of the sphere this pressure of 10 lb. per sq. in. Since

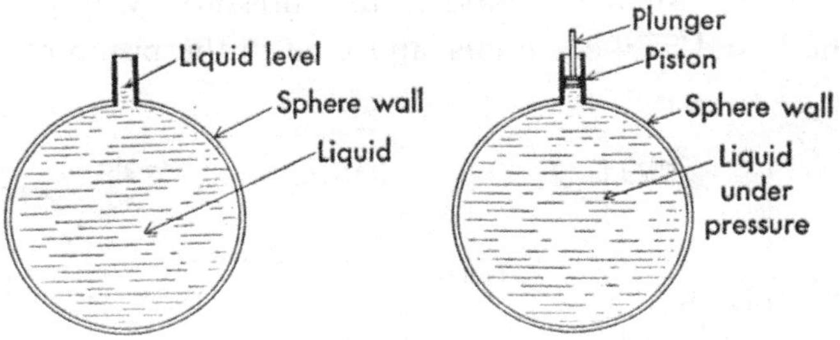

Fig. 168. Metal spheres to show distribution of hydraulic pressure.

the area of the inside of the sphere is 100 sq. in., the total pressure on the inside of the sphere would be 100 × 10 or 1000 lb. Each square inch of liquid within the sphere would press against the walls of the sphere with 10 lb. of pressure.

Force and Pressure. An understanding of the terms force and pressure is necessary. A force may be considered as a push or pull. For example, the total force acting on the inside of the sphere is equal to 1000 lb. Pressure may be considered as the push or pull exerted on a unit area, for example, 10 lb. of pressure per sq. in. The pressure multiplied by the number of units of area to which it is applied gives the total force acting on the area.

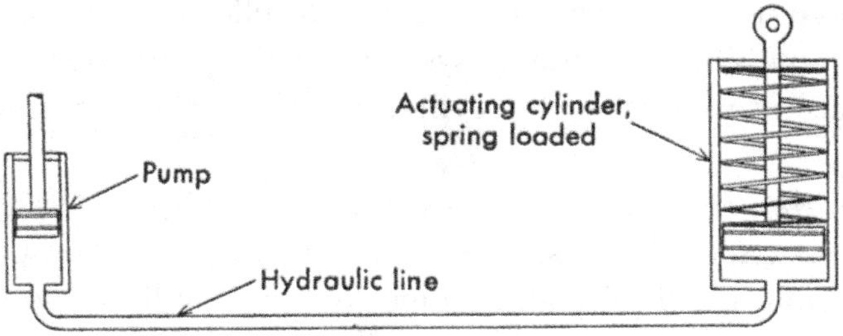

Fig. 169. A schematic drawing of a cylinder to show the hydraulic principle.

This same principle is basic in the operation of hydraulic lifts, hydraulic jacks, and hydraulic cylinders as used in the aircraft hydraulic system.

If two cylinders, each having a closely fitted piston, are connected with a pipe, the transfer of power by means of a column of liquid can be illustrated. Assume that the area of the piston in one cylinder is 1 sq. in., and the area of the piston in the second cylinder is 10 sq. in. If these cylinders are arranged as shown in Figure 169 and pressure is applied to the piston in the smaller cylinder, this pressure will be transferred through the liquid in the cylinder and pipe to the piston of the second

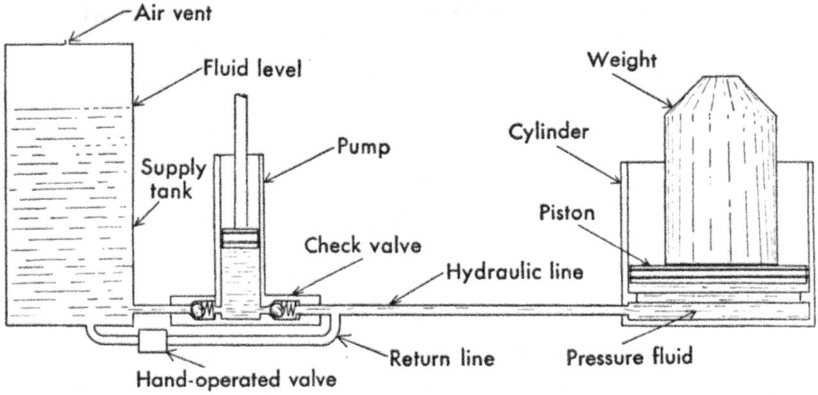

Fig. 170. A schematic drawing to show how the hydraulic principle is applied in a hydraulic jack.

cylinder. If a pressure of 100 lb. is applied to the piston by means of a lever, this pressure will be transmitted undiminished to the piston in the larger cylinder. Since the area of the larger cylinder is 10 sq. in., the pressure of 100 lb. per sq. in., when transmitted to the liquid through the piston in the larger cylinder, will lift a weight of 1000 lb. Friction losses are ignored in this example.

The above describes the action of a hydraulic jack or lift, but might illustrate also a simple hydraulic system, the first piston being the pump and the second piston being a cylinder used to move some part of an aircraft.

Simple Aircraft Hydraulic System. A simple aircraft hydraulic system must have certain essential parts. These parts consist of a pump, a supply tank to contain the liquid, and a pipe to carry the liquid to a cylinder. The cylinder must contain a piston which is fastened to the part to be moved, such as a retractable landing gear or wing flaps. As the pump is operated, liquid is drawn from the supply tank and forced through the pipe. This liquid enters the cylinder and, pressing against the end of the piston, forces it outward in the cylinder. The connection between the cylinder and the flap is such that, as the piston moves outward, the flap is opened. The flap may be stopped in any position by stopping the action of the pump.

This is the simplest form of hydraulic system. To close the flap in this simple system, it would be necessary to reverse the direction of the liquid through the pump and pump the liquid from the cylinder back to the supply tank. The fact that the liquids are incompressible and cannot be stretched makes this possible.The force exerted in a negative direction by the pump drawing the liquid from the cylinder would pull the piston back to its original position, closing the flap.

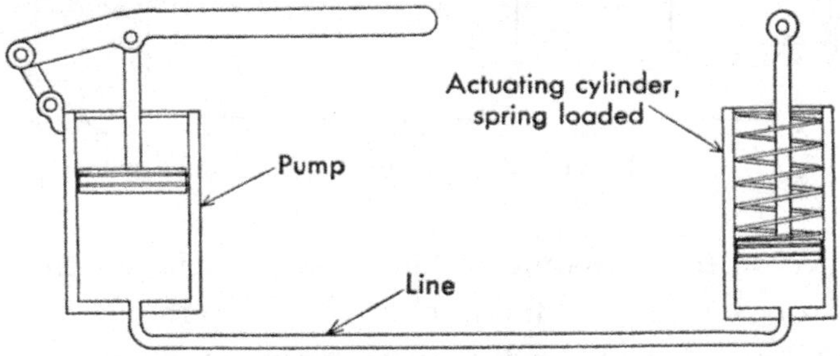

Fig. 171. A schematic drawing to show the simplest form of a hydraulic system which consists of a pump, connecting line, and cylinder.

However, this simple system would not be satisfactory for use in an aircraft because it would not be practical to reverse the direction of the pump. The following illustrations show step by step the development of a simple hydraulic system. Figure 171 shows the simplest type of what might be considered a hydraulic system.

This system consists of a cylinder in which a piston may be operated by means of a hand lever. The cylinder is connected with an actuating cylinder by means of tubing. If the system is completely filled with liquid and the piston in the master cylinder is moved back and forth, the piston in the actuating cylinder will also move back and forth. By means of this simple arrangement, it would be possible for the pilot to move flaps or landing gear by moving the lever back and forth. This system, however, would be clumsy and, because the pistons are equal in size, there would be no mechanical advantage and the pilot might not have strength enough to operate the part to be moved. In this simple system, all of the fluid used is contained within the system itself.

Figure 172 illustrates a slightly more complex system which is equipped with a reservoir to contain an extra supply of hydraulic fluid. In this simple system, the master cylinder may be much smaller than the actuating cylinder. This smaller cylinder acting through the liquid to the larger cylinder would give the pilot a mechanical advantage because a small pressure applied to the piston of the master cylinder would develop

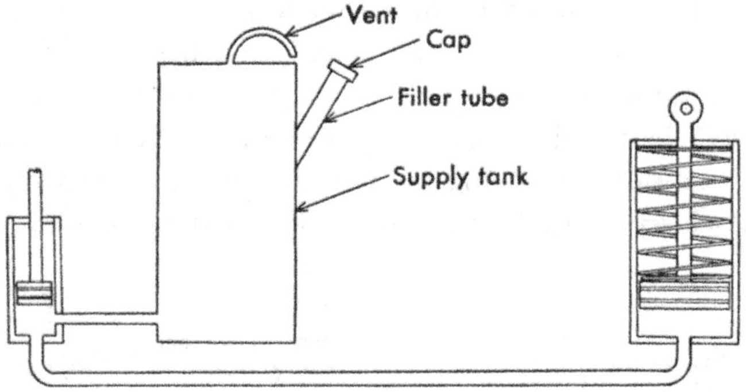

Fig. 172. A schematic drawing to show a simple hydraulic system.

large pressure on the actuating piston. However, the system illustrated would not work because the fluid, following the path of least resistance, would be pumped directly back into the reservoir.

The next step would then be the placing of check valves in the system to prevent the fluid from flowing either back into the reservoir or back into the master cylinder on the return stroke. The system with the check valves installed, as shown in Figure 173, would not be practical since

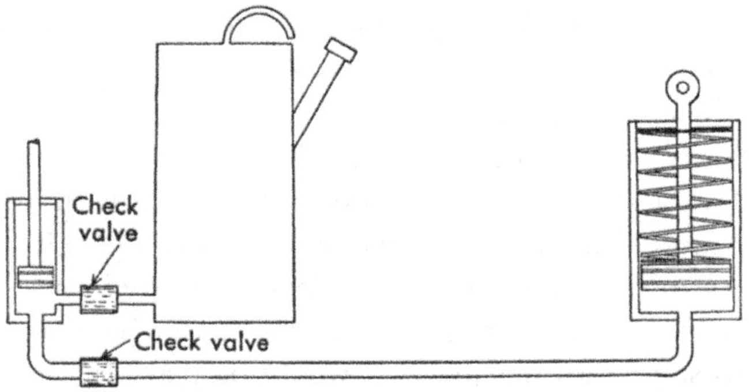

Fig. 173. A schematic drawing of a hydraulic system with check valves installed.

there is no way to allow the fluid to return from the actuating cylinder to the reservoir. Check valves operate in one direction only.

Figure 174 shows the installation of a return line connected to the reservoir and a selector valve by which the pilot may direct the flow of fluid either to the actuating cylinder or from the actuating cylinder to the reservoir. In all of these simple systems, the return of the piston in the actuating cylinder would have to be accomplished either by spring pressure or by forces acting upon the movable part which would manually force it back into its original position.

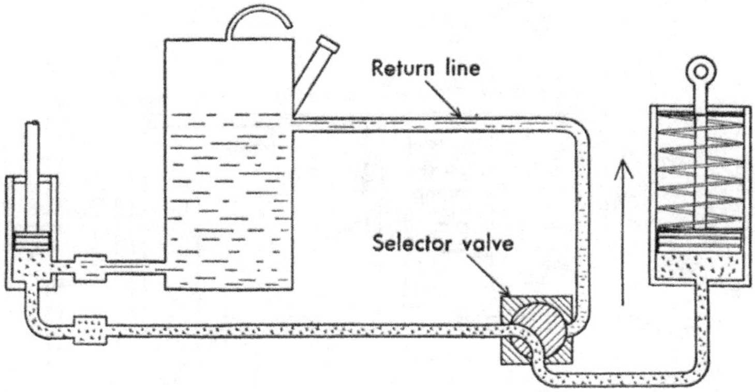

Fig. 174. A schematic drawing of a hydraulic system with return line and selector valve added.

Figure 175 shows an arrangement whereby the hydraulic fluid under pressure may, by means of a selector valve, be applied to either side of the actuating piston. The piston moving in one direction in the cylinder lowers a flap or a landing gear, but, when the direction of flow of the hydraulic fluid is reversed by means of the selector valve, the piston reverses the operation. This arrangement would constitute a practical

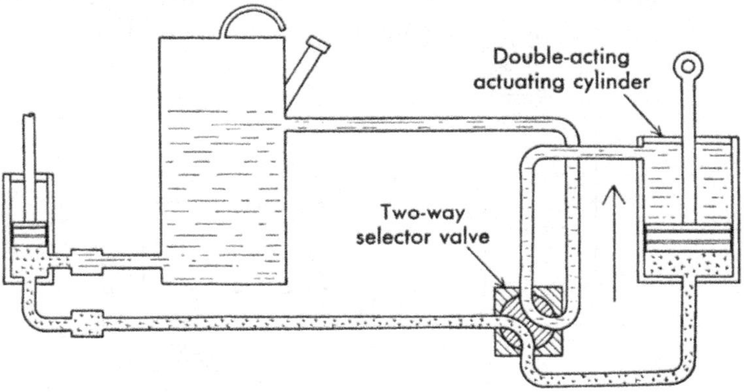

Fig. 175. A schematic drawing of a hydraulic system with a two-way selector valve and a double-acting cylinder added.

and workable hydraulic system. However, due to the mechanical advantage desired, only a small amount of fluid could be delivered by the master pump during each stroke. To relieve the pilot of the effort necessary to pump the fluid by hand, a power pump is installed.

The power pump, as is shown in Figure 176, may run continuously or operate under the control of the pilot. When running continuously, it would be necessary to install a pressure-relief valve such as that shown

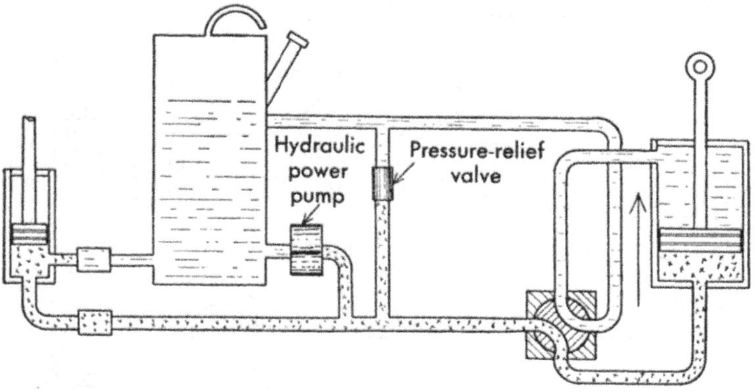

Fig. 176. A schematic drawing of a hydraulic system with a power pump and pressure relief valve installed.

in Figure 176. This valve could be set at any desired pressure. When the desired pressure in the system is reached, the fluid from the pump would be by-passed back to the reservoir. This arrangement would maintain the desired pressure in the system at all times.

Since the hydraulic fluid is incompressible, any sudden relief of pressure, such as allowing the fluid to flow into a large actuating cylinder, would cause an immediate decided drop in pressure in the system. To prevent this sudden drop, an accumulator is installed. This accumulator

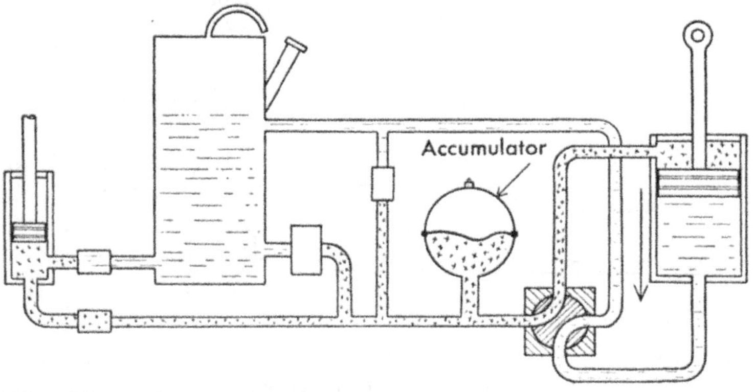

Fig. 177. A schematic drawing of an accumulator installed, forming a complete elementary hydraulic system.

is simply a device which allows a certain amount of fluid to accumulate in the system under either pneumatic or spring pressure. When fluid from the system is allowed to flow into an actuating cylinder, the fluid

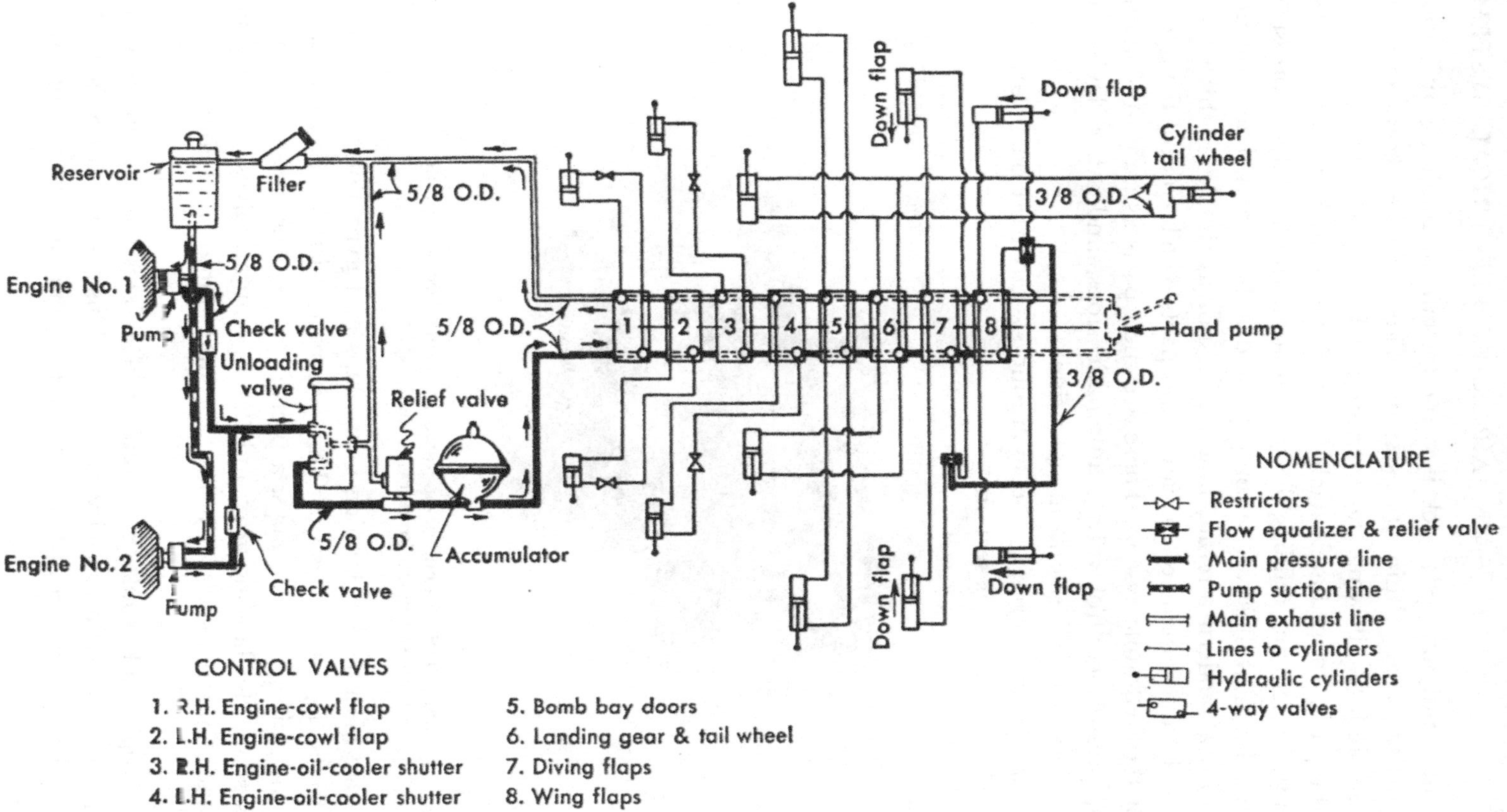

Fig. 178. A schematic layout of a complete hydraulic aircraft system. (Courtesy Pesco Products Company)

in the accumulator feeds back into the system, maintaining an almost constant pressure within the system. The pressure, however, does decrease slightly allowing the pressure-relief valve to close, and fluid from the pump immediately feeds into the system building up the pressure in the system and refilling the accumulator with fluid under pressure.

Figure 177 now illustrates a complete hydraulic system. The hydraulic system on a modern aircraft, however, is not as simple as this system because there are many other valves and parts which control the operation of the hydraulic system. There are such parts as pump-unloading valves, pressure-control valves, automatic stops, and other devices to make the system effective in its operation.

Typical Hydraulic System. The hydraulic system used on a large aircraft is described in this section.

Fig. 179. An engine-operated hydraulic pressure pump. (Courtesy Pesco Products Company)

On this aircraft, the landing gear, flaps and gyropilot are operated by a single hydraulic system. This system consists of a supply tank, hydraulic fluid pressure pumps, a pressure regulating system, the gyropilot system, the landing-gear and flap-operation valves, the landing-gear and flap-operating cylinders, a hand pump, and four pressure gauges.

The supply tank is located in the fuselage. This tank is vented to the atmosphere and has a capacity of 2½ gal. This system uses Sperry Oil. Sperry Oil is a petroleum hydraulic fluid having a viscosity somewhat less than S.A.E. 10.

The level of the hydraulic fluid in this tank is measured by means of a stick gauge. The hydraulic fluid pumps are engine driven and are mounted on the accessory drives of the engine itself. The pumps draw fluid from the supply tank and deliver it, under pressure, to the pressure-regulating system.

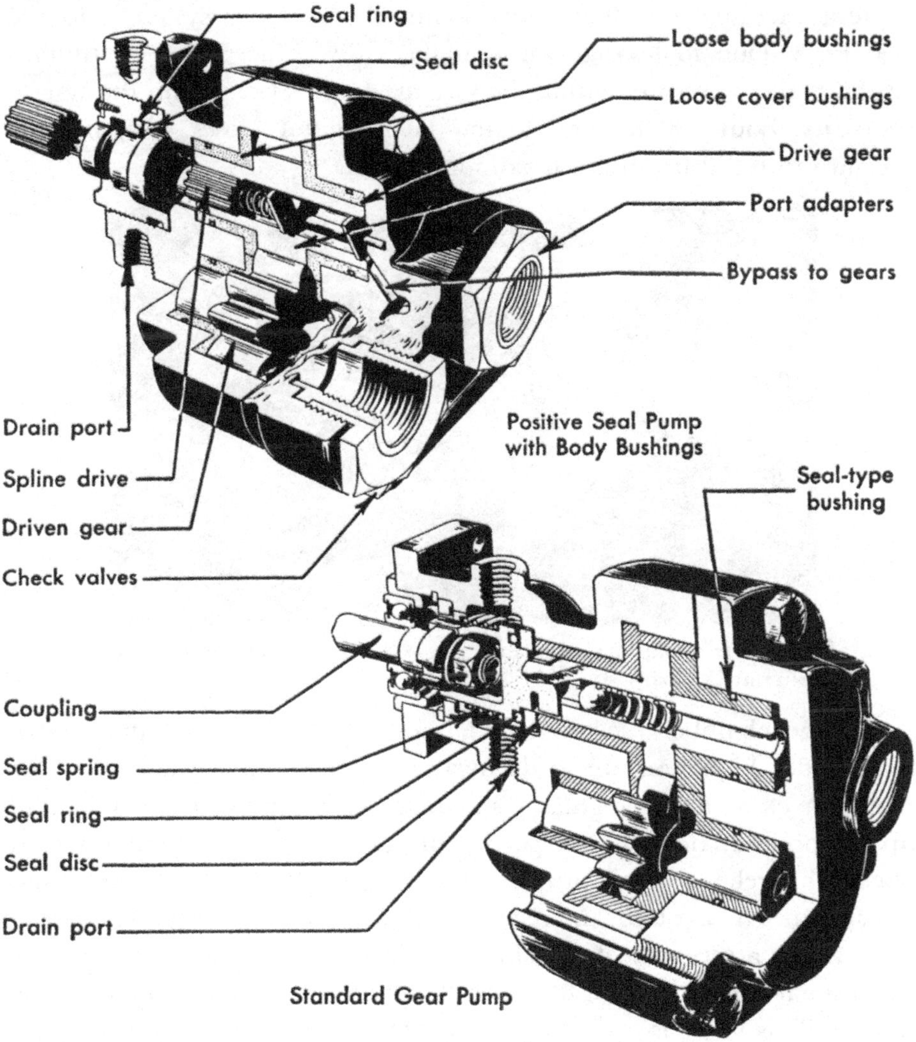

Fig. 180. A cutaway view of an engine-operated hydraulic pressure pump. (Courtesy Pesco Products Company)

The pressure-regulating system consists of one double-ball check valve, combined pressure-regulator and surge-chamber relief valves, and by-pass valves.

The pressure-regulating system is located in the control tunnel in the fuselage. The pressure regulator is set to open at approximately 700 lb. pressure per sq. in. and by-passes the fluid from the pumps directly to the supply tank. This pressure-regulating system relieves the load on the pumps. The pressure in the pump feed lines is practically zero. Because of this low pressure it is necessary to use double-ball check valves to prevent excessive backflow from within the pressure system into the feed lines. Due to leakage through the valves, there is a continuous gradual fall in pressure within the system. As pressure within the system drops to about 500 lb. per sq. in. the regulator closes, allowing the pressure to build up to the maximum again.

Fig. 181. A flap actuating hydraulic cylinder. (Courtesy Douglas Aircraft Company, Inc.)

A surge chamber (accumulator) is provided to store a quantity of fluid under high pressure to do away with the constant action of the regulator cam. If the regulator should stick, the pressure would build up to approximately 900 lb. per sq. in., and then a relief valve would open. This relief valve is provided for the protection of the pumps and the system. In case both the regulator and relief valves cease to operate properly or a serious leak develops within the system, a hand-operated by-pass valve may be used to return the fluid directly back to the engine pumps. This valve is operated by a control handle located on the floor of the pilot's compartment on the left of the control pedestal.

The landing-gear valve consists of a four-way valve and is located on the left side of the control tunnel. This valve is operated by means of the landing-gear control lever which is mounted on the control pedestal. The valve is provided with an emergency control lever which extends through the floor of the cockpit at the foot of the control pedestal. This

Fig. 182. A typical hydraulically operated landing gear. (Courtesy Douglas Aircraft Company, Inc.)

emergency lever is protected by a snap-on cover, and is built to receive a screw driver or a similar object as a handle.

The flap-operating valve is located on the right side of the control tunnel and is operated by the flap-control lever which is mounted on the control pedestal. The flap-control valve is also provided with an emergency control lever similar to the one used for operating the landing-gear-operating valves.

The landing-gear hydraulic cylinders are located in the wheel wells at the spar and are connected to the drag strut of the landing gear. When the wheels are down, the piston is protected from dirt by being entirely withdrawn into the cylinder and locked in position by a hydraulic lock.

The hydraulic lock is located at the head of the landing-gear-actuating cylinder. It consists of a spring-loaded pin which rides in a notch in the end of the piston when the piston is fully withdrawn into the cylinder. This pin locks the piston in position until the pin is withdrawn from the notch. The fluid used to extend the piston is fed through this lock and forces the pin out of the notch before the fluid is fed into the cylinder. The pin itself extends back through the end of its containing cylinder. When the pin has dropped into the notch in the piston, the end of the pin should be flush with the lock-cylinder cap.

A cushioning effect is given to the landing gear as it nears the end of its full down-travel by a cup in the end of the piston, which slides over a cylindrically shaped extension on the end of the cylinder. This extension traps a small amount of fluid in the cup, and the fluid escapes by leaking around the side of the cup, giving a cushioning effect.

XIII OPERATION
OF THE HYDRAULIC SYSTEM

A typical aircraft hydraulic system is made up of the following parts: a pump, a supply tank, a pipe for carrying the hydraulic fluid, hydraulic actuating cylinders, selector valves, by-pass valves, pressure regulators, and an accumulator.

Usually a pressure gauge is installed to show the pressure of the fluid in the system. Filters are also installed to prevent foreign matter from entering the cylinders or any part of the system.

To develop pressure in the system, a hand pump was first used which consisted of a single-action pump as illustrated in Figure 183. In this

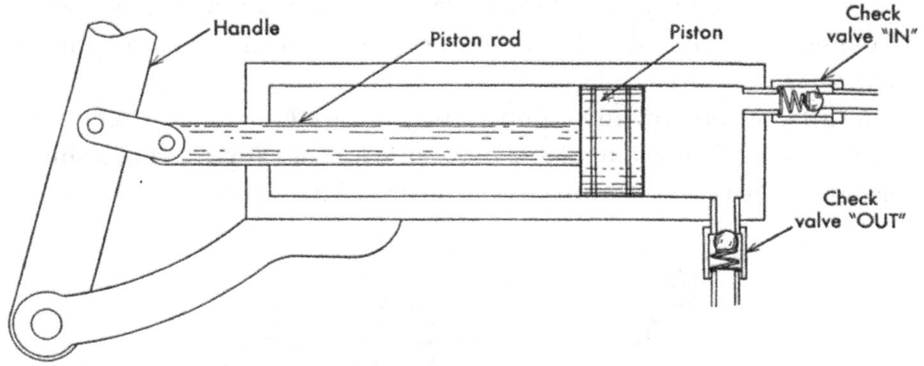

Fig. 183. A drawing to illustrate a single-acting hand pump.

pump, while the piston was drawn in one direction in the cylinder, the fluid was drawn into the cylinder from the supply tank. As the piston was moved in the opposite direction, this fluid was forced out through a valve into the pipe leading to the hydraulic system. As the piston moved back and forth, the fluid was delivered to the system in a series of spurts.

A double-action pump, as shown in Figure 184, produces a steadier flow of fluid because fluid is drawn into the cylinder on both sides of the piston, and each stroke of the piston in the cylinder delivers a quantity of fluid to the system.

In most systems, an engine-driven pump, usually of the gear type, is used, and a hand pump is also installed to be used as a stand-by in case the engine-driven pump fails. If the hand pump were connected directly

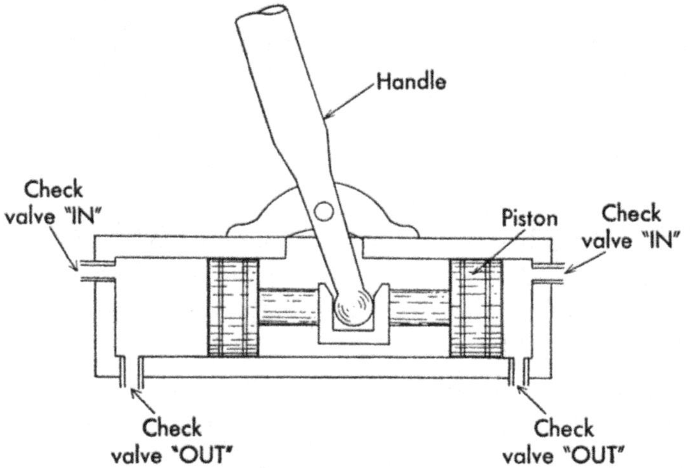

Fig. 184. A drawing to illustrate a double-acting hand pump.

to the cylinder by means of the line, the pressure would be applied to the cylinder unevenly, which would cause, in turn, an uneven operation of the cylinder, and the flap would be opened or closed in a series of short jerks.

In order to obtain smooth operation, an accumulator of some type must be used. The accumulator may consist of a cylinder containing a spring-loaded piston, shown in Figure 188, into which fluid from the

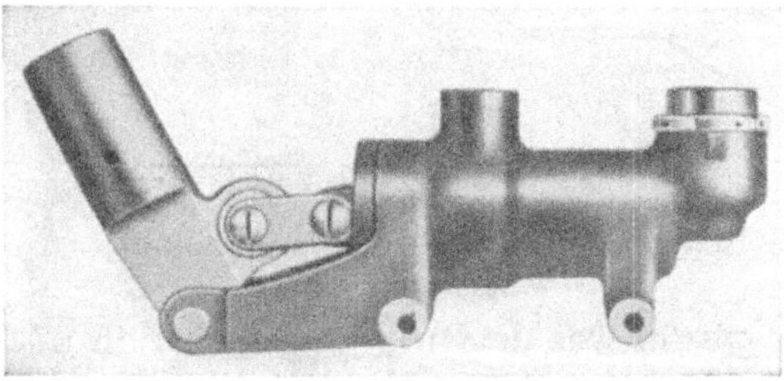

Fig. 185. A hand-operated hydraulic pump. (Courtesy Pesco Products Company)

pump is delivered, building up a pressure which remains constant as long as any fluid remains in the cylinder. However, the more common type of accumulator, illustrated by Figure 189, consists of a hollow metal sphere divided by a rubber diaphragm. Air under pressure is forced into the sphere on one side of the diaphragm causing it to fill the sphere completely. The air pressure in the accumulator is usually several

Fig. 186. An engine-driven hydraulic pump. (Courtesy Pesco Products Company)

hundred pounds per square inch. It should be approximately equal to the pressure of the liquid in the system. When the pump is started, fluid is forced into the sphere on the opposite side of the diaphragm, forcing the diaphragm to compress the air. The compressed air acts as a spring maintaining constant pressure on the diaphragm. This action causes a

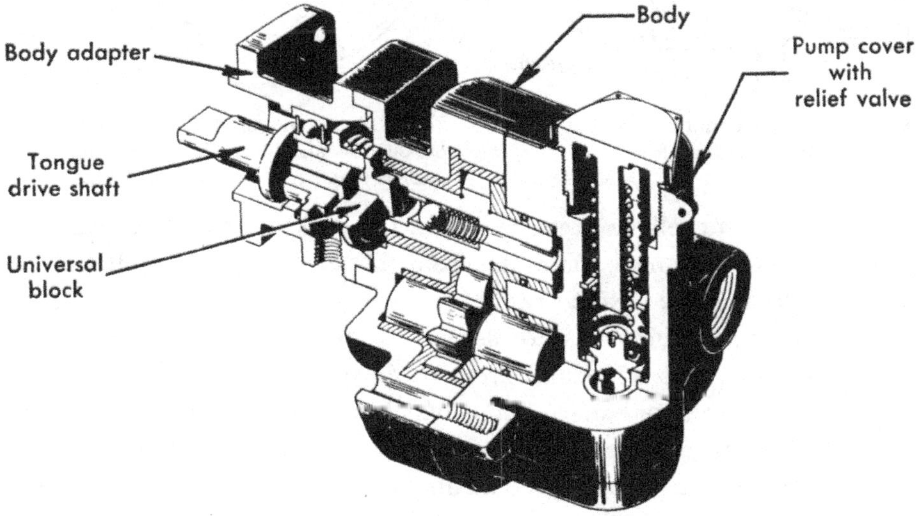

Fig. 187. A cutaway view of an engine-driven gear pump with integral relief valve. (Courtesy Pesco Products Company)

smooth flow of fluid from the accumulator to the lines of the hydraulic system.

When an engine-driven pump is used, a pressure-relief valve is installed in such a manner that, when the pressure in the system reaches a predetermined point, the relief valve opens and allows a free flow of fluid back into the supply tank. A pressure-relief-valve arrangement is shown in Figure 190.

The supply tank is usually constructed of welded aluminum and is arranged with its axis vertical. This tank must be equipped with a filler cap, drain, and vent. The vent must be installed to take care of expansion

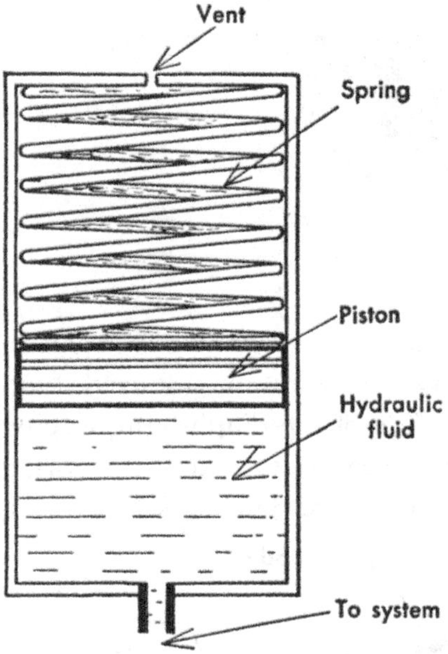

Fig. 188. A schematic drawing of a spring-loaded accumulator.

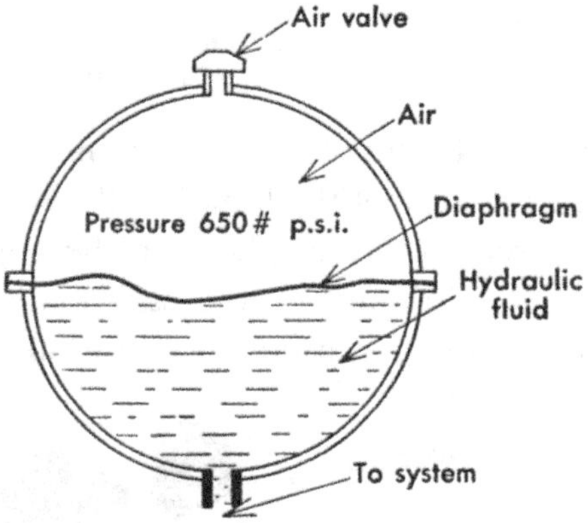

Fig. 189. A schematic drawing of an air-pressure accumulator. Pressure in accumulator is equal to pressure in system.

of the fluid or to relieve pressure within the tank in case of oil leakage. A filter is installed near the outlet from this tank. The filter should be arranged so as to be cleaned easily.

The hydraulic fluid lines are usually of aluminum tubing and must be strong enough to withstand pressures in the system. They should be installed so as to avoid sharp bends which obstruct the free flow of the fluid. The joints must be made in such a manner as to prevent leaks and cause no restriction in the pipe. The lines must be free from tensile forces. It is necessary, at times, in order to obtain flexibility in the line, to install a section of flexible, hydraulic, high-pressure hose.

The pressure gauge should be installed in the main line to show at all times the pressure within the system.

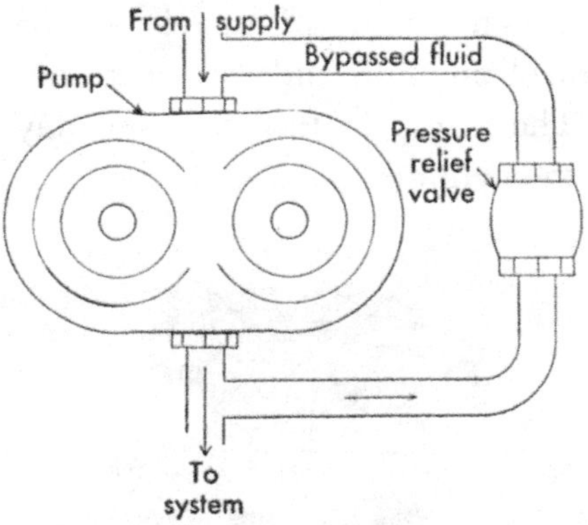

Fig. 190. A drawing to show a typical pressure relief valve installation.

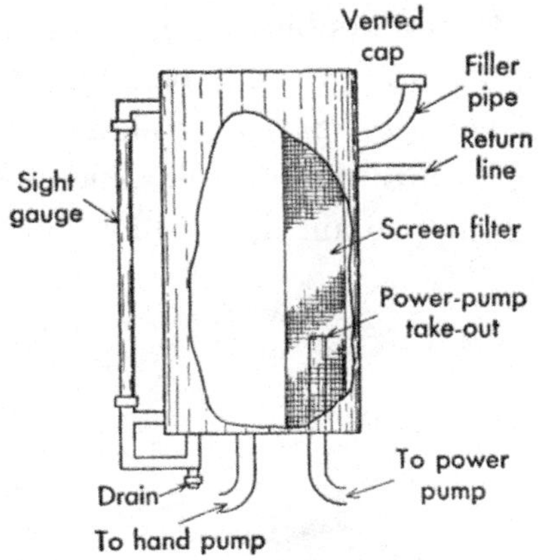

Fig. 191. A schematic drawing of a typical hydraulic storage tank installation.

The actuating cylinder is the device by which the force exerted by the liquid in the system is changed to motion. Actuating cylinders are of a number of different types. One type, called the single-acting actuating cylinder, consists of a cylinder containing a spring-loaded piston,

Fig. 192. A hydraulic pressure gauge. (Courtesy Electric Auto-Lite Company)

as shown in Figure 171. In this cylinder, the fluid under pressure acts on one end of the piston, causing it to move outward in the cylinder and, when the fluid pressure is relieved, the spring forces the piston back to its original position. Positive control is obtained in this cylinder on only one end of the piston.

In the double-acting actuating cylinder, shown in Figure 175, fluid under pressure may be admitted to either end of the piston. This type of cylinder makes possible positive control on both ends of the piston by means of hydraulic fluid pressure. The piston in this cylinder may be stopped and held in any position in the cylinder.

Fig. 193. A motor-driven hydraulic pump. (Courtesy Pesco Products Company)

The flow of hydraulic fluid to the actuating cylinders is controlled by means of a selector valve. This valve allows the pilot, at will, to admit fluid to either side of the piston in the double-acting cylinder or to the fluid side of the piston in a single-acting cylinder. The position of the piston may be regulated by opening or closing the selector valve. In a single-acting cylinder, the valve is arranged so that, in one position, fluid under pressure flows into the cylinder and, in the other position, fluid is allowed to flow from the cylinder back to the supply tank. The piston may be locked in any position by setting the selector valve to neutral. In the double-acting cylinder with the selector valve in one position, fluid under pressure is admitted to one end of the piston while fluid in the cylinder at the other end of the piston is allowed to drain back into the supply tank (see Figure 175). With the selector valve in the reverse position, fluid under pressure is admitted to the opposite end of the piston, while fluid from the other end of the cylinder flows into the supply tank. With this valve in neutral, the piston is locked in any desired position.

Separate selector valves are usually installed for each set of cylinders, such as the flap-actuating cylinders or landing-gear-actuating cylinders. Flow governors or pressure governors are usually installed in the hydraulic system. These governors are small relief valves which may be adjusted to regulate the rate of flow of fluid to the regular cylinders. For example, a flow governor may be adjusted to permit a maximum of 1 gal. of fluid per minute to flow under a pressure of 800 p.s.i. If the pump delivers 2½ gal. of fluid per minute, the by-pass valve allows the extra 1½ gal. to flow back into the supply tank.

Another important part of the hydraulic system is the brake-operating mechanism. The brake system consists of the brake pedal and a master cylinder which acts as a brake accumulator. As the brake pedal is depressed, pressure is applied to the fluid in the cylinder which is carried through a line to an expander ring that applies the brake. As the pressure is relieved on the brake pedal, the fluid in the expander ring is allowed to flow back into the main brake accumulator.

The flaps are operated by a simple double-acting cylinder. This cylinder has a slightly greater stroke than the landing-gear cylinder and is located in the fuselage near the trailing edge of the wing.

The pumps which furnish the hydraulic fluid under pressure are of such size that either one alone can operate the system in case of failure of the other.

The regular supply lines from the storage tank are taken out several inches above the bottom of the tank. The location of this take-out makes it impossible for the engine-driven pumps to empty the tank completely.

In case a line breaks, not all of the fluid will be pumped out. The hand-pump supply line is taken off at the bottom of the tank, and there is always sufficient fluid left in the tank to allow operation by the hand pump after an emergency repair or in case the break is between the engine pump and the hydraulic system. The hand pump is usually supplied by a completely independent line, both from the tank and to the hydraulic system. Cutoff valves are sometimes arranged so that one part of the system may be cut out.

The automatic pilot, which depends upon the hydraulic system, is supplied with hydraulic fluid under pressure directly from the pressure-regulating system, by-passed to a pressure regulator. A hydraulic fluid filter is provided to filter all fluid going to the hydraulic system of the automatic pilot. One pressure regulator is usually sufficient to handle the fluid necessary for this system. The one regulator will handle all of the fluid supplied by both pumps. The automatic-pilot system when in use operates the ailerons, rudder, and elevator controls.

In one typical system there are four pressure gauges connected to the hydraulic system for a check on its operation. Three of these gauges are located on the instrument panel, the left and center gauges indicating pressure being delivered by each engine pump. The righthand gauge indicates the pressure within the system available for operating the

Fig. 194. An electro-hydraulic "power package" made up of a motor and hydraulic pump. (Courtesy Pesco Products Company)

landing gear and flaps. The fourth gauge indicates the pressure in the automatic-pilot system. In case of failure of one of the engine pumps, loss of fluid, broken lines, or other such failures, the hand pump may be used for the manual operation of the landing gear and flaps.

This pump is located under the emergency control box, and a handle for the pump is clipped to the supports of the pilot's seat. Since the supply line to the hand pump is taken from the lowest point in the supply tank, the pump will operate as long as there is any fluid in the tank. After the tank has been completely emptied, the landing gear and flaps may still be operated by the fluid remaining in the lines and cylinders.

In normal operation, the righthand gauge will be found to fluctuate between 450 and 650 p.s.i. While cruising with the landing gear and flaps retracted and the automatic pilot off, the pressure on this gauge will indicate approximately 650 lb. of pressure. Pressure increases beyond this point are checked by the pressure regulator.

Oil leakages in the various parts of the system will cause a gradual drop in pressure to about 450 p.s.i. When the pressure has dropped to this point, the regulator will come into operation, and the pumps will again boost the pressure to approximately 650 p.s.i.

The left and center gauges will read "0" as long as the pressure in the system is being built up to 650 lb. again. It takes only a fraction of a second in this system for the pressure to be built up and for the by-pass valve to open. Unless there is an undue amount of leakage in the pressure system, the pressure should not fall to 450 p.s.i. more than about once every 30 min.

In cruising flight, with the landing gear and the flaps retracted and the gyropilot turned on, the left and center gauges should show a constant pressure of approximately 300 p.s.i. This is a pressure which is maintained by the gyropilot pressure regulator. The righthand gauge will drop slowly due to leakage in the pressure system.

When either the flap-operating valve or the landing-gear-operating valve is moved to raise or lower the flaps or landing gear, the pressure indicated on the righthand gauge will drop practically to 0. This pressure will build up again as soon as the pumps are able to catch up to the movement of the operating cylinder.

The load on a cylinder is fairly constant in operating the flap, and the gauge will drop only a little before going to 650 p.s.i again. This pressure should show as long as the flaps are moving. After the flap valve is returned to neutral, the pressure will again begin to drop. When the landing-gear mechanism is operated, the motion of the cylinder is at first faster than the pressure can be built up by the pumps. The pressure drop will be more severe and last longer than when operating the flaps. The pressure will, however, again return to 650 p.s.i. as soon as the cylinder has reached the end of its stroke. During the operation of the landing gear or the flaps, the left and center gauges will read approximately the same as the righthand gauge until the motion has been stopped and the righthand gauge has reached 650 p.s.i. When the by-pass valve opens, the left and center gauges will return to 0.

It will be noted from this that the righthand gauge gives a true indication of the pressure available on either the landing-gear-operating or flap-operating cylinders. The left and center gauges show the pressure of the fluid being delivered by the pumps to either the landing gear, flap-operating gear, or the gyropilot.

When operating the hand pump, the speed should be governed by watching the righthand pressure gauge. The pump should never be operated at a speed which will cause the pressure on the righthand gauge to go over about 650 p.s.i. This pressure, of course, varies in different systems.

To operate the landing gear, the automatic-pilot control lever must be in the OFF position. To raise the landing gear, the automatic pilot should be in the OFF position and the landing-gear control lever should be moved to the UP position. The landing-gear control lever should always be in either the UP or DOWN positions and never between these points. To prevent accidental raising of the landing gear when the airplane is parked, a solenoid lock is installed which locks the landing-gear lever in the DOWN position as long as there is any weight on the landing gear. If this lock fails to release after the airplane is in the air, the locking "dog" may be reached with the finger and released by pressing down. This dog is reached through a small hole provided at the side of the control column. The opening is at the level of the landing-gear control lever. In case of failure of the linkage between the control lever and the control flap, the valve may be operated directly by means of the emergency control lever previously described. A forward motion of the lever will raise the gear, and a backward motion will lower it.

To operate the flaps, the automatic-pilot control lever must be in the OFF position. The flap control lever is pushed into the DOWN position. To raise the flaps, the flap control lever is pulled into the UP position. The automatic pilot must be off while the flaps are being raised. The flaps may be stopped in any position by returning the lever to neutral. During flight, the flap-control lever should remain in the neutral position. With this lever in the neutral position, the flap-operating mechanism is cut off from the hydraulic system.

There is no chance for leakage in the flap-operating cylinders or lines when the control lever is in neutral position. In case of failure of the flap-control lever to operate properly, the flap-control valve may be operated directly in the way described for the landing-gear control valve. A forward motion will raise the flaps, while a backward motion will lower them. The automatic pilot is turned on by moving the automatic-pilot control lever to the ON position.

The proper pressure gauge should be noted to be sure that the correct pressure for operating the automatic pilot is available. This pressure should be about 100 p.s.i.

The hydraulic-system hand pump may be used at any time. The rate at which the pump is operated is usually between 50 and 55 strokes per minute. The hand pump would lower the landing gear in about 60 sec. under normal conditions. The pump action should be slow and uniform to obtain the proper pressure as indicated on the righthand pressure gauge.

Faulty operation of the hydraulic system can be detected usually only by having a thorough knowledge of the entire system. The pilot and the mechanic should both be entirely familiar with all parts of the system, the function of such parts, and the method by which they operate. Often, serious damage to parts of the hydraulic system may be avoided by an instant understanding of what is wrong and by proper methods of correcting it.

XIV FLOW-CONTROL VALVES

There are a number of different valves in the hydraulic system, the purpose of which is to control the rate of flow or the direction of flow of the hydraulic fluid.

One type of valve, such as is shown in Figure 195, controls the rate of flow by the simple method of allowing the fluid to flow through an opening of a predetermined size. This type of valve is more like a metering jet than a true valve, because it restricts and regulates the flow of fluid

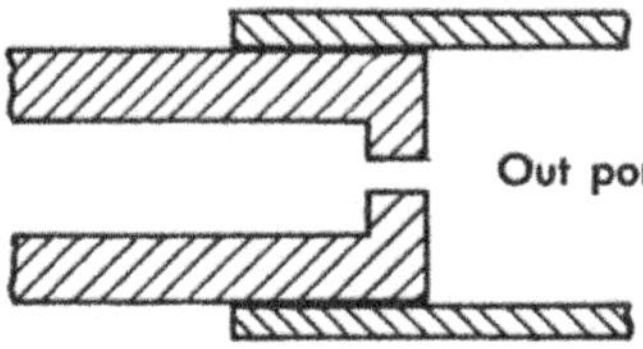

Fig. 195. A drawing of an "orifice" type flow control.

in either direction. This type of regulator is sometimes called a "restrictor." A second type of flow regulator is a jet with a variable opening. The size of the opening may be varied by a tapered pin passing through the opening or by means of an adjustable needle valve. The tapered pin or the needle valve may be of the type that is set by means of a tool or that is adjusted by means of either a lever or a handwheel.

A check valve is a valve which restricts the flow of the fluid in one direction. This type of valve may be a simple flap valve, a ball-type spring-loaded check valve, or a cone-type spring-loaded check valve, such as shown in Figure 196. A valve of this kind may have an opening through the ball or flap, or a passage by-passing the ball or flap to restrict the flow in one direction while allowing the free flow of fluid in the other direction. When a check valve has a by-pass passage, it is known as an "orifice check valve." Instead of a ball or flap, the check valve may be in the form of a cone, as shown in Figure 196. The orifice type of a cone check valve usually has the opening drilled through the cone.

A metering check valve is shown in Figure 197. In this check valve, the metering pin may be adjusted to hold the ball slightly off the valve seat. Fluid entering through port B raises the ball off its seat and flows freely out through port A. The flow in the opposite direction is restricted, and the amount of flow is regulated by the distance the metering pin holds the ball off the valve seat. Screwing the metering pin in or out increases or decreases the return flow of fluid.

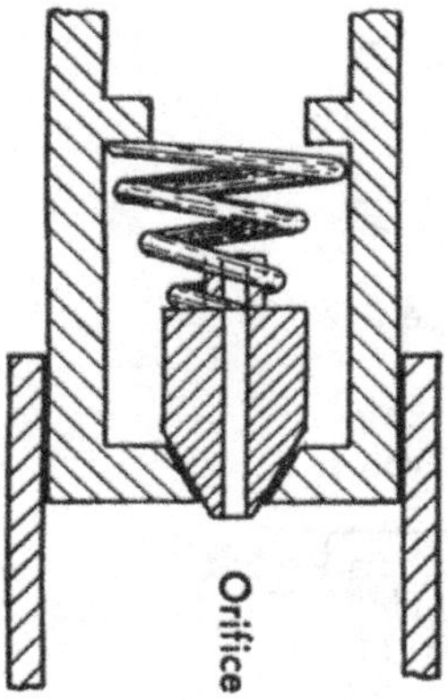

Fig. 196. A drawing of a cone-type orifice check valve.

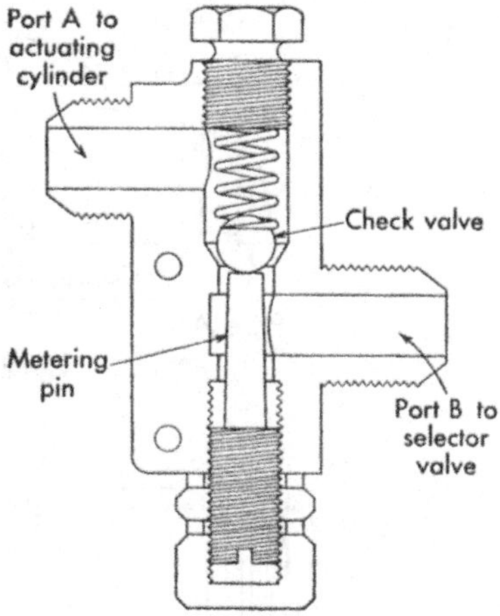

Fig. 197. A drawing of a metering pin check valve.

In order that one part of the system may be disconnected from the rest of the system without losing all of the hydraulic fluid, a line-disconnect valve may be installed. This type of valve is shown in Figure 198. When the two parts of this connection are screwed together, the pin presses the ball off the valve seat (Figure 198–A), allowing the free passage of fluid in either direction. When the connection is screwed apart (Figure 198–B), the spring presses the ball against the seat preventing the fluid from entering pipe A. A double line-disconnect valve may be installed which seals both pipes.

A by-pass check valve is usually installed between the accumulator and the hand-pump connections. This type of valve, shown in Figure 199, may be manually operated. When the valve is closed, the pressure fluid is sealed off from the hand pump. When the hand pump is to be operated, the valve is opened, thus allowing fluid from the hand pump to pass to the accumulator. The operation of the valve is accomplished by means of a plunger having a pin on the end which acts against the check valve, pushing it away from the valve seat when the plunger is pushed inward. A handle, similar to a pump handle, is used to move the plunger in and out.

A timing valve or a sequence valve is used when it is desired to have one hydraulic action follow another in a definite order. This type of valve is usually operated by some part of the moving mechanism, such as a part of a flap. As the flap reaches its full DOWN position, some part of the mechanism presses against the plunger (Figure 200). This plunger

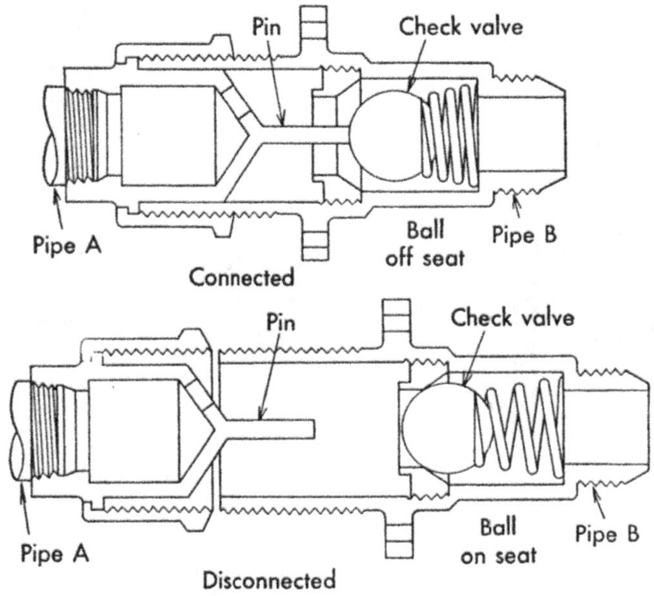

Fig. 198. A drawing of a disconnect valve.

operates a pin which lifts the check valve off the valve seat allowing the fluid to flow through the valve. This flow of fluid may be used to cut off the flow to the cylinder, operate locking mechanisms, or simply allow the fluid to by-pass the actuating cylinder. This type of valve may be used to reverse the motion of the cylinder, thus allowing the fluid to flow first into one end of the cylinder and then into the other end.

A shuttle valve may be used to direct fluid or gas to an actuating cylinder from a normal source or from an emergency source. The action of the valve is usually automatic. This type of valve may also be used to direct air from an emergency source to an actuating cylinder if the supply of hydraulic fluid is lost. In an emergency, this valve may be used to seal off a part of the system, allowing no loss of fluid even though there is a leak or break in the lines. As shown in Figure 201, there are two ports, A and C, with a common outlet port, B, midway between. A spring-loaded piston acts as a check valve when forced to either end of the containing cylinder. The piston normally closes port C. When port C is closed by the piston, the fluid is free to flow between port A and port B. During an emergency, when fluid or air enters port C

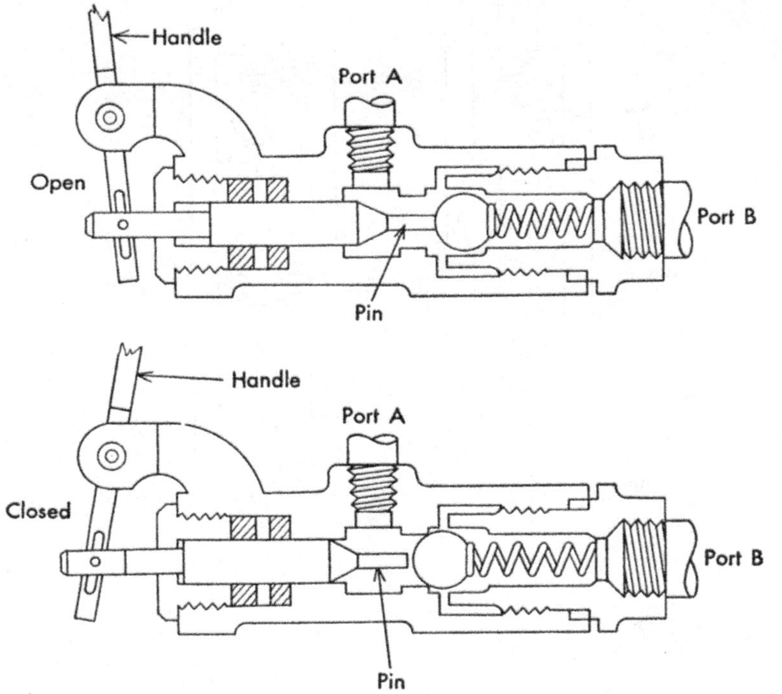

Fig. 199. A drawing to show a by-pass check valve.

forcing the piston to the opposite end of the housing, closing port A, the flow goes out through port B. Port B is connected with the actuating cylinder.

A cross-flow valve is used to by-pass fluid from the landing-gear up-line to the down-line while the gear is being extended. The weight of the landing gear is often great enough to cause it to fall so rapidly, when released, that fluid cannot fill in behind the piston in the actuating cylinder. This weight may cause pressure to build up on the opposite side of the piston. A cross-flow valve allows the gear to fall more easily and with an even motion.

One type of cross-flow valve operates automatically during the normal operation of the landing gear. This valve consists of a housing with three ports, A, B, and C, a restrictor, a check valve, and a piston (Figure 202). The piston also acts as a check valve. Fluid from the up-side of the piston enters port A, moving the piston to the right, and flows through the check valve and out of port C to the down-line. When the flow of fluid, due to the action of the pump, fills the cylinder, the pressure in the down-line closes the check valve and lets out the gear to its fully extended position.

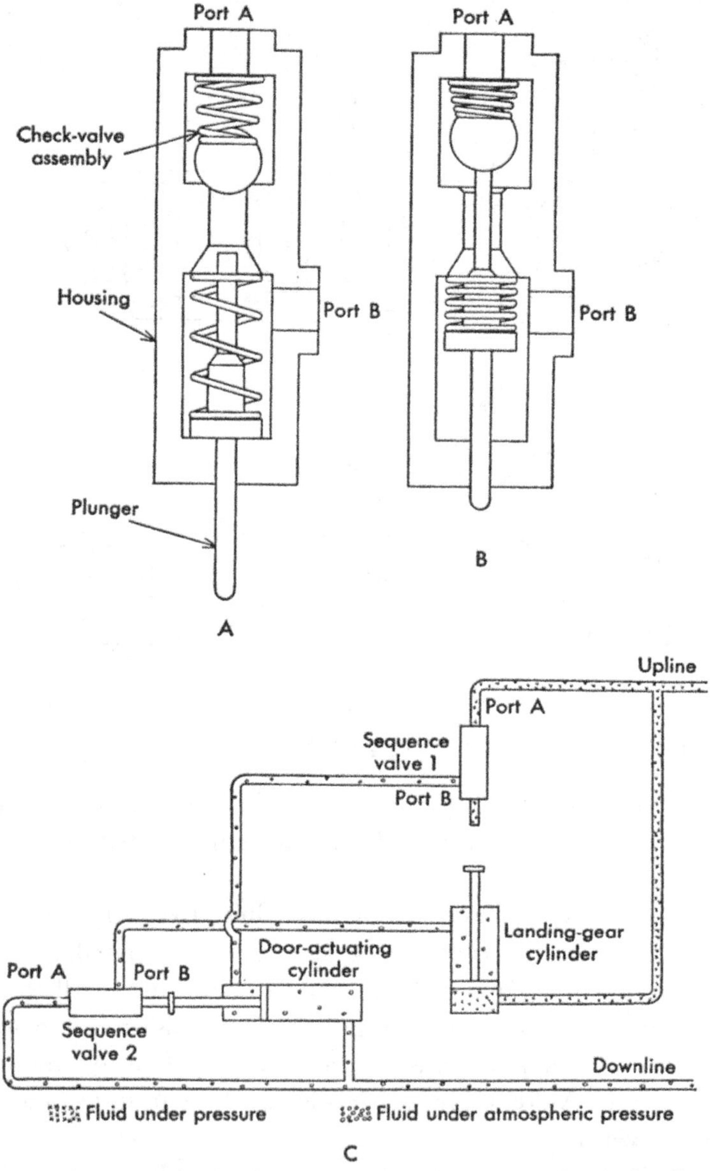

Fig. 200. A drawing to show a timing, or sequence, valve and a diagram to show its installation.

The fluid in the up-side of the cylinder will then flow through the restrictor and out of port B. When retracting the landing gear, fluid under

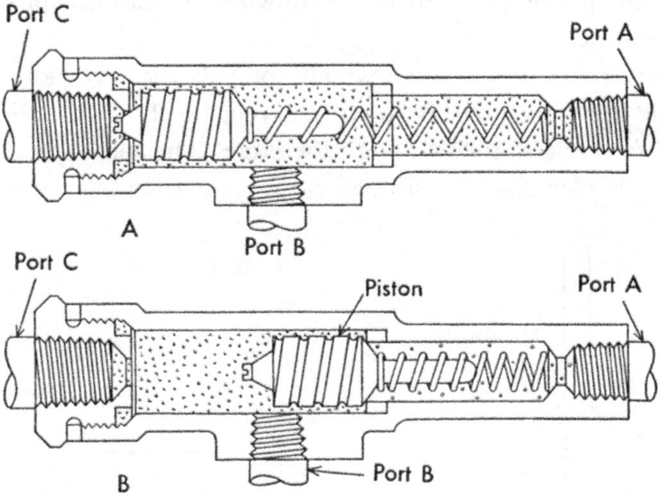

Fig. 201. A schematic drawing of a shuttle valve.

pressure entering port B moves the restrictor to the left, and flows out of port A to the actuating cylinder.

Another type of cross-flow valve is manually operated and used only during emergencies when extending the landing gear. This valve may be used when the automatic unit is not operating. By moving the operating

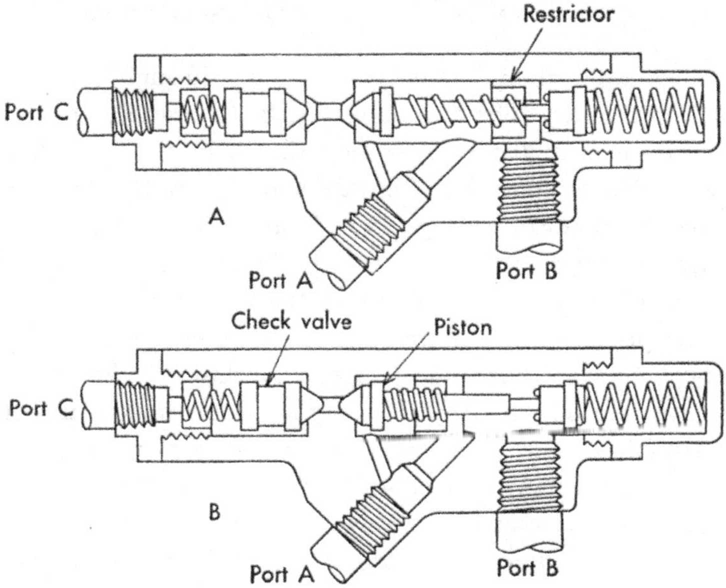

Fig. 202. A schematic drawing of a cross-flow valve.

handle, shown in Figure 203, the check valve is forced off its seat, allowing fluid to flow from the up-line into the down-line. The check valve prevents fluid under pressure in the down-line from escaping through the unit into the up-line.

The selector valve is a valve which controls the direction of flow of a hydraulic fluid and may be used to direct the flow first through one part of the system and then through the other. The rotor type of selector, as shown in Figure 204, is a simple rotating part having passages which

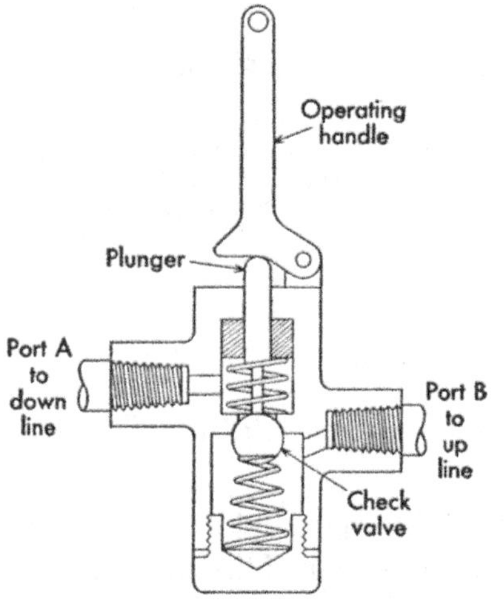

Fig. 203. A schematic drawing of a manually operated cross flow valve.

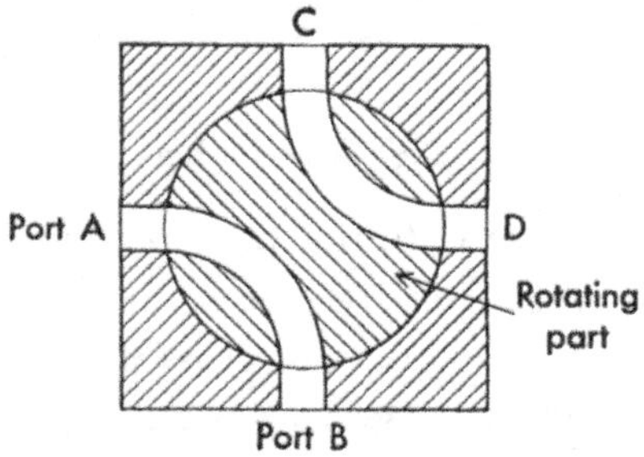

Fig. 204. A schematic drawing of a rotor-type selector valve.

may be lined up with different ports leading through the cylinder in which it is contained. The four ports, A, B, C, and D, are 90° apart. When the rotor is in the first position shown in Figure 204, pressure fluid flows in through port A and out through port B. Fluid under atmospheric pressure flows in through port D and out through port C into the storage reservoir. When rotated to the second position, pressure fluid may flow in through port A and out through port C, and fluid under atmospheric pressure flows in and out of the other two ports.

Selector valves may be of the poppet type shown in Figure 205. The poppet valves are operated by means of a cam arrangement connected with a manual control. By moving the manual control, different pairs of poppets may be opened or closed, regulating the flow of fluid in the direction desired. Selector valves may be of the piston type (Figure 206), where the control of fluid flow is brought about by a manually operated piston which opens and closes the desired openings.

The open-center selector valve is another type of valve used to direct fluid under pressure to one end of a cylinder and, at the same time, allow the fluid from the other end to flow into a return line. The open-center type of valve automatically returns to neutral when the actuating cylinder reaches the maximum stroke. When in the neutral position, this valve allows the output of fluid under pressure from the pump to flow through the unit to the storage reservoir. When the piston is moved from the neutral position, as shown in Figure 207, fluid enters through port A and goes out through port D. Port C is connected to a return line by way of port B. When the cylinder reaches the end of its stroke,

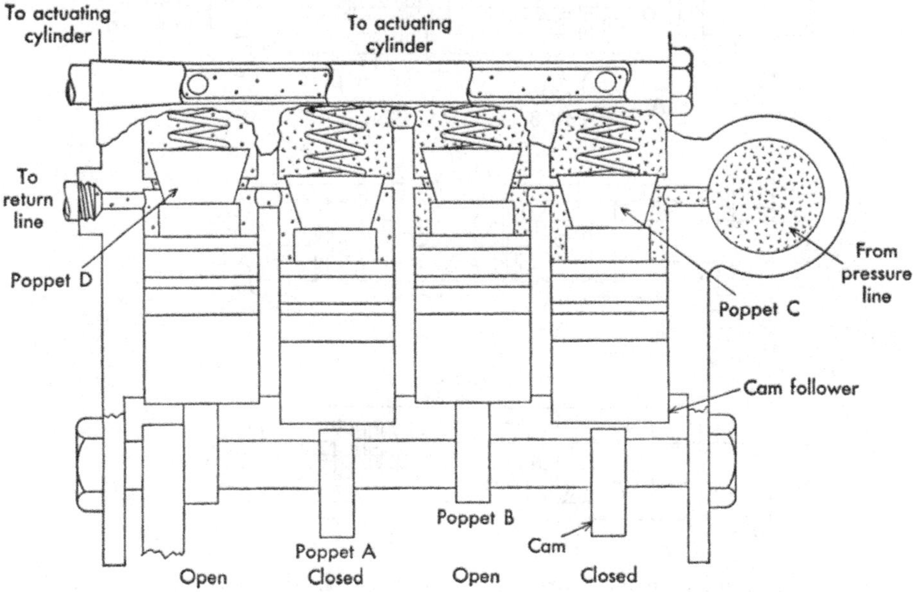

Fig. 205. A schematic drawing of a poppet-type selector valve.

the pressure will open the relief valve, E. Fluid passing through this opening flows to the right end of the piston. A metering valve, G, prevents the rapid escape of fluid from this end of the cylinder. Fluid in the other end of the cylinder is free to pass through the back-pressure-relief passage. The piston is moved towards the left by pressure which builds up on the end of the piston. When the piston is moved far enough, the neutral booster will center the piston in a neutral position. This position is shown in Figure 207–B. Fluid under pressure from port A will then be directed through the passages in the piston and housing to the right end of the piston. This pressure will quickly move the piston to the neutral position and connect ports A and B. Fluid from the pump then flows directly to the reservoir. The automatic return of the piston from the position shown in Figure 207–C is brought about by the cam and lever assembly. The operation of the valve in the opposite direction is the same as that described above, except that port C is connected to the pressure line, and the relief valve, F, and metering valve, H, act to provide pressure for the automatic return of the piston to the neutral position.

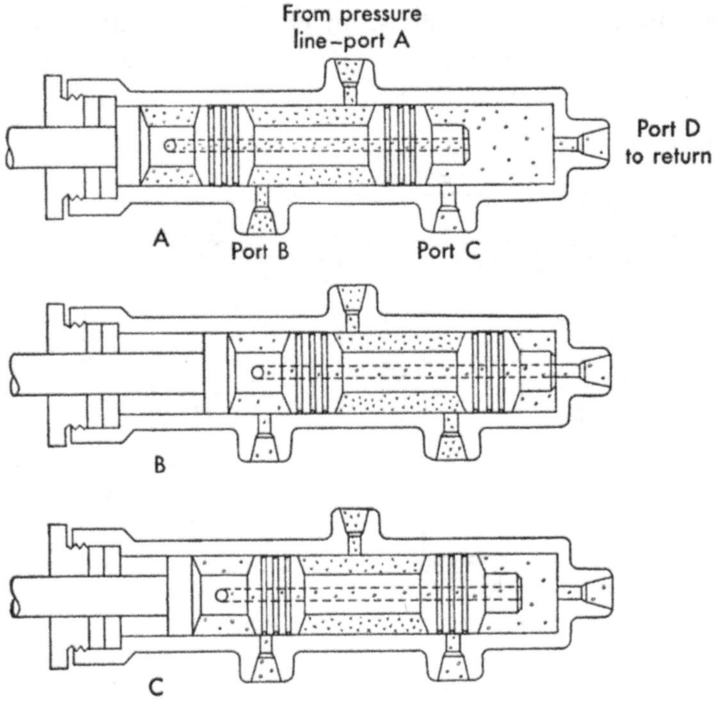

Fig. 206. A schematic drawing of a piston-type selector valve.

The surface-control booster unit is made up of a selector valve and an actuating cylinder contained in the same housing. This unit acts to multiply the distance of movement of multiple control surfaces by the exact amount necessary to compensate for the loss of distance which would normally occur in gaining mechanical advantage. This unit is made up of a three-port housing which contains a piston type of selector valve assembly, an actuating cylinder assembly, and two gust-lock valve assemblies. One port is connected to the pressure line, another port is connected to the return line, and the other port is connected to the gust-lock control valve.

The selector valve is normally in the neutral position. Moving the selector valve directs fluid to one side of the actuating cylinder piston and, at the same time, connects the other side of the piston to the return line. This unit is mounted in the control-surface system. The selector-valve piston and the actuating-cylinder piston are connected to the control through bell cranks attached to the piston rods. The housing is connected to the control surface by another bell crank. The difference

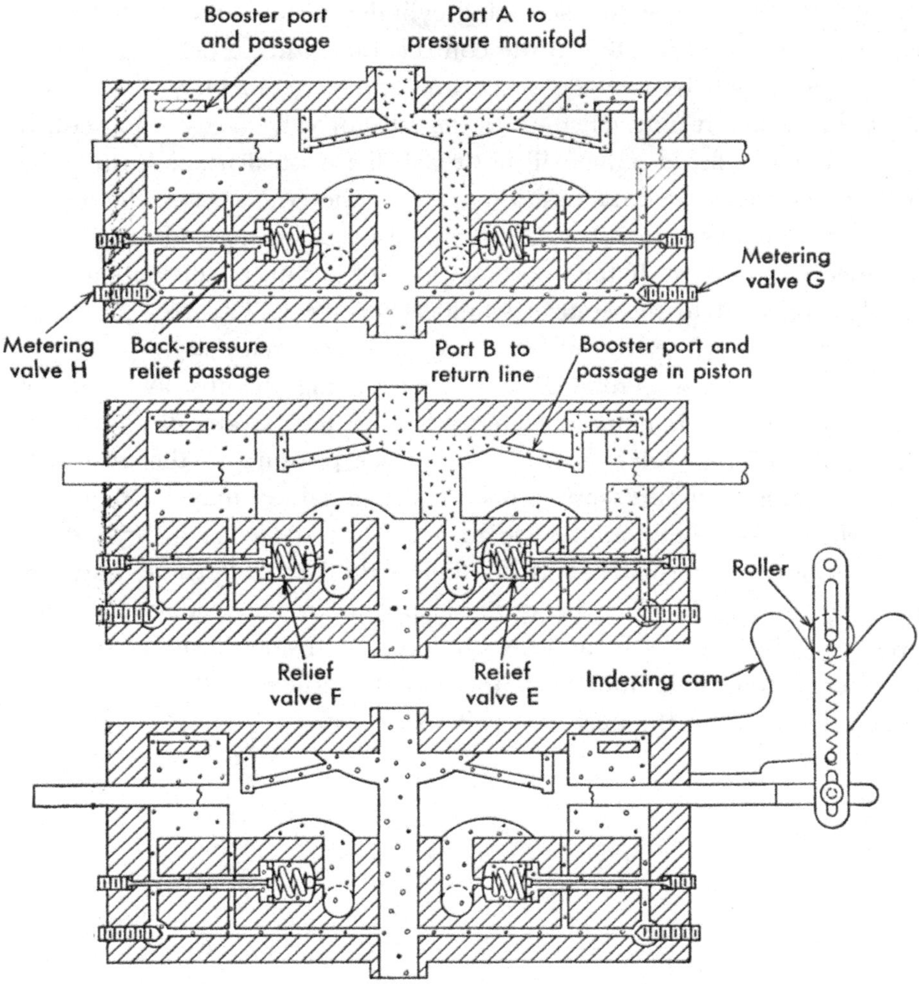

Fig. 207. A schematic drawing of an open-center-type selector valve.

in length of the bell cranks gives a mechanical advantage. Moving the control causes the actuating cylinder piston to move in since the selector valve is in neutral at the beginning of the movement. The whole unit moves because the actuating-cylinder piston and housing are locked together by the hydraulic fluid. The housing, being connected to the control surface, causes the surface to move. The difference in length of the bell crank causes the selector-valve piston to move farther than does the housing. This movement displaces the selector valve from its neutral position and allows fluid under pressure to flow in through an internal passage leading to one end of the actuating cylinder. At the same time, the passage from the other end of the actuating cylinder is opened to the return line. Since the actuating-cylinder piston and the selector-valve piston are connected to the control, they will be held stationary. The housing will be moved by the fluid pressure, moving the control surface to which it is attached, until the selector valve is again in neutral position. The selector valve will not reach the neutral position until the bell cranks are in the same position in which they were before the movement started. Even though force is gained by using the bell cranks, no distance will be lost. If the booster unit should fail, a stop on the bell crank connected to the control would strike the bell crank connected to the control surface, the surface may be moved manually.

The control surfaces may be locked by moving the gust-lock control valve. This valve is manually controlled by the parking-brake lever. Movement of this valve directs fluid under pressure to the gust-lock valves. These units are spring-loaded pistons which may be made to close the passages between the selector valve and the actuating cylinders. The pistons are normally held off their seats by the springs, and the passages are open. When pressure is applied to the pistons, they are pressed against their seats and hold fluid on both sides of the actuating-cylinder piston. This fluid locks the control surface in the position in which it was when the gust-lock valve was closed.

XV PRESSURE-CONTROL VALVES

Pressure-control valves are usually safety units placed in a hydraulic system to prevent damage by excessive pressures. The valves may be used, however, to regulate the pressure in certain parts of the system, such as the automatic pilot. Pressure-control valves may be used to retain fluid under pressure in various parts of the system.

The type of pressure-control valve used to prevent the development of excessive pressures is called a by-pass relief valve. When the pressure reaches a predetermined value, this type of valve opens and allows fluid to flow from the pressure part of the system to the storage tank return line, but because of spring action maintains the predetermined pressure in the system. A by-pass relief valve is usually installed in connection with the hydraulic pumps. When the pump has built the pressure up to the predetermined value, the valve opens, and excess fluid flows through the return lines to the storage reservoir or to the pump intake line.

Where the words UP and DOWN occur in this part of the text, they refer to the UP or DOWN position of the landing gear or flaps. Up- and down-lines refer to lines which bring about UP and DOWN positions.

One type of relief valve is shown in Figure 208. Port A is connected to the pressure line and port B is connected to the return line. When the pressure on port A becomes excessive, the valve is forced open against the spring pressure, allowing fluid to flow out into the return line. The pressure on the valve may be regulated by the pressure of the screw on the spring. The pressure in the pressure line cannot exceed that required to lift the valve off the seat against the spring. If the pressure falls, the valve is again pressed against the seat by the spring, thus allowing the pressure to be built up in the lines. Usually this type of valve dances up and down on the seat, maintaining a constant pressure in the line.

A pressure cutoff valve is usually installed in conjunction with a hydraulic motor. This valve is placed between the selector valve and the motor and is used to cut off the flow of fluid automatically when the operating mechanism has reached the end of its travel. This type of valve is shown in Figure 209 and is sometimes called a stop valve or a

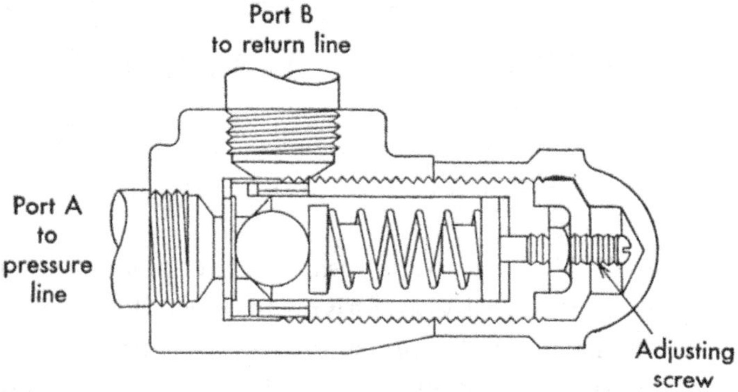

Fig. 208. A schematic drawing of a pressure relief valve.

travel-limit valve. This valve has two ports in the housing and a tapered poppet valve which is connected to an external plunger. This plunger is operated by the moving mechanism. It is so arranged that, just as

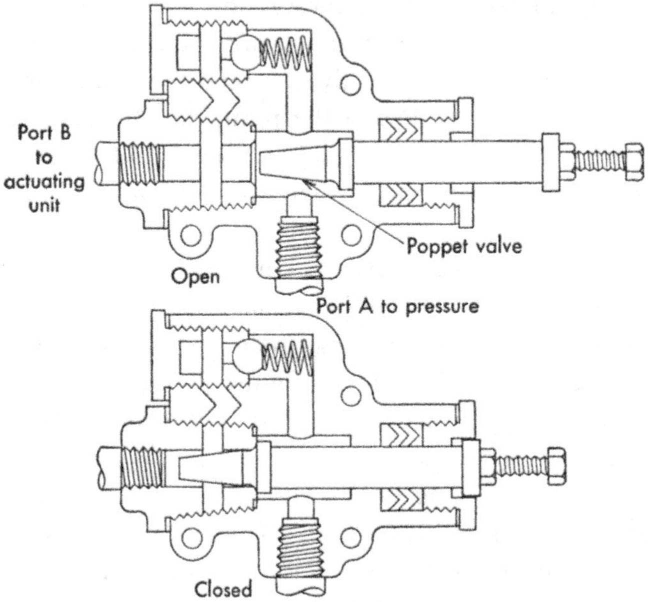

Fig. 209. A schematic drawing of a pressure cutoff valve or a "Travel Limit" or "Stop" valve.

the mechanism reaches the end of its travel, the valve is forced into the opening to the actuating unit, cutting off the flow of fluid. This type of valve allows the free flow of fluid backwards through the mechanism.

The return fluid, forcing the plunger away from the opening, allows the fluid to flow backward through the valve. A pressure cutoff valve is often installed at each end of a moving part to stop it at the end of its stroke.

A flap-overload valve, as shown in Figure 210, is often installed in the flap-operating mechanism. This valve allows the flaps to be opened but, if the pressure on the flap becomes excessive because of high air speed, the

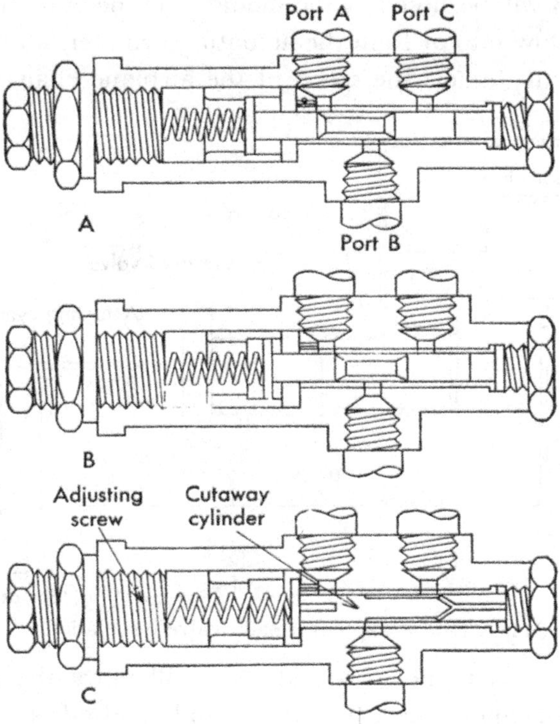

Fig. 210. A diagrammatic drawing to illustrate a flap "overload" valve.

valve allows the flaps to close partially, which prevents damage to the flap structure and keeps the flaps from being left in an open position when the air speed rises above normal. This valve consists of a three-port housing. In this housing is installed a spring-loaded, spool-shaped piston and an adjusting screw. Two external check valves are always used in conjunction with this unit. A typical installation is shown in Figure 211.

During normal operation, the unit will be in the position shown in Figure 210. Fluid from the selector valve enters port A, flows around a cutaway section of the piston and out of port B. Port B leads to the downside of the actuating cylinders and opens the flaps. As the flaps move downward into the air stream, the force resisting the downward movement increases. The hydraulic pressure in the down line and in the valve, increased by the force acting against the flap, acts against the bottom face of the piston through drilled passages in the lower end of the piston. If the pressure is excessive, it overcomes the force of the spring before the flaps reach the full down position. The piston will then be moved up, and port A will be closed by the shoulder on the bottom of the piston. No fluid can flow into or from the actuating cylinder, and the flaps will remain stationary unless the speed of the airplane changes. If there is

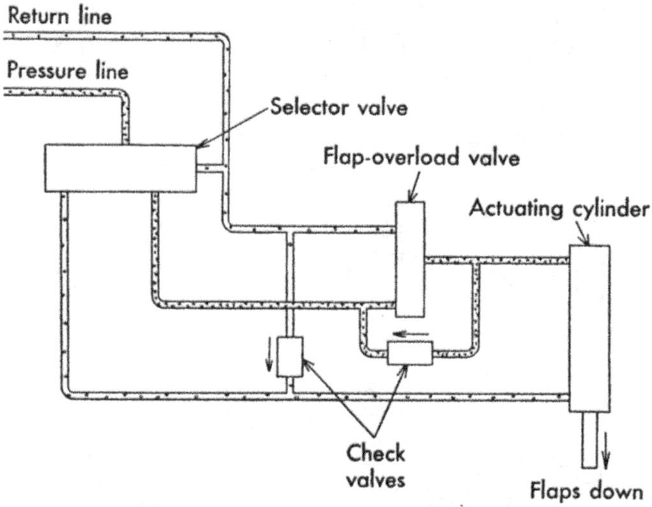

Fig. 211. A typical flap "overload" valve installation.

an increase in speed, the increased load will move the piston in the flap-actuating cylinder back into the cylinder. Fluid is trapped in the down-line, and pressure in this line and in the flap-overload valve will increase. When this pressure reaches the value for which the unit is adjusted, the cutaway section of the piston connects ports B and C, and some of the fluid on the downside of the cylinder can escape from the down-line. This allows the flaps to close enough to relieve the overload caused by the excessive speed. If the speed is decreased, the piston will be moved down by the spring, and port C will be closed. The flap will then remain stationary unless the speed is again increased or decreased.

A ratchet valve is used in the wing-flap system. When a leak type of flap-selector valve is used in the system, fluid under pressure can escape through the selector valve. A leak-type, flap-selector valve is one which uses a piston that is lapped into place rather than a piston equipped with sealing cups. A ratchet valve, or a similar valve called an auxiliary wing-flap valve, is used with this type of selector valve to retain pressure in the flap-actuating cylinder. This keeps the flaps from gradually opening when the aircraft is on the ground or from creeping to a closed position

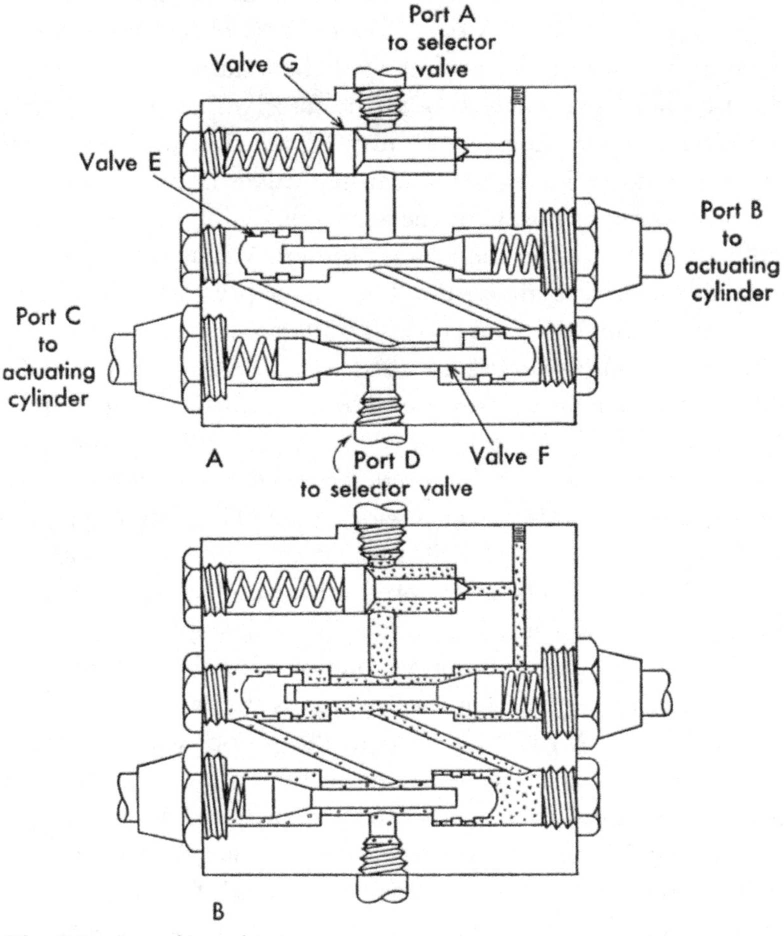

Fig. 212. A ratchet valve.

when they are down during flight. The ratchet valve is installed in the flap system between the selector valve and the actuating cylinder. The lines leading to the selector valve and those leading to the actuating cylinders all pass through this valve, the housing of which has four ports. The valve contains three spring-loaded adjustable valves. Valve G, as shown in Figure 212, is a temperature-expansion relief valve, while valves F and E are check-relief valves. When the selector valve is in the

UP position, fluid under pressure from the selector valve enters port A, pressing valves E and F off their seats. This pressure fluid then flows out of port B to the actuating cylinder and closes the flaps. Fluid from the opposite end of the actuating cylinder enters port C and flows around valve F and out port D to the return line through the selector valve. When the flaps reach the full UP position and are closed, valve E closes. If pressure applied at port A decreases because of leakage through the leak type of selector valve, pressure will be retained in the actuating cylinder by the closed valve, E, and the flaps will remain in the UP or closed position. When the selector valve is placed in the DOWN position, fluid under pressure enters port D, causing valves E and F to open. The fluid then flows out of port C to the actuating cylinder, causing the flaps to open or to be placed in the DOWN position. When the flaps are fully down, valve F closes, retaining the fluid under pressure in the actuating cylinder and holding the flaps in this position.

Hydraulic pressure switches are used to turn a switch on or off automatically with changes of pressure. The pressure-warning switch is to signal, by light or sound, or both, when oil pressures, fuel pressures, or hydraulic fluid pressures reach a predetermined point. Pressure switches may be adjusted to operate over a wide range. The simple type of pressure-warning switch, such as is illustrated in Figure 213, may be used to indicate either a high or a low pressure, or both. The pressure-warning switch consists of a housing containing a spring-loaded piston that operates the switch. The housing has one opening for the pressure fluid and an opening through which the wires pass that are connected to the parts of the switch. A rod connected to the piston, which is moved by pressure against the spring or is moved in the opposite direction by the spring when the pressure falls, operates the switch. The pressure-switch unit is connected in series with a warning device. When arranged to warn of low pressure, the fluid under pressure holds the piston up against the spring. If the fluid pressure falls, the spring forces the piston downward and operates the switch which closes the electric circuit connected with the warning device. As the pressure again increases, the switch is opened and the warning stops. This type of switch may be adjusted to switch on or off over a limited range, if desired.

A pressure switch is used when it is desired to have a large difference between the pressure at which the switch is turned on and the pressure at which the switch is turned off. The housing of the pressure switch has three ports: A and B, as shown in Figure 214, are connected to the pressure line between the power pump and the accumulator; and the return line is connected with port C. Within the housing are two pistons, D and E, and a hollow poppet valve, F. Piston D is so arranged that the end of its rod will move the ball up and close the passage when the pressure is great enough to overcome the resistance of the spring. Any increase in pressure will cause piston D to raise poppet F from its seat. Piston E is attached to an electric switch by a piston rod. Upward movement of piston E turns the switch on, while downward movement turns

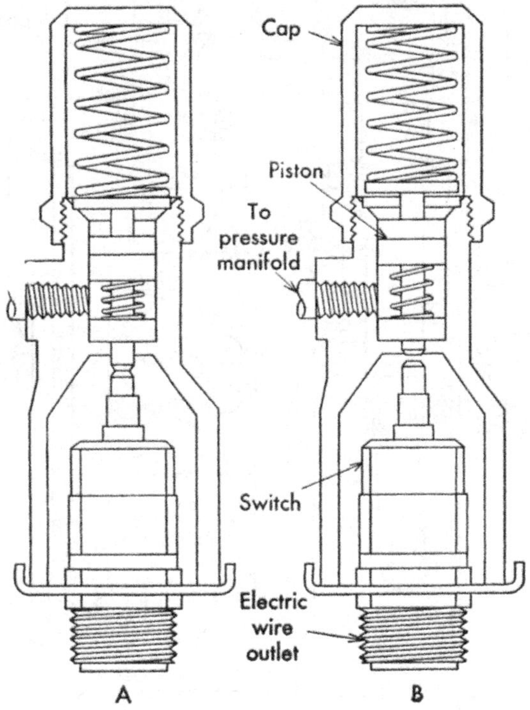

Fig. 213. A spring-loaded pressure warning switch.

the switch off. When the switch is in the OFF position, as shown in Figure 214, the pressure acting through port B holds the piston D up. When piston D is in this position, the opening in the hollow poppet valve is closed by the ball, and the poppet valve is opened. When the poppet valve is open, the upper chamber is connected to the pressure system. The pressure acting on the top of piston E is the same as the pressure acting on the bottom of piston E. However, the area of the top of piston E is greater than the area of the bottom because of the area of the piston rod. Therefore, the force holding the piston down will be greater than the force pushing it up. Because of this difference in pressure, the switch will remain in the OFF position until the system pressure is reduced. As the system pressure falls, the large spring forces D down and closes the poppet valve. Any decrease in pressure will allow the ball to move away from the opening in the hollow poppet, and the

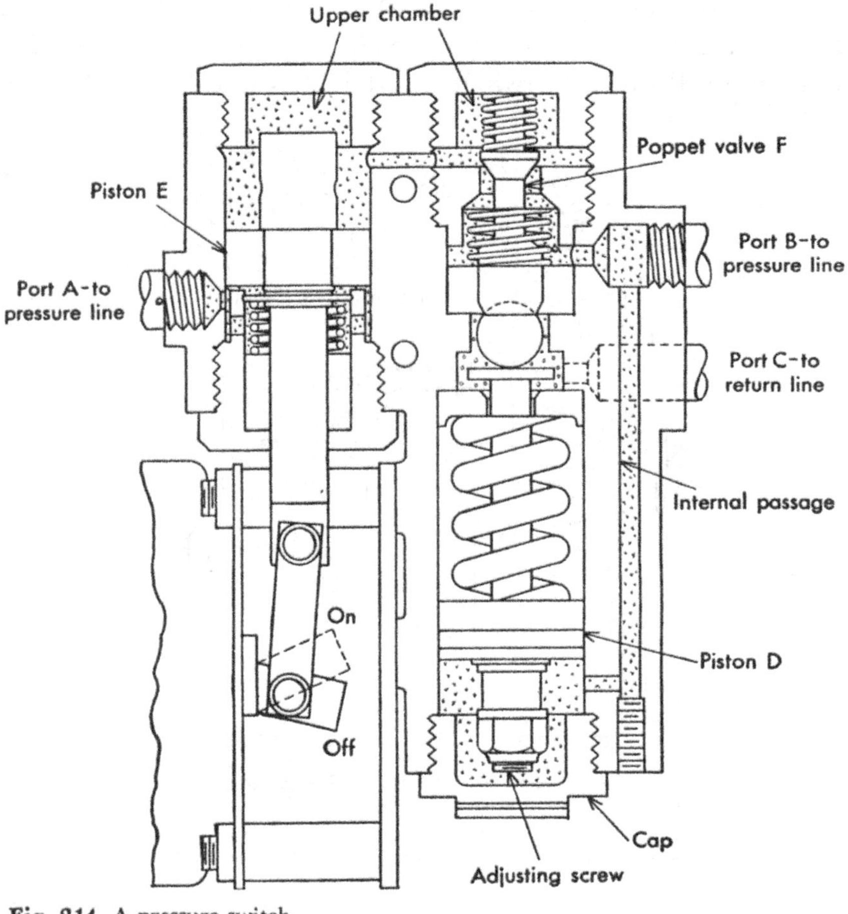

Fig. 214. A pressure switch.

pressure in the upper chamber will drop to zero. As soon as this occurs, the force holding piston E down will drop to zero, and the hydraulic pressure acting on the bottom of piston E will move it up, thus closing the switch. This will start the pump, and the hydraulic pressure in the system will increase until piston D is again moved upward far enough to open the poppet valve.

To prevent violent oscillation of the pressure-gauge pointers caused by impulses applied to the hydraulic fluid by the power-driven pumps, a pressure-gauge snubber may be installed. Violent oscillation of pressure-gauge indicating pointers makes it very difficult to obtain accurate readings.

The purpose of the pressure-gauge snubber is to dampen out these impulses and allow the instrument to give a steady reading. The diagram of a pressure-gauge snubber (Figure 215) shows a two-port housing

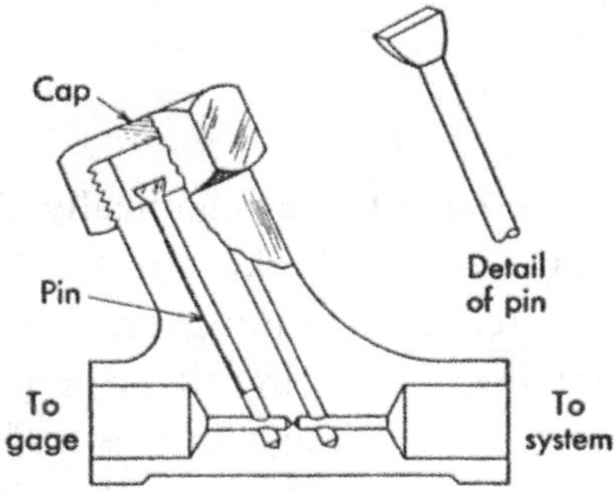

Fig. 215. Diagrammatic drawing of a pressure-gauge snubber.

which contains an orifice in which is placed a floating plunger. This plunger is free to move up and down but, because of its inertia, tends to remain stationary. The weight of the pin opposes the pressure surges in the fluid and dampens out their irregular impulses, allowing a more steady flow of fluid to the pressure gauge. The dampening out of these pressure-gauge surges allows a steady pointer indication and a more accurate reading.

XVI HYDRAULIC BRAKE SYSTEMS

Aircraft brakes may be operated mechanically, hydraulically, or by means of compressed air. Aircraft brakes are used not only to shorten the run of the aircraft after it is on the ground, but are used to steer many aircraft while taxiing. Most aircraft are equipped with parking brakes. On light aircraft, both the foot brakes and the parking brakes are usually mechanically operated. Some small aircraft, however, have a hydraulic brake system which is similar to that used on automobiles. On larger aircraft, the hydraulic brake system may be connected directly to the main hydraulic system or may be an independent system. The independent system is more common in light aircraft, which are not usually equipped with a regular hydraulic system.

Brakes operated by compressed air are usually for emergencies, being used only in case of failure of the regular hydraulic system. This type of system is not usually found on light aircraft. The master-cylinder brake system is made up of a manually operated cylinder which is supplied with hydraulic fluid from a reservoir. For each main landing wheel there is one master cylinder which is operated by means of a pedal similar to that used on an automobile. The brake pedal may be attached to the rudder-control pedals or may be a separate pedal. The brake pedals may be attached so that the rudder and brake may be used at the same time.

The reservoir is necessary to maintain a supply of fluid in the master cylinder and to allow for small losses because of leaks. The reservoir may be a part of the master cylinders or it may be separate and connected with them by means of a tube through which the hydraulic fluid flows.

It is important that sufficient fluid be maintained in the reservoir at all times. The master cylinder may be operated directly from the pedal or it may be some distance from the pedal and operated by means of rods and a linkage arrangement. The operating mechanism is equipped with a spring which brings the pedal back when the foot pressure is relieved. Each brake is independent and may be operated separately.

The lines connecting the master cylinder with the brake-operating mechanism may be a combination of rigid and flexible tubing. The cylinders that actually operate the brakes themselves are called the brake-actuating cylinders. These cylinders may act directly on the brake shoe, pressing it against the brake drum, or may cause a cam to rotate which, in turn, presses the brake shoes against the drum. Some

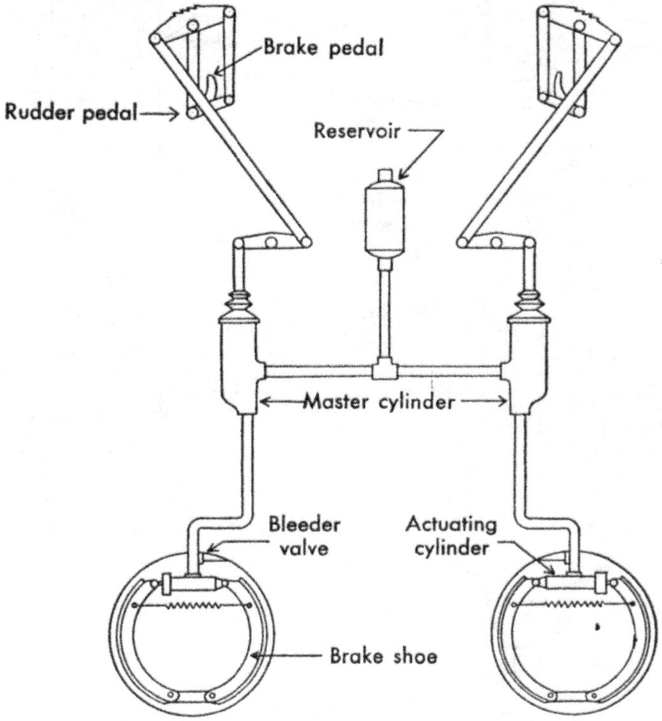

Fig. 216. A schematic drawing of a hydraulic brake system.

brakes have a baglike arrangement which is expanded by the hydraulic fluid. This bag encircles the brake drum and presses the brake shoes against the drum.

The hydraulic parking brake usually operates by trapping the hydraulic fluid in the brake-actuating cylinders or by a mechanical arrangement which locks the brake shoes against the drum. In one type of hydraulic parking-brake system, the foot brakes are depressed and then the parking-brake lever is pulled into position, locking the brakes in the ON position. To release this type of brake, the foot pedals are depressed and the parking brake released. The pedals are then released and return to the OFF position.

In most systems, the brake shoes are withdrawn from the drum by means of springs. The master cylinder is equipped with a port which is just barely uncovered when the pedals are in the OFF position. This port allows the free flow of liquid from within the cylinder backward

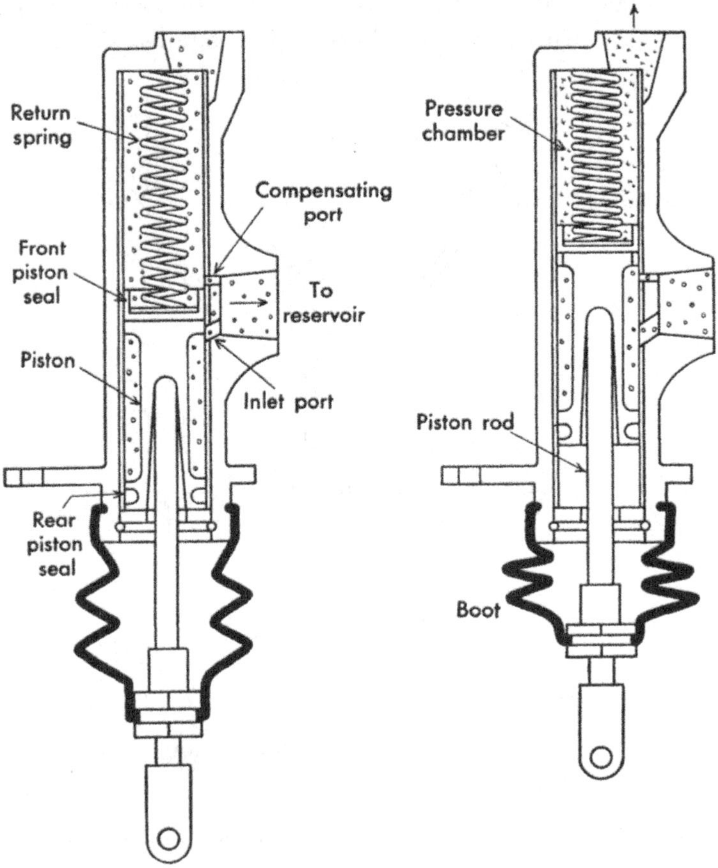

Fig. 217. A diagrammatic drawing of a master cylinder for a hydraulic brake system (*Goodyear* type).

to the reservoir to prevent pressures being built up by thermal expansion of the liquid. This port is closed as soon as the piston in the master cylinder is slightly depressed. Figure 216 shows a typical master-cylinder hydraulic-brake system using two shoes in each brake drum.

There are four common types of master cylinders in general use.

1. Fluid enters the master cylinder from a reservoir through inlet and compensating ports, as shown in Figure 217. The application of

operate the brakes. A power-brake control valve, sometimes called a "debooster," is used at times. This unit still further reduces the pressure of the power-brake control valve, allowing smoother application of the brake. A controlled amount of pressure fluid is metered by the power-brake control valve from the main hydraulic system into the brake system. The amount of fluid under pressure admitted to the brake system depends upon the amount of pressure applied to the foot pedal.

There are two common types of brake-control valves: the internal-spring type and the external-spring type. The internal-spring type consists of a push rod, a slide valve, a spring, a bullet-type valve, a cam, and a poppet valve, all enclosed in a three-port housing. Pressure fluid from the system is stopped by the poppet valve when it is resting on its seat. The push rod is moved into the housing by the actuating lever when pressure is applied to the foot pedal. This action moves the valve by compressing the spring. On the first part of the movement, the bullet valve closes the escape passage in the center of the slide valve. The bullet valve moves with the slide valve as it continues to move. The cam is rotated about its pivot by this motion and opens the poppet valve in the pressure port. The pressure fluid from the hydraulic system then flows around the poppet valve and out of the port into the brake line. The faces of the slide valve and the bullet valve are acted upon by the pressure in the brake line. The increased pressure moves the slide valve and the bullet valve against the force of the spring. As the bullet valve moves, the poppet valve closes, thus preventing additional pressure from reaching the brake line. The pressure in the brake line is governed by the amount the spring must be compressed to allow the poppet valve to seat. The distance the plunger is moved determines the pressure necessary to seat this valve. The pressure to the brakes is directly proportional to the distance the foot pedal is depressed. Any desired pressure may be metered to the brake system up to the maximum pressure allowed. However, this is governed by the thrust of the push rod, the strength of the spring, and the adjustment of the inlet-valve adjusting screws.

The external-spring type of power-brake control valve contains two control-valve assemblies, two external-spring yokes, and a four port housing. The valve assembly consists of a piston containing a floating pin and a check valve. The rod attached to the piston extends outside the housing and makes contact with one end of the spring yoke. When the check valve is on its seat, it prevents fluid under pressure in the main hydraulic system from reaching the brake line. The brake line is connected to the return line through drilled passages in the piston. When the foot pedal is pressed downward, the spring yoke rotates about its pivot; the first part of this rotation moves the piston up far enough to allow the shoulders on the pin to seat on the piston and close the return passages. Any further movement of the piston will move the pin up and unseat the check valve.

Fluid under pressure from the main system flows through a port, passes the check valve, and out another port to the wheel-brake assembly.

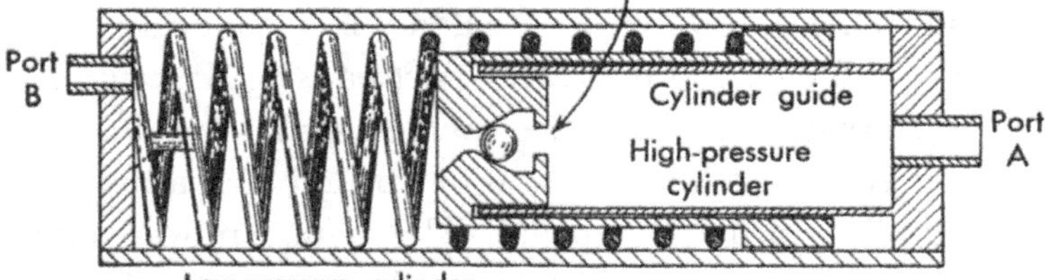

Fig. 221. A schematic drawing of a hydraulic brake debooster.

Pressure acting on the top of the piston produces a force which tends to move the piston downward. When this force is great enough to bend the spring yoke, the piston and pin will move down and allow the check valve to seat, cutting off the pressure from the brake line. The greater the rotation of the spring yoke, the greater the amount of pressure which must be exerted to bend it, and the greater the pressure built up in the brake line. Since the spring yoke is rotated by the movement of the foot pedal, the amount of fluid under pressure admitted to the brake line from the main hydraulic system depends upon the distance the brake pedal is moved. When the pressure is removed from the foot pedal, the fluid pressure on top of the piston moves it and the pin downward, and the piston continues to move down a short distance after the pin strikes the pin stop. This movement separates the shoulder on the pin and piston and opens the drill passages to the return line. The fluid then flows to the bottom of the unit and out a third port to the return line, and the brakes are released.

The brake debooster serves two purposes: (1) to reduce the pressure from the brake control valve, and (2) to insure rapid release of the brake. The brake debooster is shown in Figure 221 and consists of a cylinder barrel which is fitted with a cylinder head on each end. These heads provide connections for the control valve and the brake lines. A spring-loaded piston and a piston guide sleeve are enclosed in the cylinder. The cylinder is divided into a small-volume, high-pressure cylinder and a large-volume, low-pressure cylinder by means of the piston. A ball-type compensating valve is located in the center of the piston head. When fluid pressure from the brake control valve enters port A and acts on the inside of the piston, the piston is moved towards the left.

As the piston moves into the large-volume chamber, the return spring is compressed and fluid is forced out through port B to the brake line. This action is just the opposite of the usual hydraulic cylinder in which a small piston acts to boost the fluid pressure. The pressure in the large-volume chamber is less than that in the small-volume chamber because the pressure acting on the inside of the piston must compress the strong, piston-return spring before the piston can be moved to the left. The pressure is also less because the pressure applied to the small inside area must be distributed over the larger outside area. The pressure in the large-volume chamber is less than the pressure in the small-volume chamber in inverse proportion to their respective areas. This unit deboosts or lowers the pressure. With this arrangement, a smooth brake application can be maintained even though the maximum pressure of the system is applied to the brake control valve. As soon as the pressure from the brake control valve is diminished or released, the piston-return spring moves the piston to the right. When the piston moves to the right, the pressure is reduced in the low-pressure chamber and the brakes are unloaded. When the brakes are applied, the piston normally moves nearly the full length of the cylinder. If there is any leakage in the brake system, the piston will move toward the left until the pin located in the end of the cylinder head opens the ball compensating valve. This allows fluid to flow into the low-pressure chamber until the volume increases enough to move the piston away from the pin, allowing the compensating valve to close.

An emergency pneumatic or compressed-air brake system is sometimes installed to be used in case the hydraulic brake system fails. The brake cylinders are then moved by compressed air supplied from a high-pressure cylinder carried for this purpose. A control valve, which releases the compressed air into the brake system, is mounted within reach of the pilot and copilot.

Shuttle valves lead the compressed air to the brake-actuating cylinders and, at the same time, close the ports to the main hydraulic system to prevent loss of pressure. In some systems, an emergency bleeder valve is installed for the purpose of releasing the compressed air to the atmosphere. It is only necessary to open the control valve to apply the brakes. Compressed air from the cylinder enters the brake lines as soon as this valve is opened. As the air enters the system, a shuttle valve closes the hydraulic-fluid port, allowing the air to enter the cylinder and apply the brakes. Usually, the amount of braking can be varied by the control of the release valve. In some systems, however, opening the control

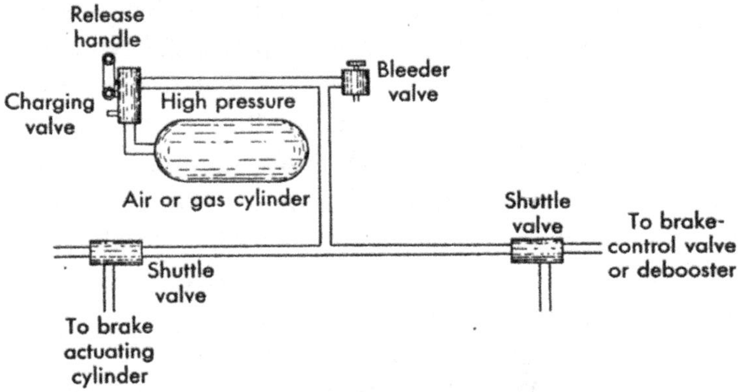

Fig. 222. A schematic drawing of an emergency pneumatic air brake system.

valve releases the full cylinder pressure into the brake-actuating cylinders to apply the brakes. To release the brakes the cylinder valve is closed and the emergency bleeder valve opened. Some systems have both these valves incorporated in one unit to simplify the operation. This enables the pilot to vary the braking pressure by rocking the valve back and forth, allowing air to escape into the system or from the system, as desired. After air has once been admitted into the hydraulic system, the system must be bled to remove all air. The control handle of the pneumatic system should be safetied in the OFF position and the cylinder recharged.

XVII TYPES OF HYDRAULIC SYSTEMS

There are five general types of hydraulic systems: (1) the power-control-valve system, (2) the pump-control-valve system, (3) the electrically driven pump system, (4) the pressure-regulator system, and (5) the open-center system.

The basic principles upon which any hydraulic system operates, as well as the basic units making up the hydraulic system, are much the same. Different systems vary only in the type of unit used, and this varies with the size of the airplane upon which it is installed and the number of parts which are to be hydraulically operated.

Power-Control-Valve System. The power-control-valve system is commonly used in smaller types of airplanes. This system operates all of the hydraulic mechanism except the brakes. As shown in Figure 223, this system consists of the following parts: reservoir, power pump, hand pump, check valve, control valve, pressure-gauge snubber, pressure gauge, relief valve, time-lag power-control valve, selector valves, orifice check valve, landing-gear actuating cylinders, ratchet valve, and wing-flap actuating cylinders.

The pump in this system is usually driven off the accessory drive shaft of the engine, and it generally operates all the time that the engine is running. This pump furnishes the hydraulic pressure to all parts of the system. The hand pump, which is usually of the double-acting type, is used for emergency operation and for checking the hydraulic system when the aircraft is on the ground.

The time-lag power-control valve is installed for unloading the pump. This valve must be closed manually, but it opens automatically after a given interval of time. A compound piston type of selector valve is used for controlling the operation of the landing gear and flaps. This type of valve does not have packing around the pistons and will therefore *not* hold pressure indefinitely. Each piston is attached to an operating handle. If the unit is installed in a two-cockpit aircraft, duplicate controls are usually installed in each cockpit for operating the selector valve.

A ratchet valve is a part of the flap system. The line operating both flaps passes through this unit, which maintains the pressure in the flap-actuating cylinder. This ratchet valve maintains the pressure even though

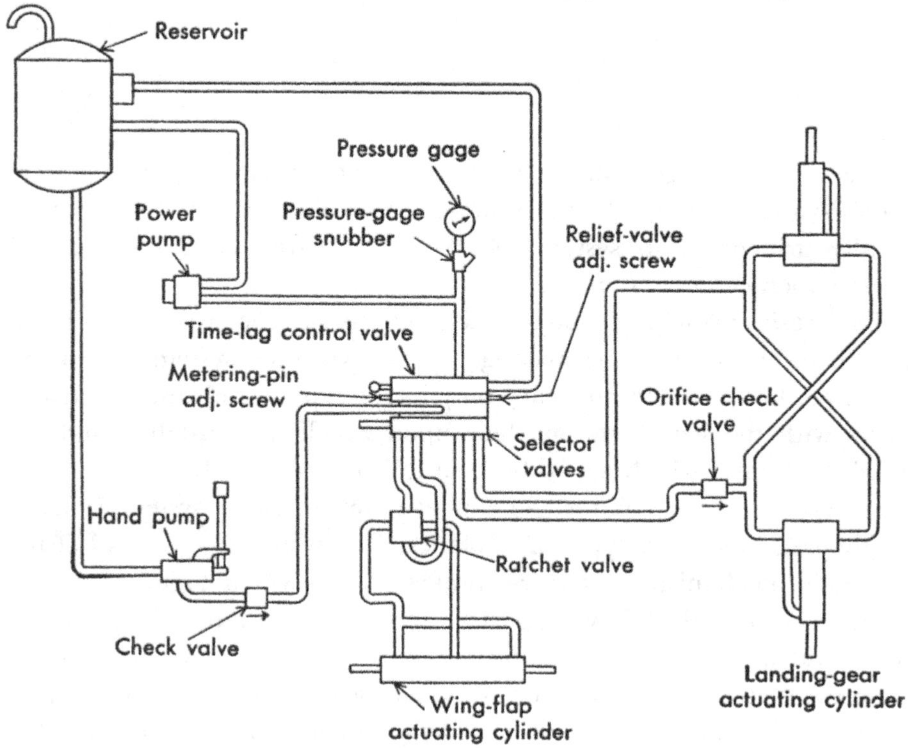

Fig. 223. A schematic drawing of a power control valve system.

the selector valve is of the leak type. A relief valve is installed to limit the pressure developed by the pump after the cylinder has reached the end of its stroke and before the time-lag valve opens.

A pressure gauge is installed to show at all times the pressure in the system.

A pressure valve is installed between the time-lag valve and the selector valve.

The pressure-gauge snubber is installed to prevent oscillation of the pressure-gauge pointer. In the upline of the landing gear is installed an orifice check valve to restrict the flow of oil from the upside of the actuating cylinder. This valve prevents the landing gear from dropping too rapidly when released.

In operating this system, the time-lag control valve is first engaged. The selector valve is next moved to the desired position. After a given length of time, the time-lag power-control valve will open. A pull on the handle will operate this valve if it fails to open automatically. If the system is equipped with a power-control valve without the time-lag feature, the selector valve is placed in the desired position and the power-control valve then engaged. When the pressure reaches the valve for which the unit is adjusted, the valve will open automatically. Pressure on the knob of the valve will open the valve in case it should fail to function.

Pump-Control-Valve System. The pump-control-valve system is similar to the power-control-valve system. The only important difference in the two systems is that a pump-control valve is used in the pump-control-valve system instead of a power-control valve. The method of operating this type of system is first to move the selector valve to the desired position and then hold the pump-control-valve handle in the DOWN position until the operation is completed. A spring-loaded mechanism is usually installed to return the pump-control-valve handle to the UP position when pressure on it is released.

Electrically Driven Pump System. On light aircraft an electrically driven pump system is generally installed. This system operates all of the hydraulic mechanism except the brakes. As shown in Figure 224, this system consists of an electric motor, an electric power pump equipped with a relief valve, a reservoir, a filter, a check valve, and a hand pump equipped with a relief valve and check valve, a selector valve, orifice valve, temperature-expansion relief valve, flap-actuating cylinder, landing-gear actuating cylinders, and a tail-wheel actuating cylinder.

A switch is installed for starting and stopping the electric motor. This motor is operated only when it is desired to operate the hydraulic equipment. The pressure in this system is developed by the power pump driven by an electric motor which is operated by current from the electrical system of the aircraft. The source of current is usually a storage battery.

The hand pump is of the double-acting type and is used only for emergency service or in checking the hydraulic system when the aircraft is on the ground. In this system, the electric pump may be operated even though the engine is not in operation and may be used for testing while the aircraft is on the ground. The operation of the landing gear, tail wheel, and flap mechanism is controlled by poppet-type selector valves. One selector valve controls the operation of the flaps, while the other

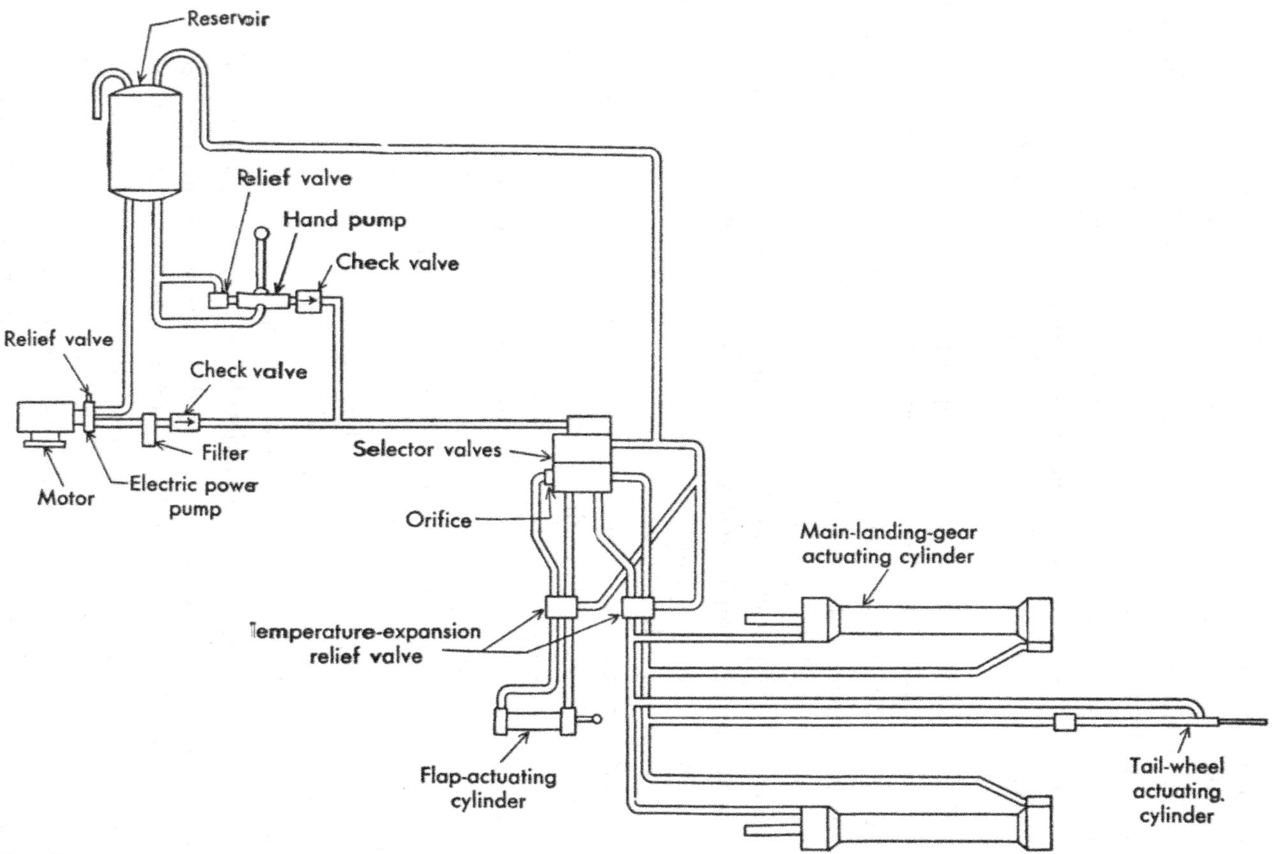

Fig. 224. A schematic drawing of an electrically driven pump system.

controls the operation of the main landing gear and the tail wheel. Both selector valves are usually built into the same unit but are independent of each other. The pressure port to each selector valve is equipped with a check valve. There are four actuating cylinders: one for each main landing wheel, one for the tail wheel, and one for the flaps. To prevent excessive pressure between the time the cylinder reaches the end of its stroke and the time the pilot turns the switch off and stops the pump, a power-relief valve is installed. The hand pump is equipped with a relief valve to keep excessive pressures from building up.

To prevent the flaps from returning to the UP position too rapidly, an orifice valve is installed.

The check valves installed in each pump line prevent the by-passing of the oil through either pump when the other is operated.

The temperature-expansion relief valves are installed to prevent excessive pressures from developing when the hydraulic fluid expands because of a rise in temperature. These relief valves are installed in both the landing-gear and flap systems.

In operating this system, the selector valve is placed in the desired position and the pump started by means of the electric switch. This switch is usually spring-loaded to return it to the OFF position and must be held on until the operation is completed. For example, the switch is held in the ON position until the landing gears are down and for several seconds more to be sure that sufficient pressure is built up in the actuating cylinders. The selector valve should be placed in the neutral position before the switch is released.

Pressure-Regulator System. The pressure-regulator system is used on larger aircraft and operates all of the hydraulic mechanism including the brakes. This system may be used on lighter aircraft, if desired. As shown in Figure 225, it includes an engine-driven gear type of power pump which operates the landing gear, wing flaps, cowl flaps, and brakes. The hand pump in this system is used in case of an emergency or to test the hydraulic system while the aircraft is on the ground. Line-disconnect valves are installed to assist in making an engine or pump change.

A rotor type of pump selector valve is used to direct fluid from either pump into the system. The fluid from the other pump is automatically directed to the gyropilot. To unload the power pump, a pressure regulator is installed. This unit also keeps the pressure within selected limits. To relieve the pressure if the regulator fails to function, a relief valve for the system is included.

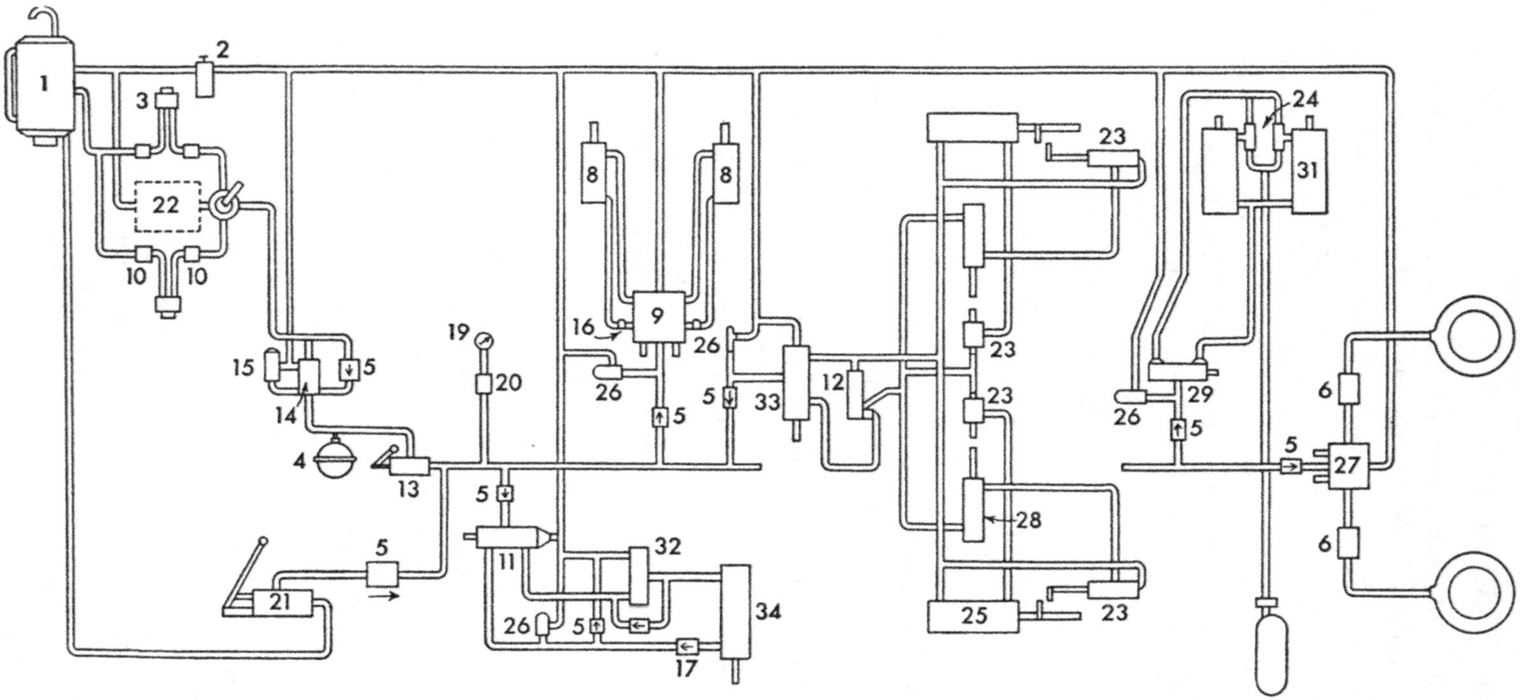

Fig. 225. A schematic drawing of a typical pressure regulator system. (1) Reservoir; (2) Cuno filter; (3) power pump; (4) accumulator; (5) check valve; (6) deboosters; (8) cowl-flap actuating cylinders; (9) cowl-flap selector valves; (10) line disconnect valves; (11) wing-flap selector valve; (12) cross flow valve; (13) by-pass check valve; (14) pressure regulator; (15) system relief valve; (16) variable restrictor; (17) orifice check valve; (19) pressure gauge; (20) orifice; (21) hand pump; (22) gyro-pilot; (23) sequence valve; (24) shuttle valve; (25) landing-gear actuating cylinder; (26) temperature expansion relief valve; (27) power-brake-control valve; (28) landing-gear door actuating cylinder; (29) bomb door selector valve; (31) bomb door actuating cylinder; (32) flap overload valve; (33) landing-gear selector valve; (34) flap actuating cylinder.

The fluid is stored under pressure in a pressure accumulator. A by-pass check valve prevents the charging of the accumulators during emergency operation of the hand pump. However, this valve permits charging the accumulators when desired. An orifice valve is installed in the line to the pressure gauge to prevent or reduce vibration of the pointer on the gauge installed to indicate the pressure in the system.

Piston-type selector valves are used to control the operation of the mechanism. In each section of the system, a relief valve is installed to prevent development of excessive pressure due to a rise in temperature. Each part to be moved is connected with an actuating cylinder. To allow the landing gear to extend evenly and rapidly, automatic cross-flow valves are installed. The flaps are kept from lowering too far at high speeds by a flap overload valve. An orifice check valve is installed in the flap down-line to prevent the flaps from going up too rapidly. Air pressure is supplied for emergency operation for various parts of the system.

Debooster valves and power-brake-control valves are installed to control the operation of the brake. Variable, resistor, fluid, flow valves are used in the cowl-flap system to restrain the rate of movement of the flaps. Sequence valves are installed to allow the landing-gear doors to open before the landing gear starts down and to ensure that it is fully up before the doors start to close.

The filter in this system is of the Cuno disk type. In this kind of system, the selector valve is moved to the desired position and, when the mechanism has reached the end of its stroke and the pressure has been built up in the pressure tank, the pressure regulator automatically opens and relieves the pump of its load.

Open-Center System. In the open-center system, the fluid circulates from the reservoir through the pump, through the selector valve in the system, and back to the reservoir. There is no pressure in the lines unless the hydraulic units are being operated. The open-center system contains an automatic, neutral selector valve or open-center valve, instead of the usual piston or poppet type of selector valve. Whenever these valves are in the neutral position, all of the fluid from the pump circulates through them to the reservoir. The selector valves in this system are connected in series instead of in parallel. In an open-center system, the selector valve is the unit which relieves the pump of its load when the actuating piston reaches the end of its stroke. The main advantage of this system is that all parts of the system are relieved of hydraulic pressure except when some part of the system is being operated.

XVIII VACUUM AND DE-ICER SYSTEMS

When aircraft are equipped with such flight instruments as the automatic pilot, turn-and-bank indicator, and artificial horizon and directional gyro for all-weather flying, they are usually equipped with a vacuum system. Vacuum systems consist of vacuum pumps and the necessary lines and valves. Venturi tubes may also be installed to create the vacuum.

The vacuum pump is usually driven from the engine. On multi-engine aircraft there is usually one pump for each engine. An arrangement is often installed whereby the carburetor of each engine may be used to create a vacuum. Each pump may be used independently of the others. In cases where the vacuum pumps fail, the Venturis or the suction created in the intake manifold or the carburetor may be used to build up sufficient vacuum to operate the suction instruments.

The entire group of gyroscopic flight instruments may be operated from any one pump or from the induction system. The automatic pilot cannot be operated from the induction system, but sufficient suction may be developed to operate the turn-and-bank indicator.

The de-icer arrangement on the wings, stabilizer, and vertical fin, or other parts of the aircraft where "rubber boots" are used, is operated by means of alternate pressure and vacuum. These rubber boots are made up of individual cells into which air may be driven to break up the ice that forms. These cells pulsate, having the air alternately pumped into them and exhausted. When the de-icer system is not in use, the cells must be kept under suction to prevent their inflation by the semi-vacuum caused by the air flow over them when they are in certain altitudes. It is necessary to have an oil separator to remove any oil from the air being pumped into the cells.

The distributor valve is connected with the electric motor which alternately supplies air to the various cells in the boots and removes air from them. Only clean, dry air should be supplied.

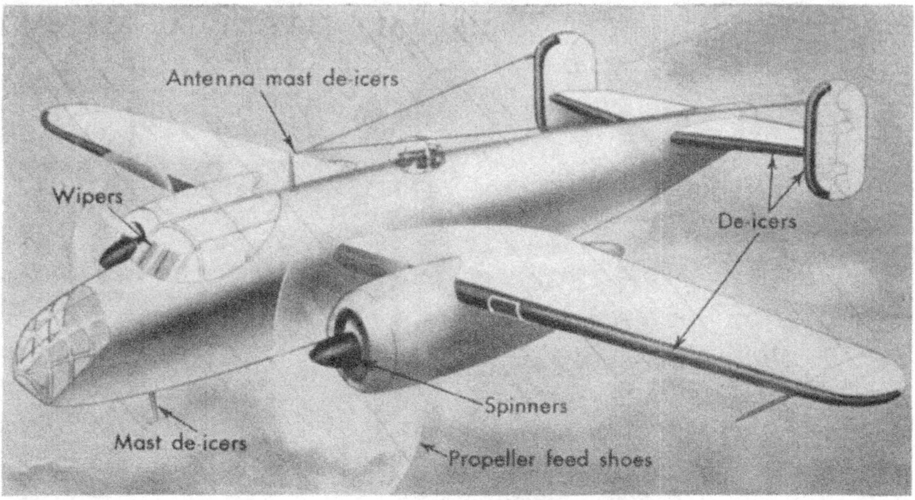

Fig. 226. Aircraft equipped with de-icers. (Courtesy The B. F. Goodrich Company.)

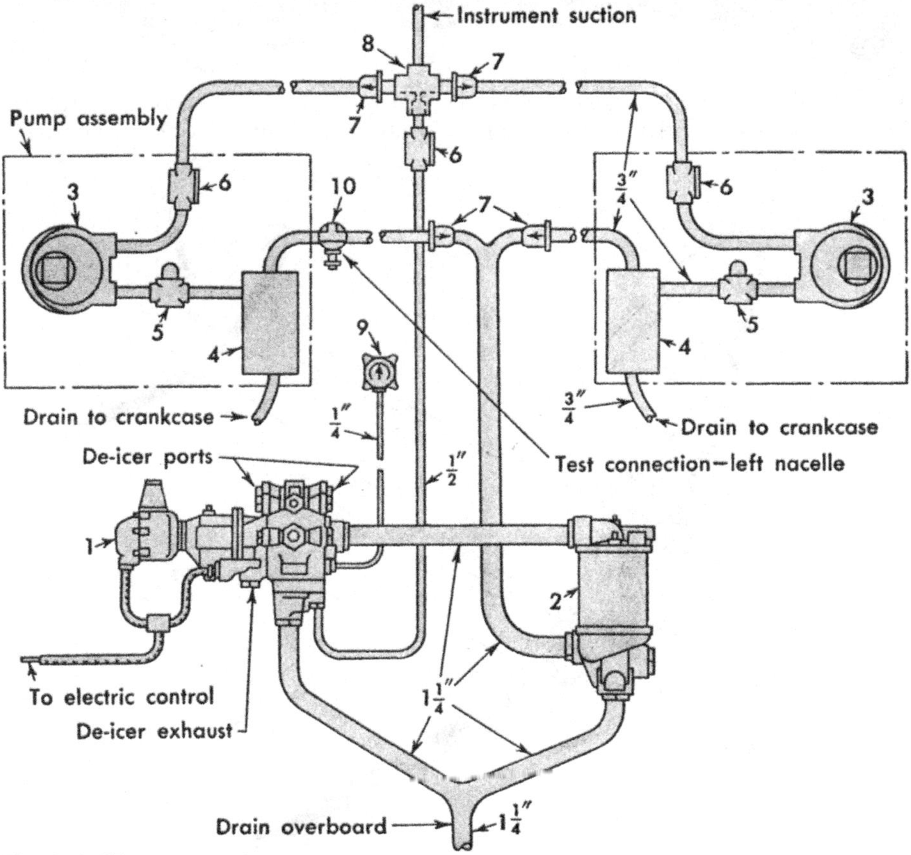

Fig. 227. The arrangement and relationship of the various accessories in a typical de-icer plumbing system. (1) Distributor valve; (2) air filter; (3) pump; (4) oil separator; (5) safety valve; (6) suction relief valve; (7) check valve; (8) tee; (9) pressure gauge; (10) pressure relief valve. (Courtesy The B. F. Goodrich Company.)

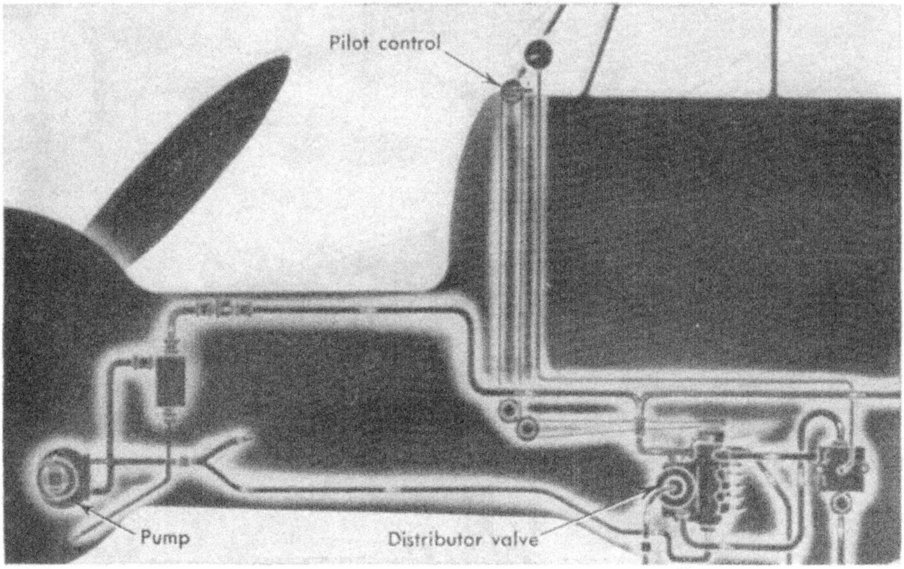

Fig. 228. An air supply system functions independently of other systems on the airplane, insuring positive inflation and deflation of de-icers. (Courtesy The B. F. Goodrich Company.)

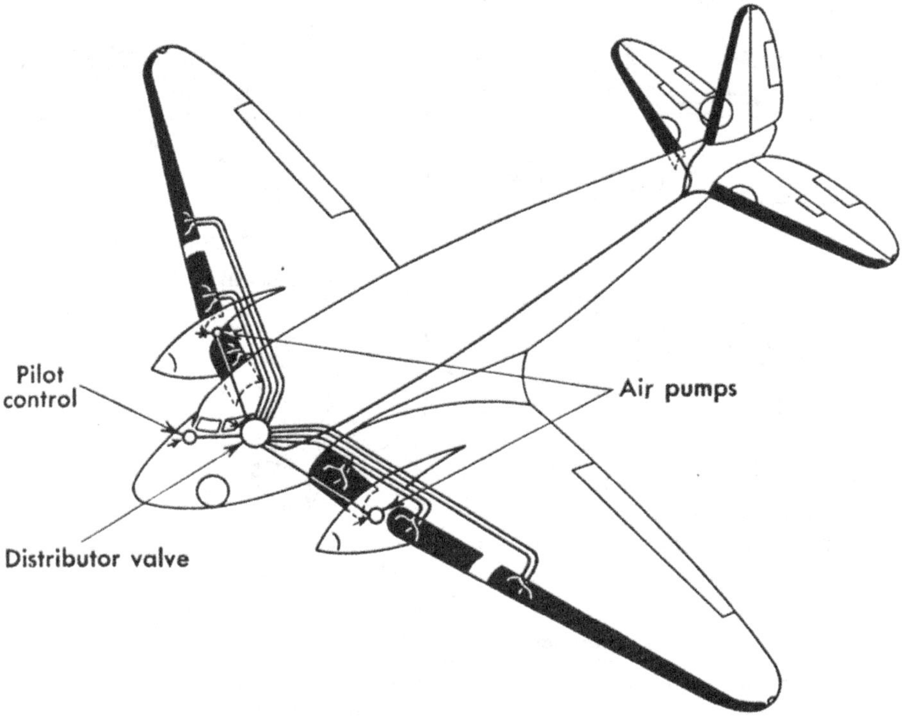

Fig. 229. A multi-engine installation provides sufficient reserve pump capacity for de-icers. (Courtesy The B. F. Goodrich Company.)

Check valves are provided to prevent the escape of air from one pump through another pump in case one of them fails. A line connecting the cells in the boot with the exhaust side of the distributor valve prevents inflation of the cells when they are not in use. A pressure of approximately 7 or 8 p.s.i. is used to operate the de-icer system.

Fig. 230. Aerodynamic "cleanliness" makes practical de-icer installations. (Courtesy The B. F. Goodrich Company.)

De-icers are made of both natural and synthetic rubber and usually have a special conductive surface which discharges static electricity. Each unit consists fundamentally of one or more tubes or cells which may be inflated at regular intervals. The tubes are built with stretch areas and reinforcing strips. The boots are usually equipped with a metal

Fig. 231. Boots are applied under tension to give smooth fit between air foil and de-icers. (Courtesy The B. F. Goodrich Company.)

strip on the edge to fasten to the wing and they are applied under tension to give a smooth fit between the air foil and the de-icer. De-icers may be applied not only to the leading edge of the wing, the vertical fins, and the horizontal stabilizers, but may be applied to the leading edge of the control surfaces or flaps and other parts of the airplane structure. Slotted wings may have a de-icer arrangement on the leading edge of the slot. Antenna mast, radio loops, and Pitot tubes may be equipped with

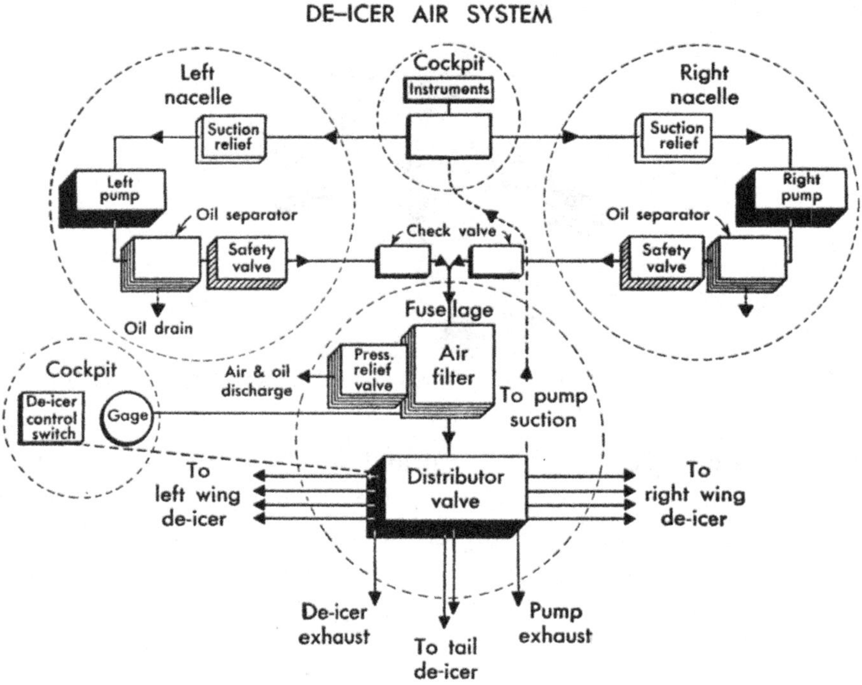

Fig. 232. A de-icer air system. (Courtesy The B. F. Goodrich Company.)

de-icers. De-icers for such parts of the airplane are usually of the boot type which slip on or may be put in place and fastened with zippers and are usually arranged so that they may be installed or removed without damage to the aircraft structure.

The typical de-icer system consists of the pumps, safety valves, oil separators, air filters, a pressure-relief valve, and a motor-driven distributor valve. The systems vary with the type of plane and the number of engines. The fundamental de-icer system is quite typical and varies only with the size of the plane which, in turn, causes a variation in the size of the pump and the type and size of the distributor valves. A vane-type, engine-driven air pump is generally used. These pumps will deliver a sufficient volume of air at approximately 7 or 8 p.s.i. of pressure. The amount of air which the de-icer tubes will hold largely determines the size of the pump.

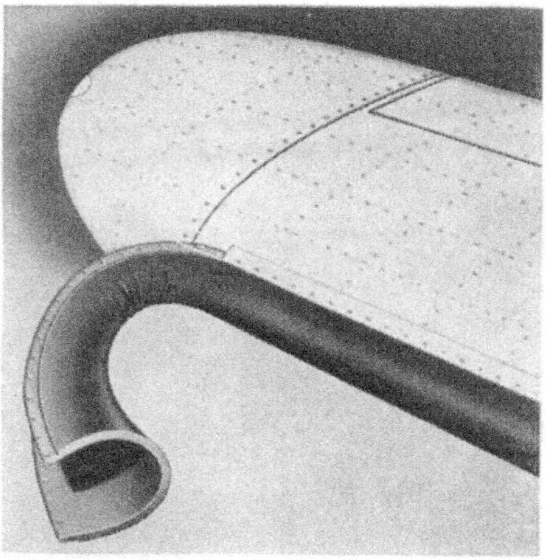

Fig. 233. De-icers may be arranged so they may be installed or removed without damage to the aircraft structure. (Courtesy The B. F. Goodrich Company.)

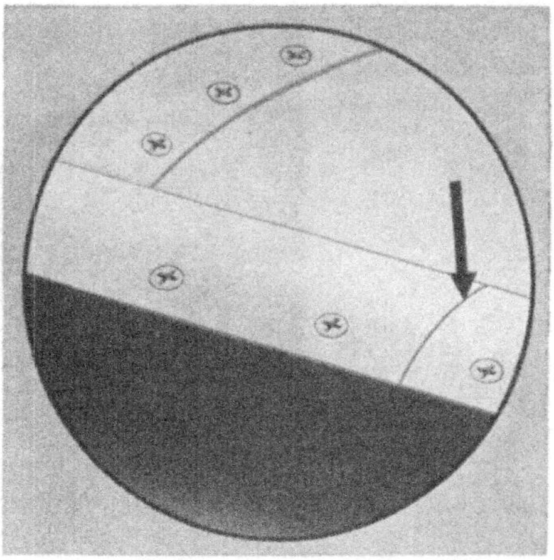

Fig. 234. Detail showing fastening of boot to aircraft structure. (Courtesy The B. F. Goodrich Company.)

As the altitude increases, the output of the pump decreases. It is, therefore, necessary to have a pump which will deliver a satisfactory amount of air at high altitudes. This pump, of course, will deliver an

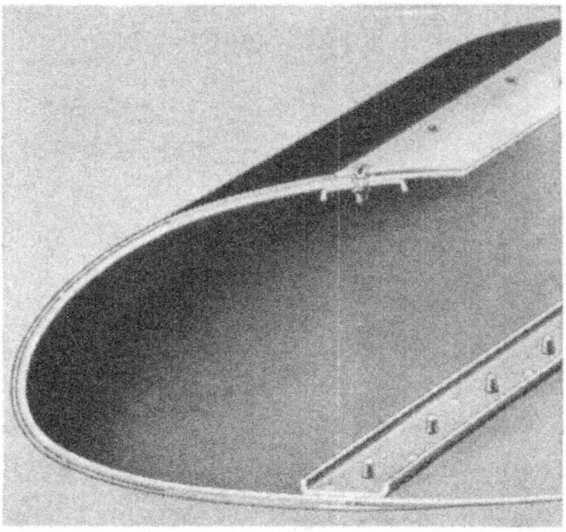

Fig. 235. The de-icer boot is applied smoothly to the leading edge of the wing. (Courtesy The B. F. Goodrich Company.)

excess of air at sea-level pressures. A suction relief valve, set at approximately 6 in. of mercury, must be installed in the system if vacuum instruments are to be operated by the de-icer air pump. This suction relief valve prevents the pump inlet from becoming starved at high engine r.p.m. Vane-type pumps are lubricated internally and, therefore, each requires an oil separator. The oil separators are usually located near

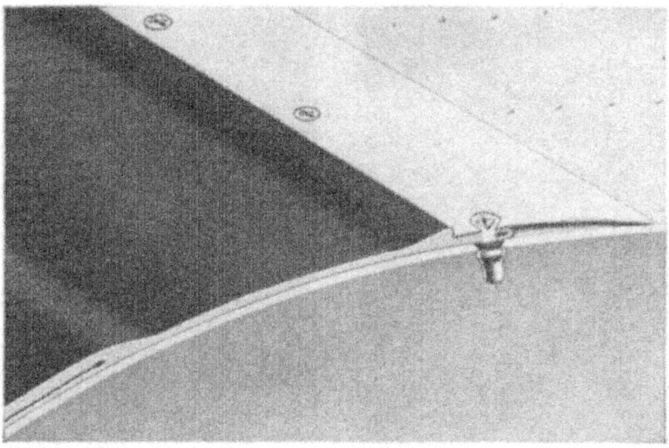

Fig. 236. Detail of de-icer boot fastening to the airplane skin. (Courtesy The B. F. Goodrich Company.)

the engine so that oil may be drained back to the engine crankcase. The oil separators should be placed where they will remain as cool as possible.

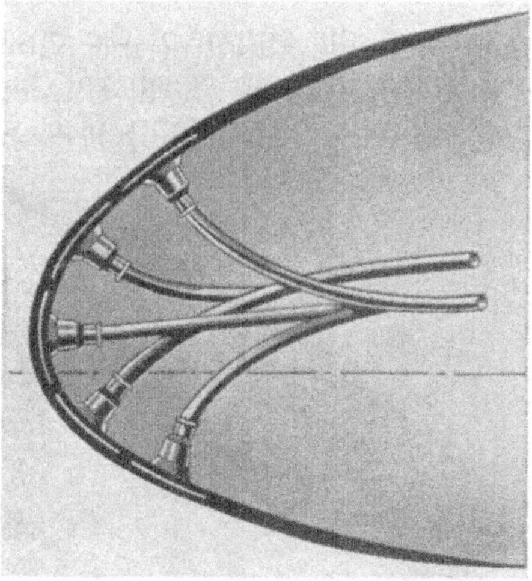

Fig. 237. Suction and pressure tubes within the airplane structure. Each tube is connected with an inflatable tube in the de-icer boot. (Courtesy The B. F. Goodrich Company.)

It is necessary to install a safety valve near each pump to by-pass excess air. These valves are usually set to by-pass the air when a pressure of approximately 9 p.s.i. is reached. The safety valve should not be used to regulate pressure in the de-icer system. To allow testing when the engine is not running, there is usually installed a test connection in the de-icer system. A three-way valve is usually placed in the de-icer system. A check valve is installed between each pump and the distributor valve. This valve is to protect vacuum instruments from damage if, for any reason, the air reverses its direction. This might be caused by engine kickback or pump failure.

An air filter with a relief valve is necessary to completely remove oil left in the air by the separator. This relief valve regulates the pressure in the de-icer system. Oil and excess air are usually discharged overboard by a line from the filter. The discharge line must be long enough to

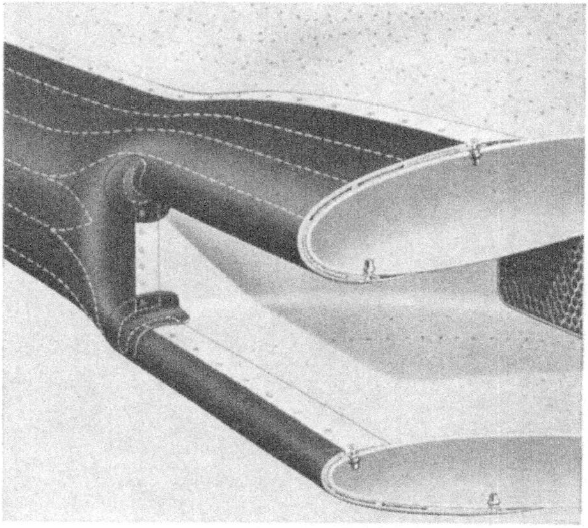

Fig. 238. De-icer boots may be applied around openings in the aircraft structure. (Courtesy The B. F. Goodrich Company.)

keep any back pressure from developing. The de-icer distributor valve is operated by an electric motor. This valve regulates the pulsating action of the cells in the de-icer boots by alternately connecting the cell with pressure and suction.

The rotor of the snap-action valve provides a three-phase cycle of 5-sec. inflation, 5-sec. deflation, and 30-sec. suction. A gear mechanism snaps the rotor valve from port to port to provide sudden inflation. A limit switch stops the rotor at the pump's exhaust port. A venting system is used to assist in preventing the boots from "lifting" by the movement of the air over the air foil.

PART III AIRCRAFT INSTRUMENTS

XIX FUNDAMENTAL PRINCIPLES OF AIRCRAFT INSTRUMENTS

The instrument board has been described as the nerve center of the airplane. The aircraft of today is equipped with many instruments. Some instrument boards contain well over 100 instruments, gauges, switches, and gadgets of various kinds. The pilot of today must depend upon these instruments to guide him safely to his destination and to keep him informed of the operating conditions of the aircraft and the engine.

Early aircraft had many forced landings because of the failure of the power plant. These failures were often due to such preventable items as

Fig. 239. An aviation magnetic compass. (Courtesy Kollsman Instrument Division, Square D Company.)

lack of fuel, lack of water, or lack of oil. The early engines were soon equipped with such simple instruments as a water-temperature gauge, oil-temperature gauge, oil-pressure gauge, and fuel gauges. As soon as the airplane became capable of flying a considerable distance, a magnetic compass and an altimeter were installed. These instruments were followed by bank-and-turn instruments to inform the pilot of the position of his aircraft while in flight. Rate-of-climb instruments assisted in telling him whether he was gaining or losing altitude and also informed him how rapidly the aircraft was climbing or diving. The aircraft instruments are divided roughly into a number of separate groups.

The first two main groups are the engine instruments and the aircraft instruments. The engine instruments keep the pilot informed of such items as oil and fuel pressure, manifold pressure, carburetor pressure,

Fig. 240. A sensitive manifold pressure gauge. (Courtesy Kollsman Instrument Division, Square D Company.)

Fig. 241. A dual engine tachometer. (Courtesy Kollsman Instrument Division Square D Company.)

engine temperature, the rate of rotation of the engine, and many other things which are important for the pilot to know while in flight. The aircraft instruments inform the pilot of such items as the position of the landing gears or flaps, hydraulic pressure, vacuum, and air temperatures.

Two other groups of instruments consist of the flight instruments and the navigation instruments. The flight instruments indicate such items as altitude, attitude, air speed, and whether or not maneuvers are being properly performed. The navigation instruments assist the pilot in finding his way from his point of departure to his destination. They give him his direction and assist him in remaining on his desired course. The clock is an important navigation instrument because it keeps the pilot informed of the time elapsed in flight.

The indication which the instrument shows may be brought about by several different means such as atmospheric pressure, the pressure of fluids and liquids, magnetic and electrical effects, changes in temperatures, speed of rotation, the rate at which fuel and oil is being used, and the gyroscope. Each instrument has its own functions to perform, but the pilot must interpret the reading of an instrument in terms of the readings of the other instruments and their relation to the performance of the aircraft as a unit. An instrument may depend upon more than one of the above for its operation.

A knowledge of the atmosphere, its composition and characteristics, is necessary in order to understand the instruments which depend upon atmospheric pressure for their operation.

Fig. 242. A vertical speed indicator operated by air pressure. (Courtesy Kollsman Instrument Division, Square D Company.)

Fig. 243. A differential pressure gauge indicating pressure in inches of water. (Courtesy Kollsman Instrument Division, Square D Company.)

The atmosphere is a gaseous envelope which surrounds the earth. It extends outward into space from the earth's surface more than 200 miles. All of the atmosphere presses against the earth's surface with a pressure which is determined by the total weight of the gases above any given point. The pressure of the atmosphere at sea level is 14.7 lb. per sq. in. The atmosphere is made up of a number of free gases. It is approximately 78% nitrogen and 21% oxygen, and the remaining 1% is made up of a number of rare gases such as carbon dioxide, argon, neon, krypton, xenon, hydrogen, and others. Since all gases are highly elastic, the lower layers of the atmosphere are compressed by the weight of the overlying layer.

The decrease in density of the atmosphere does not fall directly with the altitude. At 18,000 ft., almost exactly one half of the atmospheric air is above the pilot and one half below. Under normal conditions, the atmospheric pressure, as determined by a mercurial barometer at sea level, is 29.92 in. of mercury. This reading comes from the height of a column of mercury which is supported in the barometer tube by the pressure of the atmosphere. At an 18,000 ft. elevation, the barometer would read approximately 14.93 in. of mercury. Although 18,000 ft. is only about 3.5 miles, one half the atmosphere by weight is below this level, and almost 97% of the atmosphere is below 18 miles. Traces of the atmosphere are

Fig. 244. A sensitive altimeter. (Courtesy Kollsman Instrument Division, Square D Company.)

Fig. 245. True air speed indicator. (Courtesy Kollsman Instrument Division, Square D Company.)

thought to extend out into space about 600 miles. If the pressure decreased exactly inversely with the altitude, one half the atmosphere would be about the 300 mile level.

Air, like all other substances, has weight. This weight makes possible the flight of all aircraft and the operation of aircraft instruments which depend upon atmospheric pressure. As the altitude increases, the composition of the air remains almost constant. However, a cubic foot of air at 18,000 ft. of altitude would contain only one half as much gas as it would at sea level. This makes necessary the carrying of oxygen for the use of personnel at high altitudes and the supercharging of high-altitude aircraft engines.

There are two types of barometers commonly used to measure atmospheric pressure: the mercurial barometer and the aneroid barometer. The measurement of atmospheric pressure is simply "weighing" the amount of air above the point where the instrument is located.

Figure 246 shows a simple mercurial barometer which is made by filling a glass tube, approximately 36 in. long and closed at one end, with mercury. The open end is closed temporarily, and the tube is inverted with the open end beneath the surface of mercury in a container. The temporary closing of the open end is removed, and the mercurial barometer is complete. It is found that the mercury in the tube sinks until the top of the mercury column at sea level is approximately 30 in. above the surface of the mercury in the container. The mercury in the tube is held up by the pressure of the atmosphere on the surface of the mercury

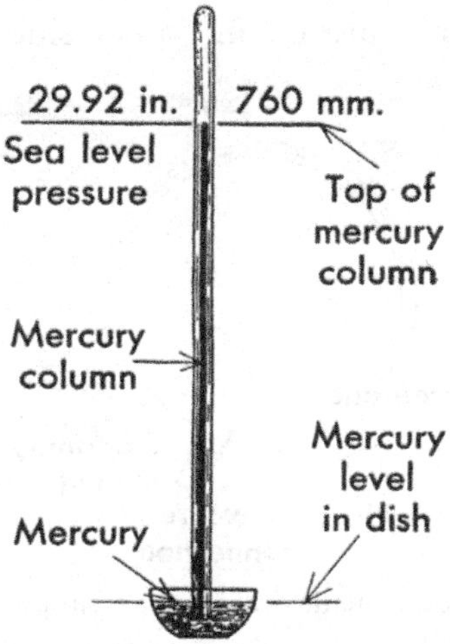

Fig. 246. A schematic drawing of a simple mercurial barometer.

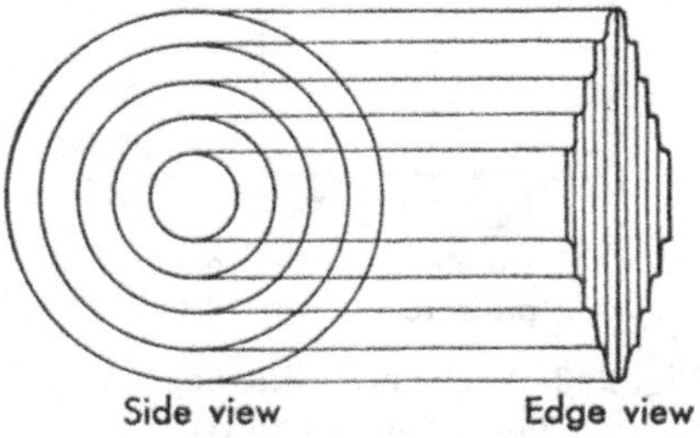

Fig. 247. A schematic drawing of a simple aneroid cell.

in the container. Any change in the pressure will cause the column of mercury to rise or fall in the tube. The pressure of the atmosphere in terms of inches of mercury depends upon the height of the mercurial column when the measurement is in inches. Many mercurial barometers are calibrated or measured in the metric system. In this system normal sea-level pressure is 76 cm. or 760 mm.

The aneroid barometer makes use of the aneroid cell, which is called the "collapsible chamber" in this type of instrument. This cell is disk-shaped and made of thin metal. The sides are corrugated to make them more flexible, and part of the air is then removed from the cell. The cell is mounted in an instrument under spring tension to prevent its collapsing completely. Any change in atmospheric pressure will cause a change in the thickness of the cell. If the atmospheric pressure increases, the additional pressure acting against the spring causes a slight collapse of the cell. The expanding and collapsing of the cell, which is the active unit in the aneroid barometer, indicate changes in atmospheric pressure.

Another kind of instrument which depends upon atmospheric pressure is the diaphragm type. This instrument is made up of a sealed cell which is partitioned off into two parts by means of a semiflexible diaphragm, as shown in Figure 248. Two tubes lead to the separate parts of the cell. One tube is exposed to static air pressure, while the other tube is exposed to dynamic air pressure or to gases under pressure. This arrangement is known as the "Pitot-static instrument." The pressure on the static side of the diaphragm is always equal to the atmospheric pressure at the point where the instrument is located. The pressure on the other side

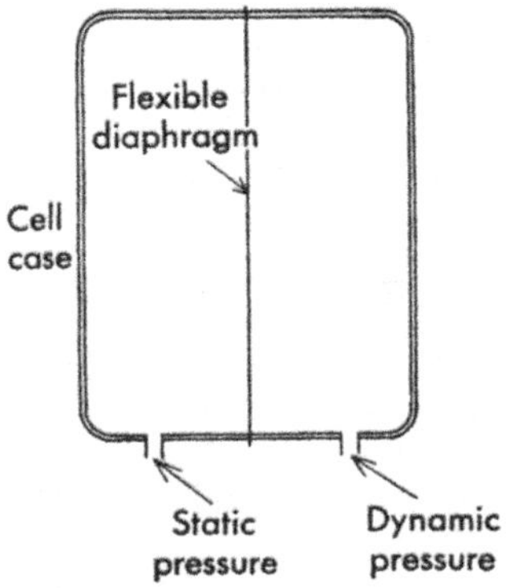

Fig. 248. A schematic drawing of a simple diaphragm-type pressure cell.

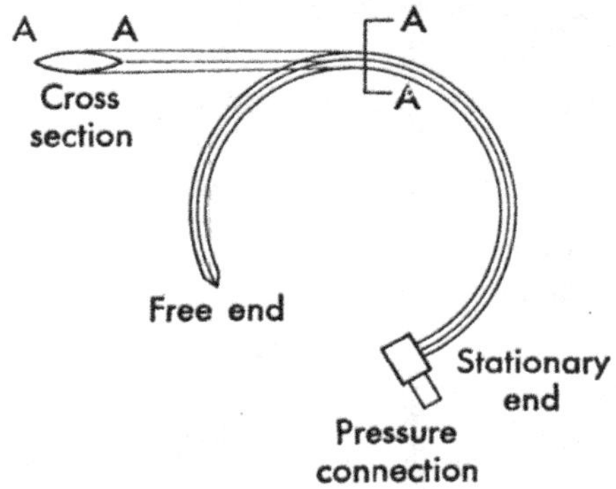

Fig. 249. A schematic drawing of a simple Bourdon tube.

of the diaphragm may be either greater or less than the atmospheric pressure. If the pressure is less than the atmospheric pressure, the instrument acts as a suction gauge and can be made to indicate decreased pressure.

Pressure Instruments. Liquids are practically incompressible, and pressure applied to any point on a confined liquid is transmitted undiminished through the liquid to the walls of the container. Confined liquid is liquid which completely fills a closed container. This property of liquids is made use of in pressure gauges. These gauges, as well as instruments measuring gases under pressure, employ a device called the Bourdon tube.

A gas is a substance in which the atoms and molecules are free to move and are not attracted closely to each other. A liquid is a substance in which the atoms and molecules are free to move, but are attracted closely to each other. Gases expand and will fill any container in which they are confined. Liquids do not expand, but settle to the bottom of a container. Both liquids and gases flow readily.

Some liquids do not flow as readily as others because they have greater viscosity. Viscosity is that property of a liquid which determines the readiness with which it flows. Fluids are substances which flow, and so liquids and gases are both classified as fluids.

Pressure is a measure of force and is usually measured as pounds pressure per square inch (p.s.i.). Pressure may be measured in any conventional unit, such as pounds or ounces or grams, which acts on one square inch of surface. A force is simply the push or pull.

Fig. 250. A manifold pressure gauge, dual indicator type. (Courtesy Kollsman Instrument Division, Square D Company)

The Bourdon tube is a flattened metal tube with an elliptical shape which has been bent into the arc of a circle. The bend, of course, does not have to be the true arc of a circle. One end of the Bourdon tube has an opening in it and is fastened solidly into the instrument case in which it is mounted. The other end is sealed and is left free to move. The opening in the fastened end of the tube is connected to a tube through which the pressure is applied. Even though the opening in the tube is small, the pressure applied is spread evenly over the inside of the Bourdon tube. As pressure is applied, the tube tends to straighten out. If the pressure is decreased or suction is applied, the tube tends to bend inward and form an arc having a smaller radius. The free end of the tube moves back and forth with changes in pressure and it is connected by suitable linkage to the indicating needle on the face of the instrument. This type of instrument is commonly known as a pressure gauge and is used to measure such quantities as engine-oil pressure and hydraulic-fluid pressure.

The Bourdon tube may be connected by means of a capillary tube with a metal cell containing a highly volatile liquid. Changes in temperature cause this liquid to develop changes in gaseous pressure which affect the Bourdon tube. When connected with a cell of this kind, the instrument may be designed so as to indicate temperature.

Magnetic and Electrical Instruments. Everyone is familiar with the common magnet. The early Greeks and Chinese were familiar with the natural magnet which they named the lodestone. This is a black iron ore now called magnetite which is strongly magnetic and sometimes possesses polarity and is then called lodestone. If a piece of this natural

Fig. 251. An electric tachometer, dual indicator type. (Courtesy Kollsman Instrument Division, Square D Company)

Fig. 252. An oil pressure gauge for indicating pressure of the hydraulic fluid in the automatic pilot system. (Courtesy Electric Auto-Lite Company)

magnet is hung by a thread so that it is free to turn, it will arrange itself so that one axis of the piece is in a north-and-south direction. This material, like ordinary magnets, will attract and pick up small particles of magnetic matter such as iron, steel, nickel, and cobalt.

A bar of iron or steel can be magnetized to form a magnet. Magnets have about them a field which exerts a force upon other magnets and magnetic particles. This field is called a magnetic field. When a conductor, which is a substance that can carry an electric current, moves through a magnetic field in such a manner that it cuts the magnetic lines of force, an electric current is set up in the conductor. A small compass placed in a magnetic field arranges itself parallel to the lines of force. There are magnetic lines of force about a bar magnet. One end of the magnet is called the north pole of the magnet, and the other end is called the south pole of the magnet. Two magnetic north poles will repel each other, as will two south poles. North and south poles, however, attract each other.

The earth itself acts as though it were a huge magnet. It has two magnetic poles, the North Magnetic Pole and the South Magnetic Pole. The North Magnetic Pole attracts the north end of a magnet, while the South Magnetic Pole attracts the south end of a magnet. A bar magnet suspended in the middle so that it is balanced and free to turn arranges itself in approximately a north-and-south direction. The magnet arranges itself parallel to the lines connecting the North and South Magnetic Poles. Neither the North nor the South Magnetic Pole is located exactly

Fig. 253. A sensitive air-speed indicator. (Courtesy Kollsman Instrument Division, Square D Company)

at the same place as the geographic North and South Poles. For this reason, a magnet does not point toward the true north and the true south at most places on the earth's surface. The amount that the magnet varies from the true north and south is called "variation." The only place that a magnet, which is a simple form of magnetic compass, points to the true north is along a line connecting the North and South Poles which has no magnetic variation. This line is the agonic line. Lines drawn between the North and South Magnetic Poles, having equal magnetic variations, are called isogonic lines.

Variation is one of the errors of the magnetic compass. Another magnetic compass error is called "deviation." Deviation is caused by magnetic substances in the vicinity of the magnetic compass which cause it to vary from its true magnetic indication. Each magnetic compass, after being mounted in an aircraft, must be tested to determine the compass errors when the aircraft is heading in different directions. A deviation card is placed near the compass which lists the amount of deviation of the compass for each 15° to 30° change in direction. Some of the compass errors may be corrected by placing small adjustable magnets near the compass needle. These magnets should be adjusted to remove as much error as possible on each heading. This process is called "compass compensating," and the small magnets are called "compensating magnets." The process of determining the amount of deviation of the compass on various headings is also called "swinging the compass."

Fig. 254. An air-speed indicator, graduated to 800 miles per hour. (Courtesy Kollsman Instrument Division, Square D Company)

When an electric current travels along a conductor, it always sets up an electric field about the conductor. The tendency of an electric current to flow along the conductor corresponds to the tendency of water to flow through a pipe. If the pressure forcing the water through the pipe is high, the tendency to flow is great. Electric pressure, which is called "difference in potential" or "electromotive force," corresponds to the water pressure in the pipes. Electric pressure or electromotive force is measured in volts, while the pressure in the water pipes is usually measured in p.s.i. The amount of water flowing through a pipe is generally measured in gallons per minute or per some unit of time. The rate of flow of electricity is measured in amperes. The quantity of water may be measured in gallons, while the quantity of electricity which flows along a conductor is measured in coulombs. A coulomb is the amount of electricity which will pass a point in a conductor in one second when a current of one ampere is flowing.

The amount of water which will flow through a pipe in a given length of time depends, not only on the pressure, but also on the size of the pipe. A small pipe offers more resistance to the flow of water than does a large pipe. A small wire or conductor offers more resistance to the flow of electricity than does a large wire or conductor.

The unit used to measure the resistance to electrical flow is the ohm. The standard ohm is the resistance offered to the flow of an electric current by a column of mercury 106.3 cm. long which has a cross section of 1 sq. mm. One thousand feet of No. 10 copper wire has a resistance of almost exactly 1 ohm.

Fig. 255. An ammeter and a voltmeter. (Courtesy AC Spark Plug Division, General Motors Corporation)

One coulomb is that amount of electricity which will deposit 0.001118 g. of silver from a chemical solution in 1 sec.

There is a close relationship between the different measurements used in electricity. If a conductor or electric circuit has a resistance of 1 ohm and the electric pressure is 1 v., 1 amp. of current will flow and 1 coulomb of electricity will pass through the circuit in 1 sec.

The rate at which electric power is used is measured in terms of watts. The number of watts of power developed when an electric current is flowing can be found by multiplying the volts by the amperes. When 1 w. is used over a period of 1 hr., 1 watt-hour of power has been used. When 1000 w. of power are used over a period of 1 hr., the amount of power or current used is called 1 kilowatt-hour.

Ohm's law states the relationship between the resistance, current, and voltage in three simple formulas. The current is equal to the voltage divided by the resistance; the resistance is equal to the voltage divided by the current; and the voltage is equal to the current multiplied by the resistance. Since current is given in amperes, voltage in volts, and resistance in ohms, the formulas become:

$$\text{Amperes} = \frac{\text{Volts}}{\text{Ohms}}$$

$$\text{Ohms} = \frac{\text{Volts}}{\text{Amperes}}$$

$$\text{Volts} = \text{Amperes} \times \text{Ohms}$$

The conventional symbol for amperes or current is I, that for volts is E, and for resistance is R. So, Ohm's three fundamental formulas are:

$$I = \frac{E}{R}$$

$$R = \frac{E}{I}$$

$$E = R \times I$$

The amount of current flowing through a circuit determines the strength of the magnetic field about the conductor. The greater the current, the greater the strength of the magnetic field. If the conductor is in the form of a wire coil, the coil acts like a magnet. The coil has north and south magnetic poles, and an iron core placed within the coil becomes a strong magnet.

If a small coil of wire having an iron core is placed between the poles of a magnet, it will arrange itself so that the lines of force between the poles of the magnet will pass through the core and the coil. If the coil of wire and its iron core are suspended under spring tension and an electric current is passed through the coil, the core will turn against the spring tension a distance which varies with the amount of electric current flowing through the coil. If the coil is connected to an indicating needle, this device may be used to measure the amount of electric current flowing or as an indicator to detect electric currents in the coil. The arrangement is called a "galvanometer."

The voltage meter (voltmeter) for measuring volts and the ammeter for measuring an electric current both depend upon this sort of arrangement for their operation.

There are two kinds of electric particles. The positive particle is called the "proton," and the negative particle is called the "electron." The proton is approximately 1845 times as heavy as the electron. Electric current is caused by the attraction between the proton and the electron which brings about the movement of electric particles through a conductor. Since the electrons are much lighter, and the electrons in a conductor are more free to move than are the protons, the flow of electric current is from negative to positive.

The first experimenters with electricity assumed that the flow was from positive to negative and, conventionally, most workers with electricity assume this to be true. Actually, an electric current is a flow of electricity along a conductor from the negative to the positive pole.

Two kinds of electric current are in common use: alternating current and direct current. Both, of course, are made up of electricity flowing in a conductor, but alternating current reverses itself and flows in the opposite direction several times a second. These changes in direction are called "cycles." In a 60-cycle current, the alternations occur 60 times a second. In a 400-cycle current, these alternations occur 400 times each second. A complete wave or cycle is described as having 360 electrical degrees. Phase is a term used to describe any point in the cycle. More than one alternating current may be made to flow along a single conductor, differing only in phase. A single alternating current having 60 cycles can be flowing along a wire, and two other 60-cycle alternating currents can be made to flow along the same wire at the same time. The alternations occur at different times, being 120 degrees apart. This would be called a 60-cycle, 3-phase alternating current. Phase is generally used to point out the separation between two or more waves or cycles.

A direct current flows in one direction only. Direct current is the type of current obtained from a wet or dry cell or storage battery. Ordinary house current is usually of the alternating type, and the generator used to generate this kind of current is called an "alternator." The generator that is used to generate direct current is similar to the alternator, but has a commutator which causes the alternating current from the generator to flow out along the wires in one direction only.

An electric generator is made up of a revolving part called the armature, which carries coils of wire through a magnetic field, thus generating electric current. An electric motor is similar to a generator, except that the electric current flows through the armature causing it to turn in an electric field.

A number of aircraft instruments depend upon the magnetic field or upon the effects of magnetic fields set up by electric currents.

Remote Indicating Instruments. On the early, small aircraft it was comparatively easy to connect various indicating instruments directly with the operating mechanism. It was also comparatively easy for the pilot to see the position of the landing gear and other parts of the aircraft. As aircraft became larger and as retractable landing gears, landing flaps, oil-cooler shutters, and cowl flaps became common, it was no longer possible for the pilot to see the position of these parts of the aircraft.

Fig. 256. Directional indicator. (Courtesy Kollsman Instrument Division, Square D Company)

In multiengine aircraft the engines are located at a comparatively great distance from the instrument board, and the fuel and oil tanks are located in various parts of the aircraft structure. For these large planes, it became necessary to develop remote indicating instruments because it was impractical to connect the instruments with the instrument board by means of flexible cables or a linkage of rods or tubes.

An electric, remote indicating instrument has two units, the transmitter and the indicator which are connected by electric wires. The transmitter is located at the point at which the action takes place, and the indicator is located on the instrument board where it is easily seen by the pilot. In one type of instrument the transmitting unit consists of a resistance strip in a more-or-less circular form. A movable contact arm, which is properly attached to the landing-gear flaps or another movable part of the airplane whose position is to be indicated, makes contact with this resistance strip. Any change in the position of this movable control member will vary the resistance to the flow of electric current through the indicator which is usually of the voltmeter or galvanometer type.

The resistance strip is connected in series with the coil which is located in the indicator. When the movable contact arm changes its position because of the movement of the landing gear or flaps, the amount of

Fig. 257. A transmitter for remote type indicating instrument. (Courtesy Kollsman Instrument Division, Square D Company)

electric current flowing through the coil in the indicator changes. This instrument has a three-wire circuit. Any change in the amount of electric current flowing causes a movement of the indicating element which can be arranged to show the position of the landing gears, flaps, and other moving parts. The indicator may contain several indicating elements. For example, one instrument shows the position of the nose wheel, the main landing gear, and the flaps.

Another type of remote indicating instrument contains synchronous motors. The transmitting element contains a rotor and a stator which make up the synchronous motor in the transmitting unit. The indicator also contains a similar arrangement of rotor and stator which makes up the synchronous motor in the indicating unit. The motors in the transmitter and the indicator are almost alike. The rotor in each motor is the part which is free to turn, while the stator is fastened to the motor case. The basic principle of this type of instrument is that the rotor in the indicating unit exactly matches any movement of the rotor in the transmitting unit.

The rotor in the indicating unit follows every motion of the rotor in the transmitting unit and matches its position exactly. The transmitter and the indicator are connected by electric wires. Each stator consists of a closed, three-phase, two-pole winding connected in a Y. The rotor has two poles, but has a simple single-phase winding. Alternating current is supplied to the rotor windings which are connected in parallel circuits. When the current is applied, both rotors are energized. The stators are not connected to the current supply, but to each other in parallel by means of three wires. The magnetic field set up in the rotor winding creates a voltage in the stator winding. A different voltage is set up in each of the three phases of the stator winding. In any given position of the rotor, these three voltage values are faced and correspond to one position only. As the transmitter rotor turns, the voltage in one phase of the stator winding becomes larger than that in the other, which becomes less. A stator voltage value is thereby set up for every possible position of the rotor. Since the stators of the transmitter and indicator are connected electrically to each other, the same voltage exists in each stator winding. The rotor of the indicator will take a position inside its stator that exactly matches the position of the transmitter rotor. The indicator rotor is attached to an indicating needle or other device used on the dial of the indicator face.

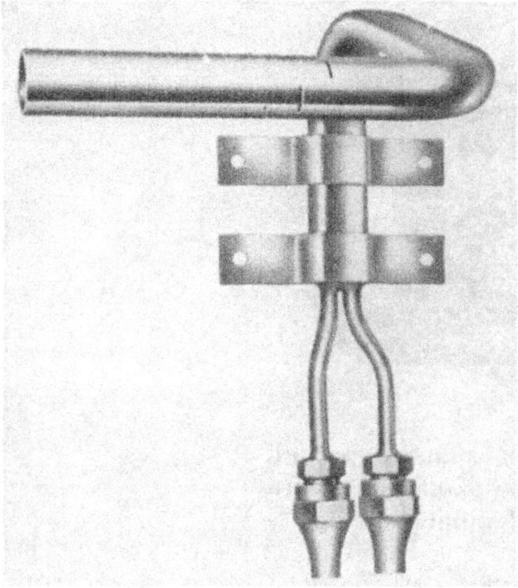

Fig. 258. A Pitot static tube. (Courtesy Kollsman Instrument Division, Square D Company)

Fig. 259. An electric tachometer generator. (Courtesy Kollsman Instrument Division, Square D Company)

Gyroscopes. A gyroscope is a small, heavy flywheel which may be rotated at extremely high speeds. While the gyroscope in the form of a toy has been known for many years, it is only in recent times that it has been used in instruments. A number of aircraft instruments make use of the gyroscope and its reaction, such as the automatic pilot, the turn

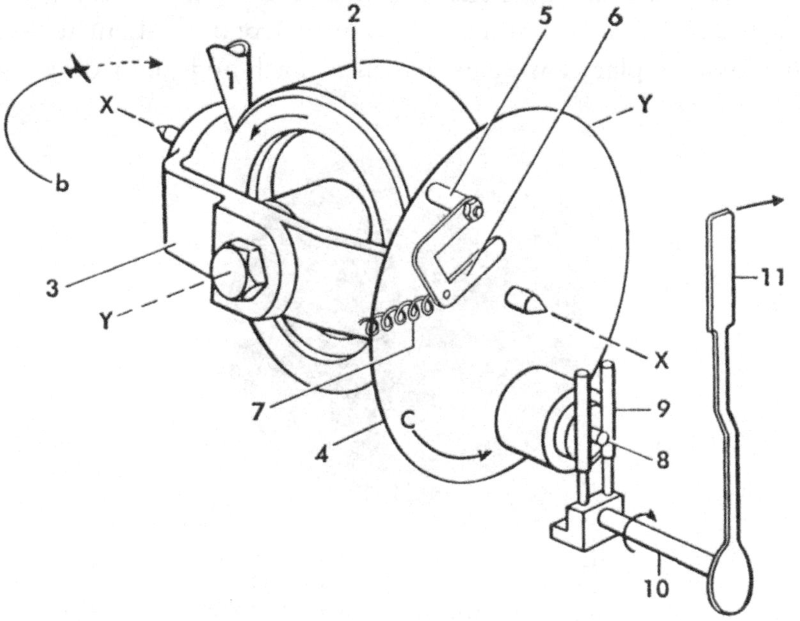

Fig. 260. A diagrammatic drawing to show the construction of a turn indicator. (1) The air jet; (2) gyro wheel; (3) gimbal frame; (4) balance plate; (5) lever; (6) U-arm; (7) centralizing spring; (8) pin; (9) fork; (10) hand staff assembly; (11) the pointer; (b) direction of turn; (c) balance plate; (x) gyro pivot; (y) gyro axis. (Courtesy Eclipse-Pioneer Division, Bendix Aviation Corporation)

indicator, the directional gyro, the horizon indicator, and the attitude-indicating instrument. A well-made gyroscope, as used in the laboratory, is usually mounted on gimbal rings so that the whole gyro is free to turn in any direction. A gyroscope mounted in this way is called a completely free gyroscope or, as the gyroscope is more commonly called, gyro.

The gyro has two fundamental properties which are made use of in all gyroscopic instruments. One of these properties is known as "rigidity in space" and the other is known as "precession." The axis of a rapidly rotating gyroscope tends to remain pointing in the same direction with respect to space. The gyroscope does not always remain pointing in the same compass direction because the rotation of the earth affects the direction in which the gyro points and the earth is constantly changing its relative position in space. If a continuously spinning gyro is mounted at the earth's equator with its spinning axis in an east-west direction, the gyro will apparently rotate about a horizontal axis at right angles to the spinning axis once each 24 hours. During this time, the spinning axis of the gyro has remained in the same direction relative to space, but the earth has made a complete rotation about its axis. If the free gyro is placed with its axis in a north-and-south direction, it

Fig. 261. Navy type air-speed indicator calibrated in knots. (Courtesy Kollsman Instrument Division, Square D Company)

will again maintain the direction of its plane of rotation in space. While the earth rotates under the gyro, the gyro appears to rotate about a vertical axis while spinning on its horizontal axis. This property of rigidity in space is made use of in the directional gyroscope and in the horizon indicator which is also called the "artificial horizon."

When a gyro is mounted in an airplane as a free gyro, it does not follow the movement of the airplane about any of its three axes but remains fixed in relation to space. If the proper indicators are attached to the gyro mounting, these indicators will show the motion of the airplane about the gyro. The indicator then shows the position of the airplane in relation to space.

A spinning gyro can be moved from place to place or tipped forward and backward in the direction of its spin without disturbing the gyro. If, however, an attempt is made to change the plane of rotation of the gyro, it resists this motion. If a gyro is rotating on a horizontal axis and a weight is hung on one end of its axis, instead of tipping downward, the gyro remains with its axis horizontal but turns in a horizontal plane about its vertical axis. This turning in the horizontal plane about its vertical axis will continue as long as the weight pulls down on the axle

Fig. 262. Horsepower indicator. (Courtesy Kollsman Instrument Division, Square D Company)

of the gyro. The gyro turns in a direction which is at right angles to the force which tends to change its plane of rotation. This is the action of the gyro called precession, and this property is used in the turn indicator. The indicating needle of the turn indicator shows whether an airplane is changing its direction of flight and, if so, how rapidly the change is taking place.

XX FLIGHT INSTRUMENTS

Pitot Static Tube. A Pitot static tube is an instrument used to operate the air-speed indicator. The Pitot static tube supplies the air-speed indicator with both impact (dynamic) and static air pressures. The Pitot static tube (Fig. 263) is placed out away from the fuselage of the airplane where it strikes air undisturbed by the propeller. This tube is

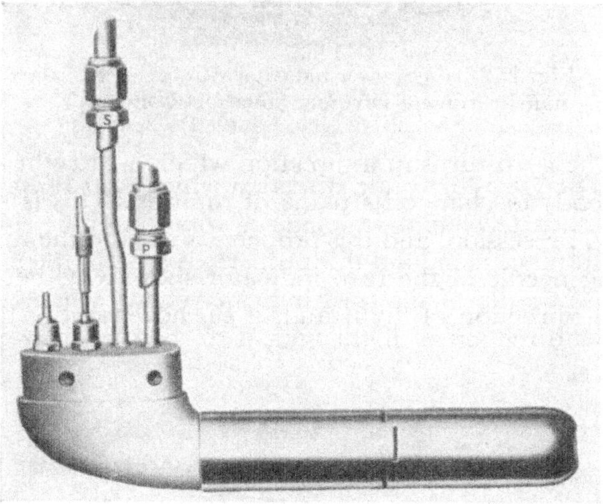

Fig. 263. An electrically heated Pitot static tube. (Courtesy Eclipse-Pioneer Division, Bendix Aviation Corporation)

usually attached to the leading edge of the wing toward the outer end. On multiengine aircraft, the Pitot static tube may be located below the nose of the airplane. The Pitot static tube, as shown in Fig. 264, is made up of two separate tubes, one inside the other. The inner tube, which is the Pitot tube, is open to the air stream, receives the impact of the air, and carries this impact pressure to the indicating instrument. This tube is open in the direction of flight. The other tube, which contains the static pressure and is called the static tube, is closed at its forward end but has a number of openings back from the end of the tube through which the air is free to circulate. These openings maintain, within the outer tube, an air pressure equal to the surrounding atmosphere.

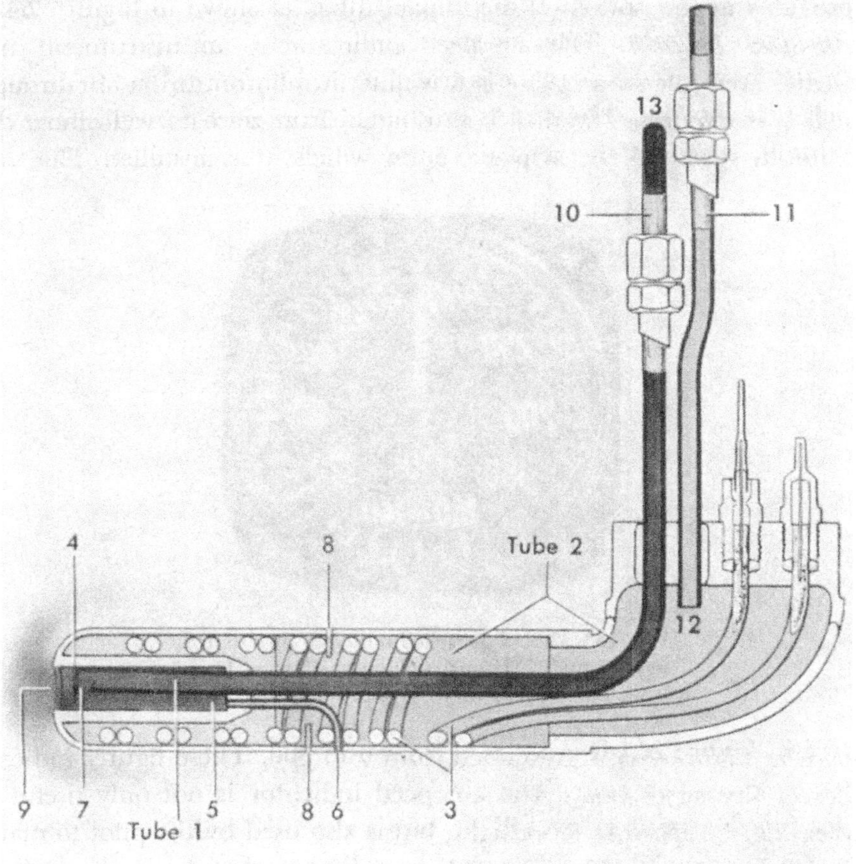

Fig. 264. A diagrammatic drawing to show the construction of an electrically heated Pitot tube. (1) Impact pressure tube; (2) static pressure tube; (3) electric heater coil; (4) baffle plate; (5) pressure chamber; (6) drainage vent; (7) pressure vent; (8) static pressure slots; (9) impact pressure opening; (10) impact pressure line; (11) static line; (12) opening to static line; (13) discharge opening from pressure line. (Courtesy Eclipse-Pioneer Division, Bendix Aviation Corporation)

The speed of the airplane drives the open end of the Pitot tube against the air. The tube and the air collide head on as the airplane speeds through the air, air pressure is built up in the tube, which is opened toward the front, and is carried by this tube to one side of a diaphragm in the air-speed indicator. The normal or static pressure is maintained by the static tube on the other side of the diaphragm in this instrument. The Pitot static tube must be protected against ice accumulations, and an arrangement must be made to drain out any water which may collect in the tube from moisture in the air. To melt any ice which may gather on the tube or in the opening, an electric heater coil is installed. A drain is provided at the bottom of the impact tube, as shown in Figure 264.

Air-Speed Indicator. The air-speed indicator is an instrument that indicates how fast an airplane is traveling in relation to the air through which it is moving. The dial is graduated from zero to well above the maximum speed of the airplane upon which it is installed. The dial

Fig. 265. Dial of air-speed indicator, graduated from 0 to 300. (Courtesy Eclipse-Pioneer Division, Bendix Aviation Corporation)

shown in Figure 265 is graduated from 0 to 300. These figures indicate miles or knots per hour. The air-speed indicator is not only useful in indicating the speed in level flight, but is also used by the pilot to maintain safe speeds while in a dive or to keep the speed of the airplane above the stalling point when landing or climbing.

Each airplane has a rated maximum air speed above which it is not safe to operate. Each airplane is also rated as to its normal cruising speed. The air-speed indicator is used by the pilot to check the speed of the aircraft. Usually, rated cruising speed should not be exceeded in level flight. The air-speed indicator is also used in navigation where the time element is considered. Time multiplied by air speed is equal to distance traveled. The air-speed indicator gives the true air speed only at normal, sea-level pressure at 59° F.

At higher altitudes, or other than normal barometric pressures, a correction must be made to the air-speed-indicator reading. An approximate correction of plus 2 per cent of the indicated reading for each thousand feet above sea level is generally used. The correction for effect of temperature must be made separately but is not usually great enough to interfere with ordinary flights.

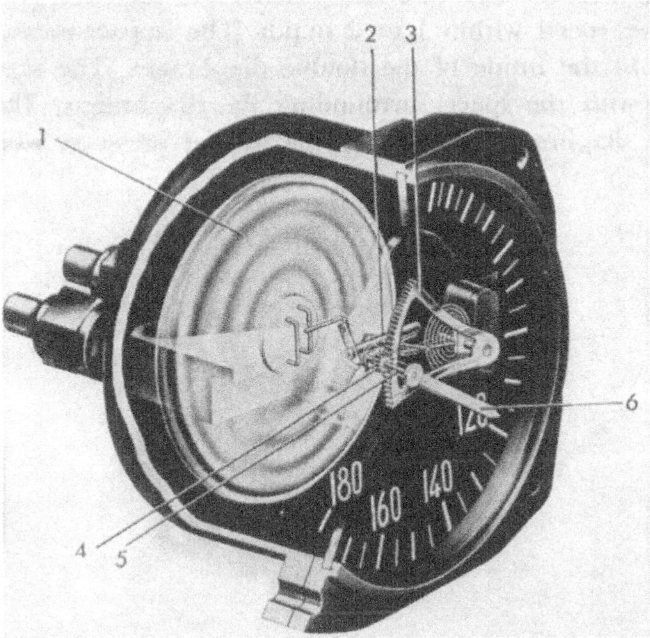

Fig. 266. A cutaway view of an air-speed indicator. (1) Diaphragm assembly; (2) rocking shaft; (3) sector; (4) pinion; (5) tapered staff; (6) pointer. (Courtesy Eclipse-Pioneer Division, Bendix Aviation Corporation)

An airplane flying at a 10,000-ft. elevation with an air-speed-indicator reading of 200 m.p.h. is actually flying at approximately 240 m.p.h. When very accurate figures for precise navigation are necessary, tables, charts, and calculators are used in making the corrections for such items as altitude, humidity, and temperature. An air-speed indicator is especially important in blind or instrument flight.

The air-speed indicator acts on the double-diaphragm or aneroid principle. An aneroid cell or a diaphragm is connected by a linkage system which converts the linear motion of the diaphragm into a rotating motion that turns the needle over the face of the dial. The air-speed indicator is simply a sensitive pressure gauge which measures the difference between the impact air pressure and the static air pressure. The impact pressure from the Pitot static tube is led to the instrument where it causes the diaphragm to expand. As the airplane's speed increases, this pressure becomes greater and the corresponding movement of the diaphragm becomes greater. While the motion of the diaphragm is quite small, this motion is multiplied by the linkage arrangement and rotates the needle over the dial a sufficient distance to enable the pilot to determine his air speed within 1 or 2 m.p.h. The impact-pressure tube is connected to the inside of the double diaphragm. The static tube is connected with the space surrounding the diaphragm. The pressure within the diaphragm is equal to the impact pressure, while the air

Fig. 267. An altimeter is simply an aneroid barometer. (Courtesy Eclipse-Pioneer Division, Bendix Aviation Corporation)

surrounding the diaphragm is equal to the static pressure. The difference between these two pressures indicates the air speed on the face of the dial. As the air speed decreases, the pressure within the diaphragm decreases and the indicating needle backs downward over the dial. A hairspring keeps the linkage under tension and causes the movement of the needle to follow closely the movements of the diaphragm.

Altimeter. The altimeter is the instrument used to indicate the altitude of an aircraft above sea level or some other known elevation. The ordinary mercurial barometer is used to indicate atmospheric pressure and so could be used to indicate altitude. However, this instrument would be extremely unwieldy in an airplane, so the aneroid principle is used to measure barometric pressure. An altimeter is simply an aneroid barometer. Instead of being graduated in units of pressure, it is calibrated in feet of altitude. The ordinary barometer is calibrated in inches of mercury, that is, the numbers around the face of an aneroid barometer indicate the atmospheric pressure necessary to support a column of mercury of a height equal to the number on the face of the aneroid barometer in inches. Normal, sea-level, atmospheric pressure is 29.92 in. of mercury at 59° F. The atmospheric pressure at 5000 ft. is approximately

Fig. 268. A cutaway view of an altimeter. (1) Diaphragm assembly; (2) a link; (3) rocking shaft; (4) a sector; (5) pinion; (6) tapered hand staff; (7) pointer; (8) hairspring. (Courtesy Eclipse Pioneer Division, Bendix Aviation Corporation)

24.89 in. of mercury, while at 10,000 ft. it is approximately 20.58 in. of mercury. Atmospheric pressure does not vary at the same rate as altitude. For example, from the above figures, the difference in inches of mercury between zero and 5000 ft. is 5.03 in. of mercury, while the difference between 5000 ft. of elevation and 10,000 ft. of elevation is 4.30 in. of mercury.

An altimeter in an aircraft on the ground indicates changes in elevation with any change in barometric pressure. A fall of about ½ in. on the barometer will cause the altimeter to indicate between 450 and 500 ft. of elevation even though the aircraft is on the ground at sea level.

Unless the altimeter is set at zero before the airplane is taken off, the altimeter will be in error approximately 475 ft. and, at a true altitude of 1000 ft., would indicate between 1450 and 1500 ft. If the pilot came in to land with the altimeter this much in error, he would fly into the ground while the altimeter indicated that he had over 400 ft. of altitude.

Most altimeters are equipped with a second scale which is calibrated in inches of mercury. By setting this scale with an adjusting knob to equal the barometric pressure at the airport at which the pilot wishes to land, the correct reading of the altimeter may be obtained. This operation is called "altimeter setting."

Sensitive altimeters have two or three pointers. One pointer indicates thousands of feet, the other indicates hundreds of feet, and a red pointer, when it is used, indicates tens of thousands of feet for flying at extremely high altitudes. The pointer indicating hundreds of feet moves around the dial ten times for each revolution of the pointer which indicates thousands of feet. Each graduation mark for the needle indicating hundreds of feet represents ten feet. This altimeter indicates very small changes in altitude.

The simple altimeter has a single indicating needle, and altitude may be read only within large fractions of hundreds of feet. One graduation on the scale usually equals 100 ft. of elevation.

The aneroid barometer operates on the aneroid or double-diaphragm principle. The diaphragm is an aneroid cell from which air has been exhausted and the cell sealed. Since approximately all of the air has been removed from the inside of the diaphragm, the diaphragm maintains a constant internal pressure of approximately zero. The cell is usually mounted under spring tension which makes it extremely sensitive to changes of pressure around the cell.

The static pressure of the atmosphere is allowed to enter the case containing the altimeter mechanism, thus causing the cell to collapse partially. At sea level under normal conditions of temperature and pressure, the barometer is calibrated so that the indicator on the dial registers zero. There is always an adjustment knob on the altimeter by which the hand may be set to any point desired.

If the altitude of the airport at which an airplane is located is 1200 ft. and the pilot wishes to land at another airport which has an elevation of 500 ft., the pilot may set the altimeter at 1200 ft. before taking off. The altimeter will then read 500 ft. when he lands at the other airport or he may set the altimeter at 700 ft. the difference in elevation between the two airports, and the altimeter would then indicate zero when landing on the airport at the lower elevation.

The diaphragm is connected with the aneroid cell by a linkage mechanism which is kept under spring tension, forcing the needle to follow the movement of the diaphragm.

Bank-and-Turn Indicator. The bank-and-turn indicator is two instruments in one. One instrument indicates the degree of bank, while the other indicates the rate of turn. When the airplane is flying straight ahead with the wings level, the ball in the bank indicator is in the center and the pointer which indicates the rate of turn is also in the center position and pointing straight up directly over the ball. As the airplane turns toward the left, the pointer swings to the left. Two points are often located on each side of the center mark on the dial which are known as "reference marks." When the pointer lines up with one or the other, it is turning in the direction of the mark with which it is aligned at exactly 180° per min. This is what is called a "precision turn." In other words, if the pointer is lined up with either of the reference marks, the airplane will complete, regardless of its speed, a turn of 180° in 1 min. As soon as the turn is stopped, the pointer returns to the center position.

Fig. 269. A bank-and-turn indicator. (Courtesy Eclipse-Pioneer Division, Bendix Aviation Corporation)

The bank indicator is simply a curved glass tube in which is placed a ball. The tube is filled with liquid and sealed at each end. On each side of the center of the curved tube is a wire or other mark just the right distance from the center so that when the ball is in the center, it lines up with the wire or mark on each side. These wires indicate the centered position of the ball. The tube is mounted in rubber in the front of the instrument case toward the bottom of the bank-and-turn instrument. The liquid in the tube is to dampen rapid movements of the ball, thus causing it to move smoothly and accurately from side to side. During level flight, the force of gravity acting upon the ball keeps the ball centered in the glass tube. If either wing is lower than it should be, the ball rolls toward the low wing. As the wings approach the proper position, the ball returns toward the center of the tube.

When making a bank turn, the ball is acted upon, not only by gravity, but also by a centrifugal force. The centrifugal force tends to cause the ball to roll away from the center of the turn, while gravity tends to pull it backward to the lowest part of the tube. When a perfect bank turn

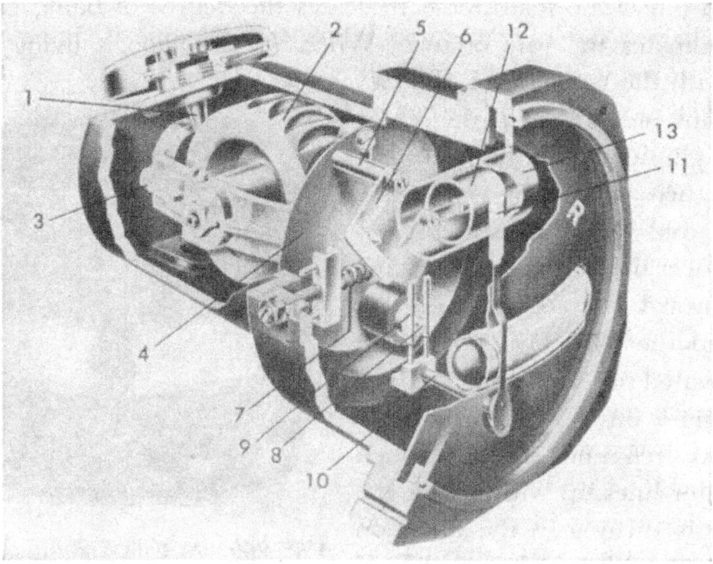

Fig. 270. A cutaway view of a bank-and-turn indicator. (1) Air intake jet; (2) gyro wheel; (3) gimbal frame; (4) balance plate; (5) lever; (6) U-arm; (7) centralizing spring; (8) pin; (9) fork; (10) hand staff assembly; (11) pointer; (12) piston; (13) cylinder. (Courtesy Eclipse-Pioneer Division, Bendix Aviation Corporation)

is made, the ball remains in the center of the tube. If one wing is lower than it should be to make the turn properly, the ball rolls toward that wing. In a steep bank turn, if the upper wing is too low, the ball rolls up in the tube toward the high wing. This is an indication that the airplane is skidding outward on the turn. If the lower wing in the turn is too low, the ball rolls downward toward the lower wing, indicating that the airplane is slipping inward on the turn.

The turn indicator operates upon the gyroscopic principle. A small gyroscope is mounted with its axis parallel to the lateral axis of the aircraft and it rotates in the vertical plane passing through the longitudinal axis. That is, the wheel rotates fore and aft with its axle parallel to the surface of the earth and pointing toward the wing tips.

The gyroscope is simply a wheel having most of its weight located near the outer rim. A gyroscope, when rotated rapidly, possesses a peculiar property called precession, described in Chapter XIX, and that property is the basis of the turn indicator.

When the airplane containing the rapidly rotating gyroscope flies in a straight line, the gyro wheel, spinning about its horizontal axis, has its line of travel in the same direction as that of the airplane. As long as the airplane continues in a straight line, the gyroscope spins in the same direction about its axis which coincides with the lateral axis of the aircraft. The gyroscope is mounted in a gimbal frame which is free to rotate about the longitudinal axis of the airplane. As the airplane turns, the gyroscope, instead of following the turn, tends to keep on spinning about its horizontal axis in the same straight line as before. The inertia of the wheel, which has a tendency to resist a change of direction, starts the precessing action of the gyroscope which rotates the gimbal about its long axis. The whole assembly, that is the gyro wheel and the gimbal frame, tends to turn over on its side a distance which corresponds to the rate of turn. The gyro wheel, which has been spinning on a horizontal axis, tends to rotate more about the vertical axis of the airplane. When the airplane turns to the left, the gyro assembly rotates to the right. When the airplane turns to the right, the gyroscope rotates toward the left. That is, the bottom of the wheel swings in the direction of the turn, while the top of the wheel swings in the opposite direction in relation to the aircraft.

In addition to the gyro wheel and the gimbal frame, the turn indicator is made up of a balance plate, a lever, a U-arm, a centralizing spring, a pin, a fork, a handstaff assembly, and a pointer. There is also in this instrument a piston and cylinder assembly. The gyroscope rotates at about 9000 r.p.m. and is driven by an air stream which is directed against little cup-shaped cavities in the rim of the wheel. The air stream is produced by an engine-driven vacuum pump or a Venturi tube. In some airplanes, a Venturi tube is the only source of suction, while in others a Venturi is used only when the vacuum pump fails.

The balance plate is attached to the gimbal frame. The rotation of the gimbal frame turns the balance plate and operates a lever and a U-arm which is attached to it.

A centralizing spring is attached to the U-arm at one side and to the instrument case at the other. The spring is balanced against the force of the precessing wheel and allows the balance plate and attached units to move only an amount proportional to the rate at which the turn is being made. Fixed to the bottom of the balance plate is a pin which fits in a fork at one end of the handstaff assembly. This pin rotates with the balance plate and carries the handstaff with it in the direction of the turn. The pointer which is attached to the handstaff moves over the face of the dial an amount that is directly proportional to the movement of the pin on the balance plate.

As soon as the airplane stops turning, the force causing the gyroscope to precess also stops and the centralizing spring returns the entire mechanism to the neutral position.

The piston and cylinder assembly is a dampening device which controls the precession motion. It causes the indicator to move smoothly and dampens out sudden movements of the pointer. The following sketches show the position of the bank-and-turn indicator under various flight attitudes.

Fig. 271. Indications of a bank-and-turn indicator under various flight attitudes: (1) straight and level; (2) perfect bank, left turn; (3) straight flight, right wing low; (4) left turn, right wing low, "skid"; (5) left turn, left wing low, "slip."

The first shows both the pointer, which is called the turn indicator, and the ball centered. The second sketch shows the position of the bank-and-turn indicator when a perfect bank is being made. The turn indicator indicates the turn toward the left, with the ball in the center position showing that the wings are in the proper position for the bank. The third sketch shows that, although the airplane is flying straight ahead, the right wing is low and there is a displacement of the ball to the right. The fourth figure shows that the airplane is turning to the left, but that the right wing is too low because the ball is displaced from the center towards the right. This position of the turn indicator and the ball indicates a skid. The fifth sketch shows that, while the airplane is turning to the left, the left wing is too low as the ball is displaced to the left. This condition of the bank-and-turn indicator indicates a slip.

Rate-of-Climb Indicator. The rate-of-climb instrument indicates the rate at which an airplane is gaining or losing altitude. It indicates only changes vertically. Rate-of-climb indicators are sometimes called "vertical-speed indicators." The rate of climb does not indicate forward speed.

Fig. 272. A vertical speed indicator. (Courtesy Eclipse-Pioneer Division, Bendix Aviation Corporation)

The rate-of-climb indicator is an aid in maintaining level flight, and this instrument is particularly valuable in flying under conditions of low visibility, such as fog, or in the clouds, or at night. As long as the pointer remains at zero, the altitude of the airplane remains unchanged. The numbers on the dial indicate thousands of feet per minute. Should the pointer on the rate-of-climb indicator move off zero, the other instruments should be checked to determine the cause of the altitude change. By means of this instrument, predetermined angles of climb or descent can be accomplished.

The rate-of-climb indicator consists of an airtight case and a diaphragm assembly, a capillary tube, a calibrated dial, and a linkage system. The linkage system consists of an arm, a rocking shaft, a geared sector, a pinion, and a pointer. In this instrument, in addition to the airtight case, there is an air-enclosing sensitive diaphragm. The diaphragm is connected directly to the static pressure, while the airtight case is vented with the outside air by means of a capillary tube. The capillary tube or a thin-walled porcelain tube permits air to seep through its pores at a relatively slow rate and permits the slow escape of air. The diaphragm,

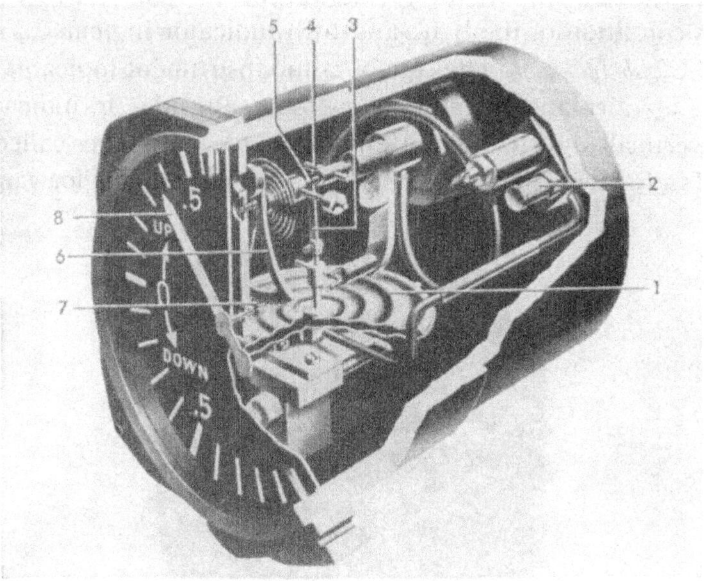

Fig. 273. A cutaway view of a vertical speed indicator. (1) Diaphragm assembly; (2) capillary tube; (3) a link; (4) an arm; (5) rocking shaft; (6) sector; (7) pinion; (8) pointer. (Courtesy Eclipse-Pioneer Division, Bendix Aviation Corporation)

being vented directly to the atmosphere, reacts almost instantly to changes in atmospheric pressure. The pressure inside the case increases or decreases very slowly, since air escapes from, or is admitted to, the case very slowly by its passage through the capillary or porous porcelain tube. The pressure, however, equalizes in a very short time after level flight at any altitude is resumed. As the airplane climbs and atmospheric pressure is reduced, the diaphragm loses pressure. The pressure within the diaphragm, however, remains practically the same as the pressure before the climb started, because the volume inside the case cannot change rapidly. When altitude is lost, the opposite condition is built up. The pressure inside the diaphragm becomes greater at a more rapid rate, and the pressure within the case changes slowly. That is, the differ-

A phase adapter may be used to convert the 115-volt, single-phase, 400-cycle power to 115-volt, three-phase, 400-cycle power, making it possible to operate three-phase aircraft instruments from a single-phase power supply. An inverter may be used with this instrument to change a 12-volt direct current or a 24-volt direct current to a 115-volt, three-phase, 400-cycle alternating current.

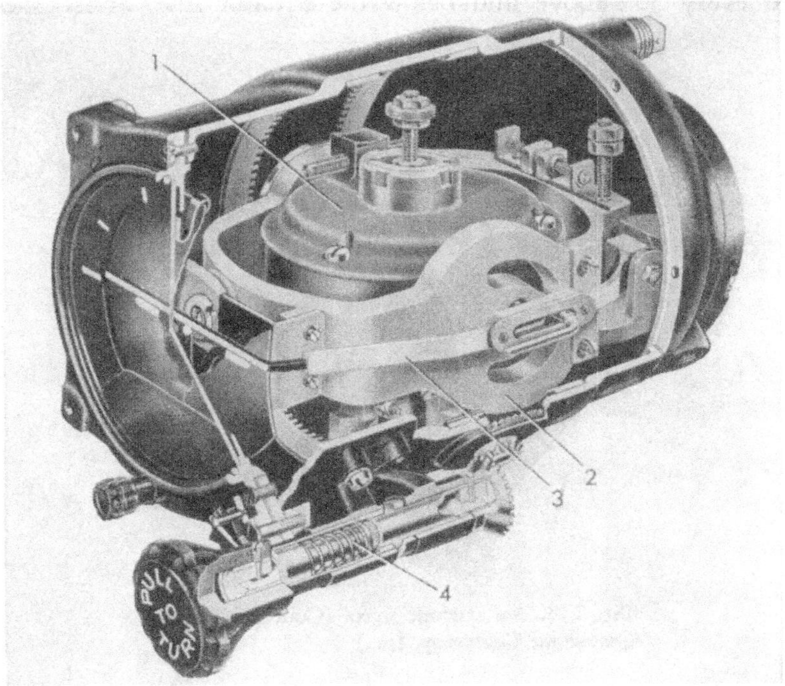

Fig. 277. A cutaway view of an electrically driven gyro horizon: (1) Gyro assembly; (2) gimbal ring; (3) horizon bar; (4) caging mechanism. (Courtesy Sperry Gyroscope Company, Inc.)

Attitude Gyro. The attitude gyro combines the functions of a gyro horizon with a number of added features which make it an outstanding instrument in the flight group. This is the only instrument which gives the pilot a continuous indication of the attitude of the aircraft in relation to a horizontal plane throughout all aircraft maneuvers. This instrument has no angular limits. It indicates through the full 360° about all the axes of the airplane.

The sensitive gyroscopic element consists of a stabilized sphere, the polar axis of which always points toward the center of the earth. The upper half of this sphere is dark, while the lower half is light-colored. When in level flight, the pilot sees both the upper half and the lower half of the sphere as though it were viewed at the equator. A "lubber line" represents the equator. The lubber line is alternately marked dark and light. The center of the lubber line is in the form of a small circle which is used as a pitch index in steep turns.

The sphere is divided vertically into equal halves by a meridian line which passes through both poles of the sphere. Latitude circles are spaced every 30° above and below the equator line. Short lines are

Fig. 278. An attitude gyro. (Courtesy Sperry Gyroscope Company, Inc.)

drawn across the meridian line every 10° between the latitude circles. The 90° markings are in the form of solid circles at each of the poles of the sphere. These markings make the various lines stand out from each other, so that the pilot can quickly interpret the attitude of the airplane and control it under instrument conditions for long periods of time without tiring. The pattern markings on the sphere appear the same under ultraviolet light at night as they do in daylight.

The gyro sphere has a universal mounting without limit stops. Because of this free movement, the instrument shows continuous indications of attitude for any angular position of the airplane. By watching this instrument throughout any maneuver, the pilot can readily picture the attitude of the airplane in relation to the earth's surface. After maneuvers, the pilot can immediately go "on instruments" without first having to level out and uncage. No caging device or other adjustment is required. Recovery from any position can be accomplished with speed and accuracy.

In any maneuver in which the nose of the airplane is vertical, the gyro may pass through the position in which one of the three degrees of freedom of the gyro is momentarily lost. If the aircraft passes near

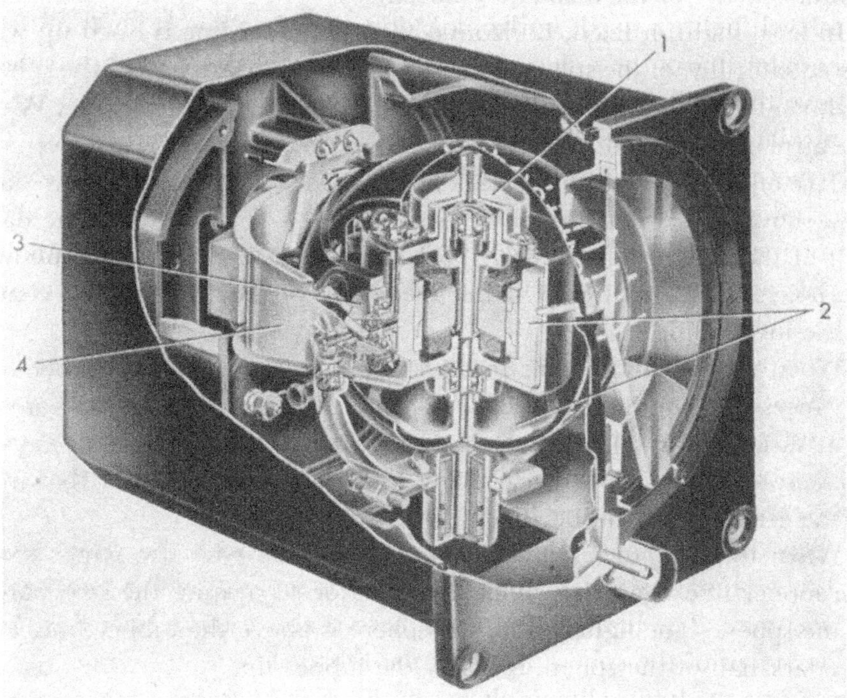

Fig. 279. A cutaway view of an attitude gyro. (1) Gyro housing; (2) rotor and erection cone; (3) gimbal ring; (4) erecting magnet and erector gimbal ring. (Courtesy Sperry Gyroscope Company, Inc.)

this position, the sphere will rapidly rotate 180° about its polar axis. If the airplane passes exactly through this position, only a slight wobbly motion occurs. The instrument has been so designed that errors resulting from these conditions are small and are automatically corrected upon return to level flight.

This instrument is always in operation when the aircraft engine is operated. Closing the switch to start the engine starts the gyro spinning, and the instrument continues to function until the engine is shut off.

The operation of this instrument should not be confused with the indications shown by the regular gyro horizon. The pilot should keep in mind that this instrument is a stabilized sphere whose polar axis is always vertical in relation to the earth's surface. The airplane maneuvers around the stabilized sphere. The apparent motion of the sphere is the result of the motion of the airplane and the instrument case about the sphere. This movement produces the change in pattern seen through the transparent opening in the front of the case and gives the pilot a mental picture of the airplane's movement.

In level flight, a fixed, horizontal, dotted lubber line is lined up with the equator line of the sphere. In this position, the dark half of the sphere is above the lubber line with the light half below the lubber line. When the airplane is placed in a dive, the lubber line moves upward over the surface of the sphere into the dark area. As the dive steepens, more dark area shows and, when the plane has reached a 90° dive, only the dark half of the sphere may be seen, with the contrasting circle surrounding the pole appearing directly under the center of the circle in the center of the lubber line.

When an airplane begins to climb from level flight, the lubber line advances into the light half of the sphere. As the climb steepens, more light surface than dark surface is seen. In a 90° climb, the bottom pole of the sphere is in the center of the field of view directly under the circle in the center of the lubber line.

When the airplane is in the inverted position with the wings level, the lubber line again lies along the equator separating the two halves of the sphere. The light half of the sphere is above the lubber line, and the dark half of the sphere is below the lubber line.

In a dive the lubber line is always in the dark half of the sphere, while it is always in the light half of the sphere when the aircraft is climbing. This is true whether the aircraft is in normal flight or in inverted flight.

Various degrees of bank can be read at the angle between the vertical index on the case and the meridian on the sphere. During a right bank, the horizontal reference line and the vertical index on the case rotate clockwise over the surface of the sphere. In a left bank, the vertical index rotates counterclockwise over the surface of the sphere. This movement holds true whether the aircraft is in level flight, dives, or climbs.

The meridian line always supplies a vertical reference, and the vertical index always moves from the meridian in the direction of the bank. To recover from any flight attitude, the pilot has only to line up the vertical index with the meridian and the lubber line with the equator of the sphere so that the dark half of the sphere is up and the light half of the sphere down.

A rotor and a stator make up the rotating and stationary parts of a three-phase induction motor. The rotor spins on bearings in the top and bottom of the gyro housing. The gyro housing, which is enclosed in the sphere, is supported on horizontal bearings at each side by the gyro gimbal ring. The sphere and all parts within have full 360° freedom of movement about the lateral axis of the airplane.

Fig. 280. A panel installation: attitude gyro, left; gyrosyn compass, right. (Courtesy Sperry Gyroscope Company, Inc.)

The gyro gimbal ring is supported on horizontal bearings by the back plate of the instrument. The gyro gimbal ring and the sphere which it supports have full 360° freedom of movement about the longitudinal axis of the airplane.

The gyroscopic effect of the spinning rotor stabilizes the sphere in space with its spinning axis vertical. The spinning axis forms the polar axis of the sphere. If left to itself, the gyro, affected by friction and other factors, would depart from its vertical position. To maintain the vertical position an electromagnetic erection system is installed. This system consists of a cup-shaped erection cone attached to the bottom of the gyro rotor and an erecting magnet below the erection cone.

The attitude gyro operates on a 115-volt, three-phase, 400-cycle alternating current. The power consumption is 15 w. with a back factor of 0.50.

A phase adapter may be used to convert 115-volt, single-phase, 400-cycle power to 115-volt, three-phase, 400-cycle power, making it possible to operate three-phase instruments from a single-phase power supply. An inverter may be used to change a 12- or 24-volt d-c supply to 115-volt, three-phase, 400-cycle alternating current for the operation of electrically driven instruments.

XXI AUTOMATIC PILOTS

The automatic pilot is a mechanism which can relieve a pilot of the strain of flying an airplane for long periods of time on long flights. It also allows the pilot to devote more time to navigation, engine operation, and other flight problems. The automatic pilot controls the attitude of the airplane in flight without assistance from the pilot. It also furnishes visual indication of the position of the airplane in yaw, roll, and pitch. These indications are similar to the indications shown on the flight indicator (artificial horizon) and the turn indicator (directional gyro). The automatic pilot can be set, not only to maintain straight, level flight, but also to execute normal maneuvers, such as turns, spirals, climb, or descent. Automatically, these maneuvers are performed by the automatic pilot with greater precision than by the human pilot.

Sperry-Automatic Pilot. One type of automatic pilot, the Sperry, consists of a combination of gyroscopes and hydraulically and pneumatically operated mechanisms. The functions of these parts are so coordinated that they resemble the operation of the human muscles, brain, and nerves. The mounting unit is a frame to which balanced oil valves, follow-up pulleys, air relays, vacuum lines, exhaust and drain manifolds, and electrical contacts are attached.

There are two control boxes, consisting of the directional-gyro unit and the bank-and-climb-gyro unit. When the control boxes are in place, all necessary connections are made automatically.

The directional-gyro control unit contains one gyro which maintains the directional attitude, the directional-gyro rudder knob, the directional-gyro card, the follow-up card, the caging and setting knob. This unit is located on the left side of the mounting unit. The bank-and-climb control unit contains one gyro which maintains the lateral and longitudinal attitude, indication, and control. This unit also contains the follow-up system, the caging knob, the control knob, the air pick-off, a vacuum gauge, and a level-flight control. This unit is located on the right side of the mounting unit. To engage or disengage the automatic pilot from the control surfaces of the airplane, a lever is set to ON or OFF.

The oil sump is an oil reservoir for the automatic pilot's hydraulic system and usually has a capacity of about 1½ gal. An engine-driven oil pump provides the necessary pressure and flow of oil to operate the automatic pilot. The pump requires about ½ hp. and has a capacity of about 3 gal. of oil per minute.

The oil-pressure regulator is a spring-loaded ball type of regulating valve connected on the pressure side of the pump. An oil-pressure gauge

Fig. 281. A gyropilot. (Courtesy Sperry Gyroscope Company, Inc.)

is provided which shows the pressure of the oil being supplied to the automatic pilot. A by-pass valve allows the oil output of the hydraulic pump either to be supplied to the main system or to be returned to the intake side of the pump. There is a speed-control valve for each surface control, to regulate the amount of hydraulic fluid actuating the controls. The oil relays are five port valves, each provided with a four-step piston. One passage of the valve provides a channel for the passage of oil under pressure from the pump. Two of the ports lead to the servo unit. One port is used for drainage, and the other for exhaust. The piston is balanced by means of a spring located at one end. The spring is so arranged that sensitive adjustments may be made. The hydraulic surface controls complete the servo unit. They are mounted permanently and are connected in or to the airplane control system.

The servo unit consists of three hydraulic cylinders, one for each control surface. The speed at which the pistons move may be regulated by the speed-control valves. A servo is made up of a piston within a cylinder. The piston is attached to the center of a piston rod which extends out of the cylinder at each end. To each end of the rod is attached the cable controlling the movement of a control surface. These cables are equipped with turnbuckles so that accurate adjustment may be made. Oil pressure acts on one or the other side of the piston. This oil pressure is controlled by the balanced oil valves in the mount assembly and it regulates the movement of the control cables. A servo unit contains a relief valve which is normally set for a pressure of from 25 to 40 lb. pressure per square inch more than the tension of the airplane control cables.

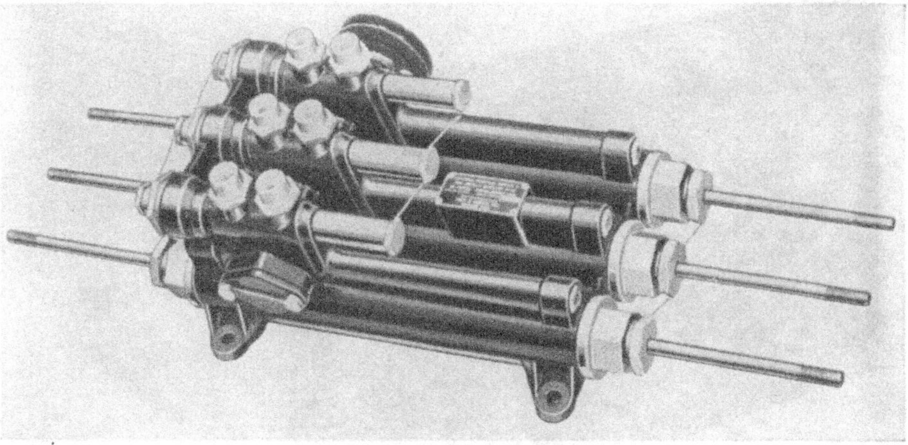

Fig. 282. A typical servo unit. (Courtesy Jack & Heintz, Inc.)

If, for any reason, the automatic pilot cannot be disengaged, the pilot can operate the controls by using excessive pressure, thus overpowering the automatic pilot and taking over the control of the airplane.

The oil-drain trap is connected between the drain manifold on the mount assembly and the suction line to the oil pump. On an alternate installation which omits the drain manifold, the oil-drain trap is connected between the oil-drain outlets on the balanced oil valves and the suction line of the oil pump.

The oil filter is connected with the main oil-pressure line. The drain manifold is connected between the drain outlets of the balanced oil valves on the mount assembly and the oil-drain trap.

In some installations, the drain outlets are connected directly to the oil-drain trap. The exhaust manifold is connected between the exhaust outlets of the balanced oil valves on the mount assembly on the oil-pump side.

A vacuum pump is necessary to this system and is engine driven. This pump furnishes the source of suction which operates the gyros and the air pick-offs.

The suction or vacuum gauge is a part of the bank-and-climb-gyro control unit. This gauge shows the amount of vacuum in the control system and is usually graduated over a range of from 0 to 10 in. of mercury.

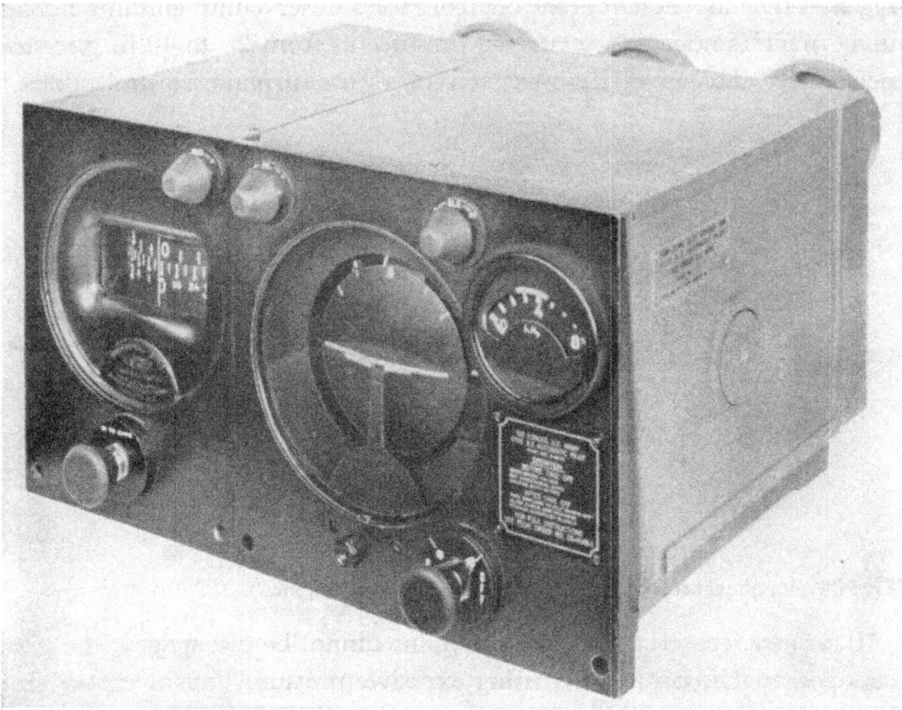

Fig. 283. Three-quarters front view of a gyropilot. (Courtesy Sperry Gyroscope Company, Inc.)

The suction regulator is connected with the air line between the vacuum pump and the mounting unit.

Each air pick-off consists of the air nozzle and the air-nozzle plate. On the directional-gyro control unit, the nozzle plate is secured to, and is a part of, the vertical ring. Since the vertical ring is held in position by the gyroscope, the airplane will remain in the same plane unless the entire mechanism is turned manually. The air nozzle is pivoted above the vertical ring, but is not a part of it.

The air nozzle can be rotated, either manually or through the follow-up system, about the same axis of the airplane in relation to the control surface associated with it. The air nozzle has a small slot approximately ½ in. long and 1/64 in. wide and is supported above the air-nozzle plate. The distance between the two surfaces is between 0.0002 and 0.0003 in. This very thin opening limits the flow of air between the air nozzle and the air-nozzle plate to a very small amount.

The air relay consists of two convex disks operated by a metallic-reinforced leather diaphragm. The two convex portions are held together with screws, allowing the leather diaphragm to be suspended between

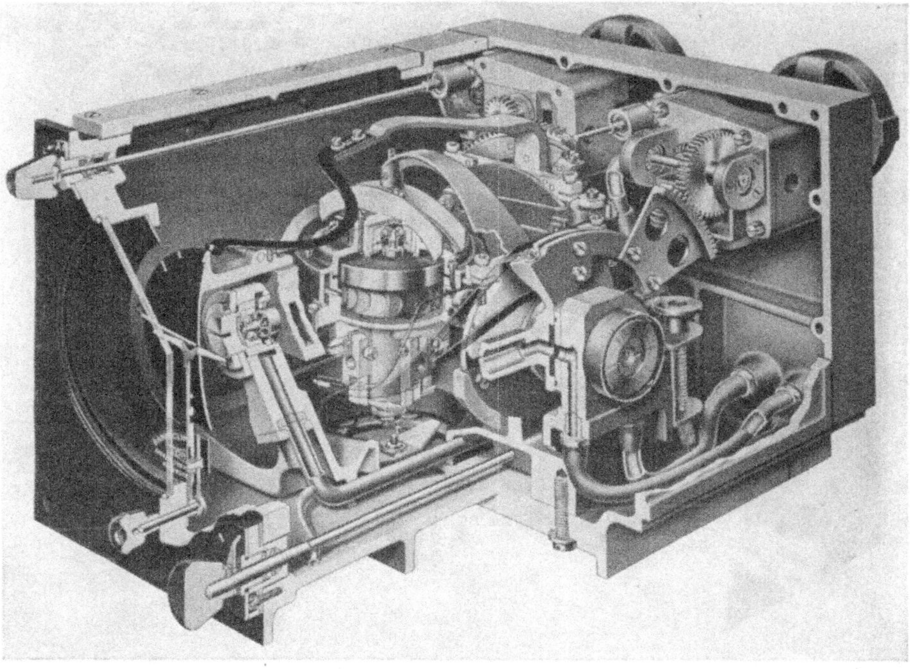

Fig. 284. A cutaway view of a gyro horizon. (Courtesy Sperry Gyroscope Company, Inc.)

them. On top of the air relay are two screens, and at the bottom are the suction nipples which lead into the area of low pressure of the control boxes. The leather diaphragm is very sensitive, and a vacuum of approximately 0.4 in. of mercury is enough to cause the diaphragm to be displaced from a balanced position to an extended position, moving with it the corresponding balanced oil valve.

The follow-up cables coordinate the air pick-off system with the hydraulic system. There is one cable for each control.

The follow-up clutch is a cork-lined clutch which automatically makes contact with the clutch plate when the control units are in the mount assembly. The follow-up clutch spring is used to keep the follow-up cable tight and to take up slack when it is returning.

The hydraulic oil and vacuum pumps start as soon as the engine is started. A partial vacuum is created, and oil pressure is built up in the system. The operation of the automatic pilot is dependent upon the

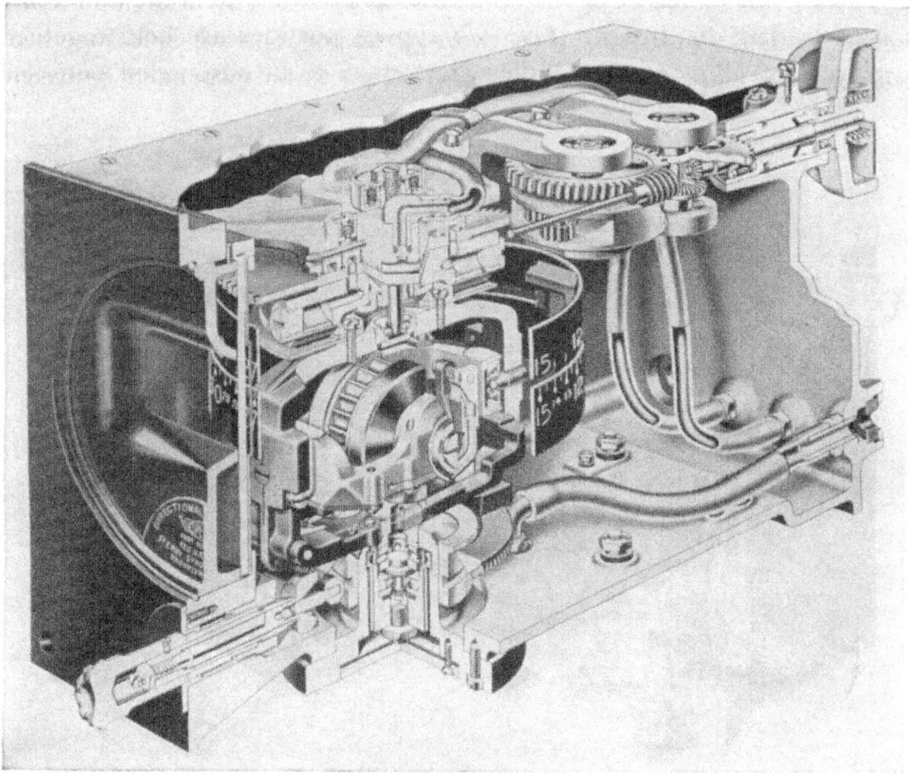

Fig. 285. A cutaway view of a directional gyro. (Courtesy Jack & Heintz, Inc.)

principle that a rapidly rotating gyroscope maintains its axis in a position which is fixed with reference to space. The gyroscopes used in the automatic pilot are small air-driven rotors which are universally mounted.

The spinning axis of the rotor in the directional-gyro control unit is always in the horizontal plane. (The axis of the rotor in the bank-and-climb-gyro control unit is always in the vertical plane.)

The rotor in the directional-gyro control unit is mounted in a vertical ring on pivots which are, in turn, pivoted to the case proper. This type of rotor suspension allows the rotor to set up a force which is transferred to the vertical rings, causing the entire assembly to remain rigidly in position regardless of any outside torque imposed upon it. The gyro thus holds a fixed position, allowing all associated units to move about it.

When the air nozzle and air-nozzle plates are exactly in balance, each air port in the air nozzle is partially opened and is cutting the edge of the air-nozzle plate by the same amount. When the air nozzle is in this position in relation to the plate, the same amount of air will escape from each nozzle. Any change of condition which tends to change the direction of flight, such as a cross-wind pressing on the rudder control, would tend to cause the airplane to leave its course. The spinning rotor, however, retains its position in space and holds the nozzle plate in position, while the air nozzle, moving with the aircraft, rotates about it. This causes the air port on one side to open wider and the one on the other side to close up, with the result that the air pressure becomes unequal. Any difference of pressure at the nozzles immediately causes a displacement of the diaphragm in the air relay. Any movement of the air relay causes a movement of the piston in the balanced oil valve. The movement of the oil valve unbalances the pressure on the two ports leading to the servo cylinder. Unequal pressure on the opposite end of the cylinder will cause the piston to move. Since the piston is a continuous part of a control cable, the control surface will be moved, thus changing the position of the airplane control surface. This change will cause the airplane to return to its normal, straight, and level flight. While all this takes time in explanation, the entire sequence of events takes place almost instantaneously. Actually, the movement of the nozzles and other parts is so small that the eye can hardly discern it.

Overcontrol is prevented by the follow-up system. One end of the follow-up cable is attached to the servo piston, and the other is wound around the follow-up pulley. The follow-up pulley clutch and a differential gear box are arranged so that any movement of the follow-up clutch will be taken up in the differential gear box. Follow-up action will cause the air nozzles to turn in relation to the air-nozzle plate, following the plate until the signal is cut off. The system is so arranged that the signal will be stopped and the control surface returned to neutral as the aircraft comes back to its proper attitude.

The purpose of the differential in the control mechanism is to permit the human pilot also to displace the air nozzles by means of the turn knob, whenever he desires to put the aircraft in a turn or to change course.

While the explanation given above pertains particularly to the directional control, the principle of the bank-and-climb follow-up system is the same, and its application very similar.

Jahco Automatic Pilot. The Jahco automatic pilot, like other automatic pilots, makes possible the automatic control of an airplane in flight. The automatic pilot maintains the attitude of the airplane which has been selected by the pilot. Visual indications of the attitude of the airplane are provided by indicators attached to the gyroscopic components which are a part of the automatic pilot. This automatic pilot may be provided with accessories for automatic and manual remote control.

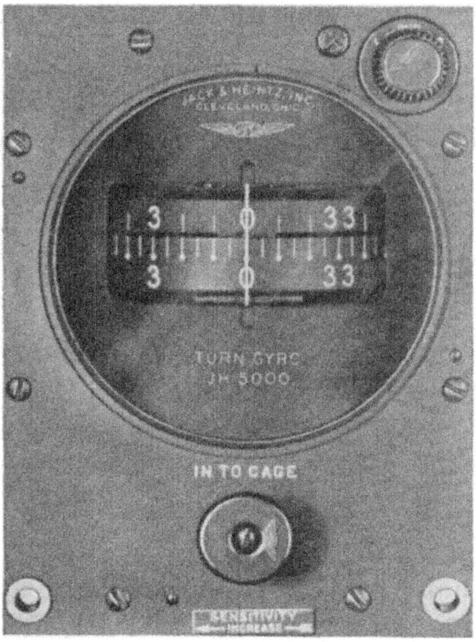

Fig. 286. A turn-gyro-control unit. (Courtesy Jack & Heintz, Inc.)

Fig. 287. Rear view of a turn-gyro-control unit. (Courtesy Jack & Heintz, Inc.)

A directional-gyro or turn-gyro control, with its card and lubber-line, indicates the heading of the aircraft at all times. The directional gyro is provided with a caging device which should be used during maneuvers that may exceed the operating limit of the instrument. The operating limits of the directional gyro are 55 degrees from the vertical in any maneuver. The gyro may be set to any heading by turning the same knob used for caging the instrument.

The turn-gyro-control unit has two index cards graduated from 0 to 360 degrees. The lower card, which is the turn-gyro card, may be set to any heading by pushing in and turning the caging knob. The upper card, which is the turn-gyro follow-up card, may be set by turning the rudder-control knob. The follow-up card is attached directly to the air valves in the instrument. The air valve is in neutral position when the readings of the upper and lower cards are the same. In order to have automatic control on any desired heading, the readings of the two cards must be matched before engaging the automatic pilot.

The bank-climb-gyro control unit contains the gyro which controls the movements of the ailerons and the elevator. This unit controls the movements of the airplane about the lateral and the longitudinal axes. This unit has a caging device for locking the gyro during maneuvers that may exceed its operating limits, which are 50° from the vertical.

Fig. 288. Three-quarters front view of a turn-gyro-control unit. (Courtesy Jack & Heintz, Inc.)

The horizontal dial is attached to the gimbal ring of the gyro and provides a horizontal reference as the airplane banks. This action depends upon the principle of the gyro known as rigidity in space. The horizontal bar in front of the horizontal dial is moved through linkage by a pin in the side of the gyro-rotor housing. The elevator follow-up index and the aileron follow-up index provide a visual indication of the air-valve setting for these controls. By use of the aileron knob, the air valve for lateral control may be set for any desired lateral attitude. It is set to maintain the position in which the airplane is flying by matching the aileron follow-up index with the bank index.

Fig. 289. Bank-climb-gyro control unit, caged. (Courtesy Jack & Heintz, Inc.)

Fig. 290. Rear view of a bank-climb-gyro control unit. (Courtesy Jack & Heintz, Inc.)

The air valves for longitudinal control may be set for the desired longitudinal attitude by the use of the elevator knob. This knob is set in the same manner as the aileron control to maintain the attitude in which the airplane is flying. The aileron and elevator controls should be set before the automatic pilot is engaged. The indices for each control knob should match.

There are two types of mount assemblies, depending upon the model of instrument used. One mount, which has fixed air-relay bleeds, requires the use of speed-control valves. The other mount includes adjustable

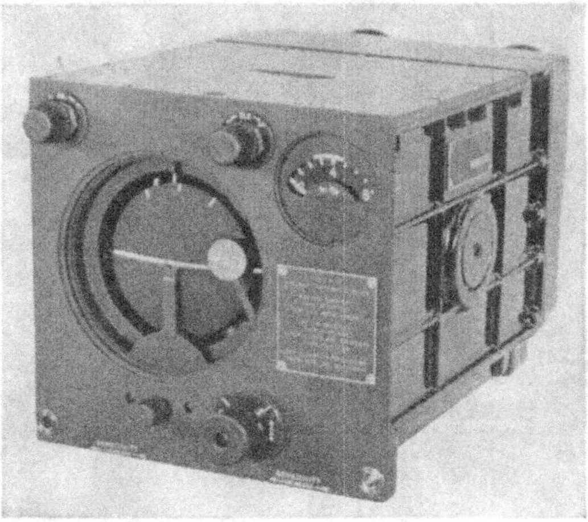

Fig. 291. Three-quarters front view of a bank-climb control unit, caged. (Courtesy Jack & Heintz, Inc.)

air-relay bleeds which do away with the need for speed-control valves. The adjustable air-bleed device makes it possible to change the size of the air-relay bleeds. This change in the size of the air-relay bleeds controls the sensitivity and speed of reaction of the automatic pilot.

The gyro-control-mount assembly consists of a shock-mounted frame to which air relays, balanced oil valves, and the follow-up mechanism are attached. The gyro-control-mount assembly is the support for the control units.

The control units slide on tracks into the mount. All mechanism and air connections located on the rear of the control units are automatically made when the control units are bolted into place.

The follow-up pulleys are provided with clutch disks which carry their motion to the control unit. The follow-up cables are attached to the follow-up pulleys. A spring which keeps the cable tight during the follow-up action is located inside the follow-up pulley drum.

The pressure and drain manifolds are connected to the three balanced oil valves. These manifolds are located in the bottom of the gyro-control-mount assembly. The manifold distributes oil pressure to the balanced

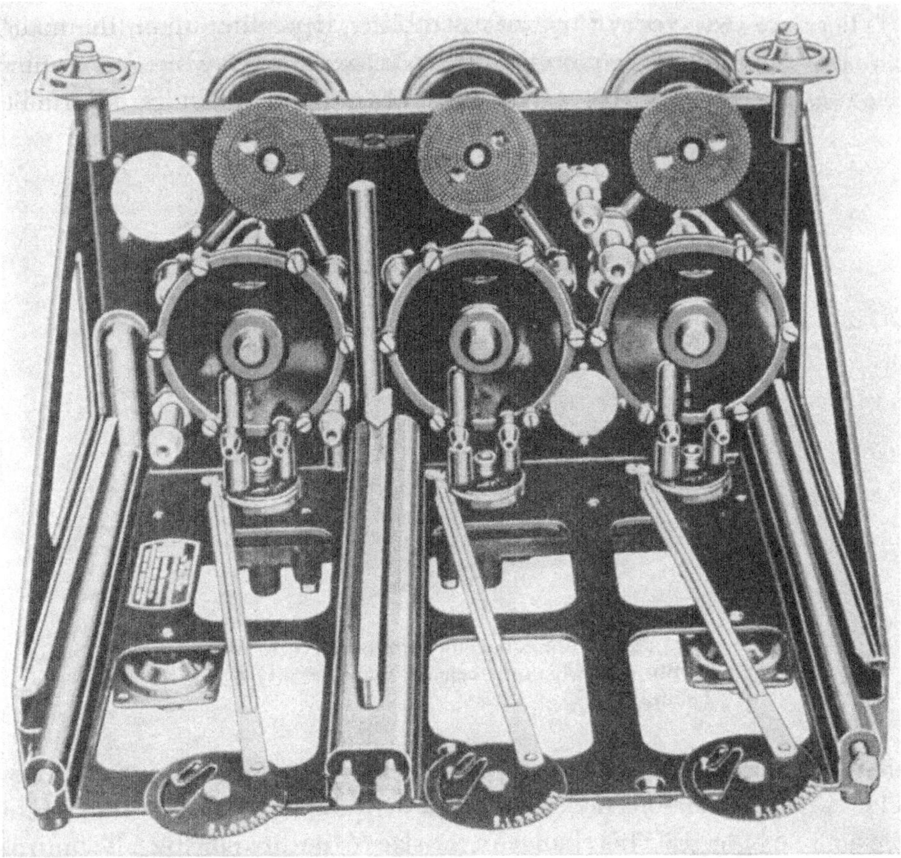

Fig. 292. A gyro-control-mount assembly. (Courtesy Jack & Heintz, Inc.)

oil valves. The manifold also collects the drain oil from these valves. The air-intake connections for the control units are connected to an intake manifold which permits the entire system to be connected to an air valve.

When speed-control valves are used, they are connected in the return line from the hydraulic surface control to the sump. These valves regulate the rate of oil flow from each hydraulic surface control. This regulation of the rate of oil flow controls the rate of speed with which the rudder, ailerons, and elevator are operated by the automatic pilot. The speed-control valve is made up of three identical units and is equipped with adjustment knobs for the rudder, aileron, and elevator. The dials above the knobs provide a reference for the valve adjustment. When the speed-control valves are not used, a return manifold is used to join the return

Fig. 293. A speed-control valve. (Courtesy Jack & Heintz, Inc.)

connections from the balanced oil valves to the main return line. This function is performed by the speed-control valve when it is used.

The oil-pressure regulator automatically regulates the oil pressure to the automatic pilot. It is adjustable to maintain the oil pressure at the value best suited for any particular installation. The pressure regulator

Fig. 294. An automatic-pilot return manifold. (Courtesy Jack & Heintz, Inc.)

has three connections: one for the line to the automatic pilot, one for the line from the pump, and one for the drain line. The drain line returns excess oil to the sump.

An oil filter is required to ensure a flow of clean hydraulic oil to the automatic-pilot system.

An air filter is required on the intake line to the automatic-pilot control units which ensures a flow of clean air to the gyros and to the air signal system.

The hydraulic surface control consists of three servo cylinders in one unit. These cylinders have piston rods extending from the cylinder at each end of the unit. This unit is equipped with a manually operated

Fig. 295. An oil-pressure regulator. (Courtesy Jack & Heintz, Inc.)

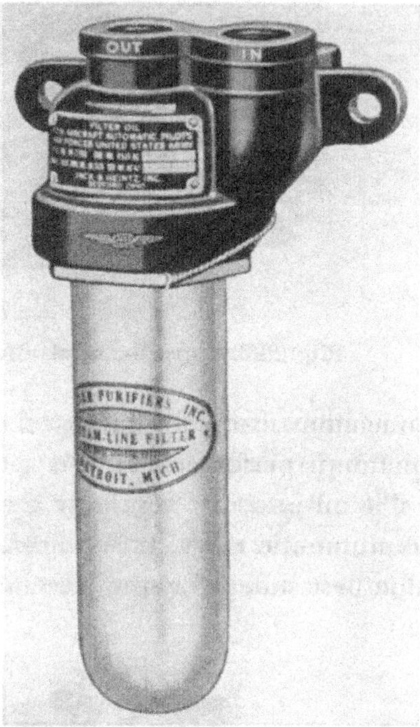

Fig. 296. An automatic-pilot oil filter.

by-pass valve affecting all cylinders at the same time for engaging or disengaging the automatic pilot. The hydraulic surface control is equipped with spring-loaded relief valves which permit the human pilot to overpower the automatic pilot.

Each of the three pistons operates one of the three main controls of the airplane. A drain trap is usually installed for installations which do not permit gravity drain from the mount assembly to the oil sump or reservoir.

A drain trap collects the oil drainage from the mount assembly. The drain trap valves this oil periodically by means of a float control valve into the suction line of the oil pump.

Three types of manifolds are used to assist in joining metal tubes to flexible hoses in the hydraulic system.

Fig. 297. An automatic-pilot air filter. (Courtesy Jack & Heintz, Inc.)

Figure 301 is a drawing showing the installation of the automatic-pilot equipment. The operation of this automatic pilot is similar to that of any gyroscopic automatic pilot. The two gyroscopes used are of a standard type: one is the turn gyro and the other is the bank-climb control unit. The first gyro operates the rudder, while the second gyro operates both the ailerons and elevator.

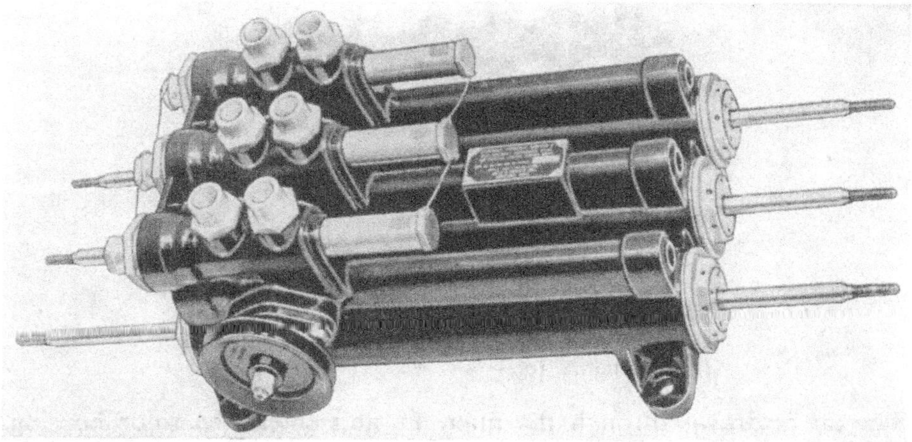

Fig. 298. An automatic-pilot servo unit for hydraulic surface control. (Courtesy Jack & Heintz, Inc.)

The bank-climb gyro spins with its axis vertical, as shown in Figure 302. This gyro is precessed erect by the action of pendulum-like vanes. When the airplane is flying level, as shown in the figure, the horizontal bar in

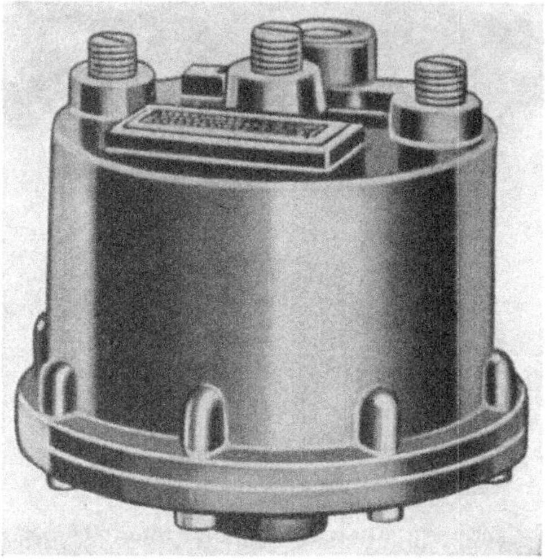

Fig. 299. An automatic-pilot drain trap. (Courtesy Jack & Heintz, Inc.)

the bank-climb-gyro control unit is level and the miniature airplane is parallel to it. Because of its rigidity in space, this rotor remains fixed in space as the airplane banks. The direction of movement is indicated on the bank scale of the automatic pilot. The rotor, A, is supported in a gimbal ring in the case. The vacuum pump, B, draws air from the box.

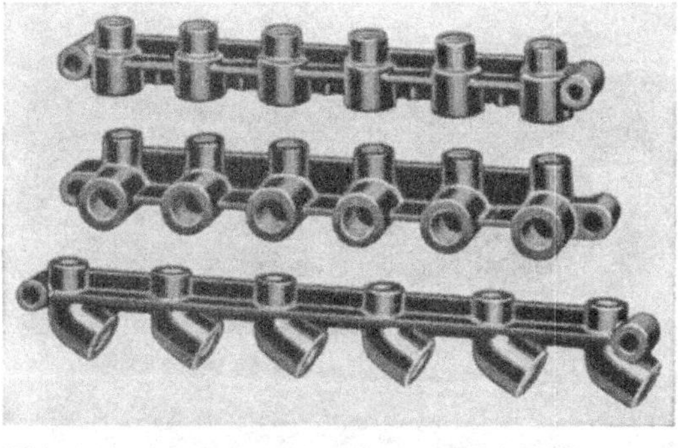

Fig. 300. An automatic-pilot manifold. (Courtesy Jack & Heintz, Inc.)

The air is drawn through the filter, C, and enters the rotor housing through a jet. This jet is directed against the circumference of the gyro rotor and causes it to spin at high speed.

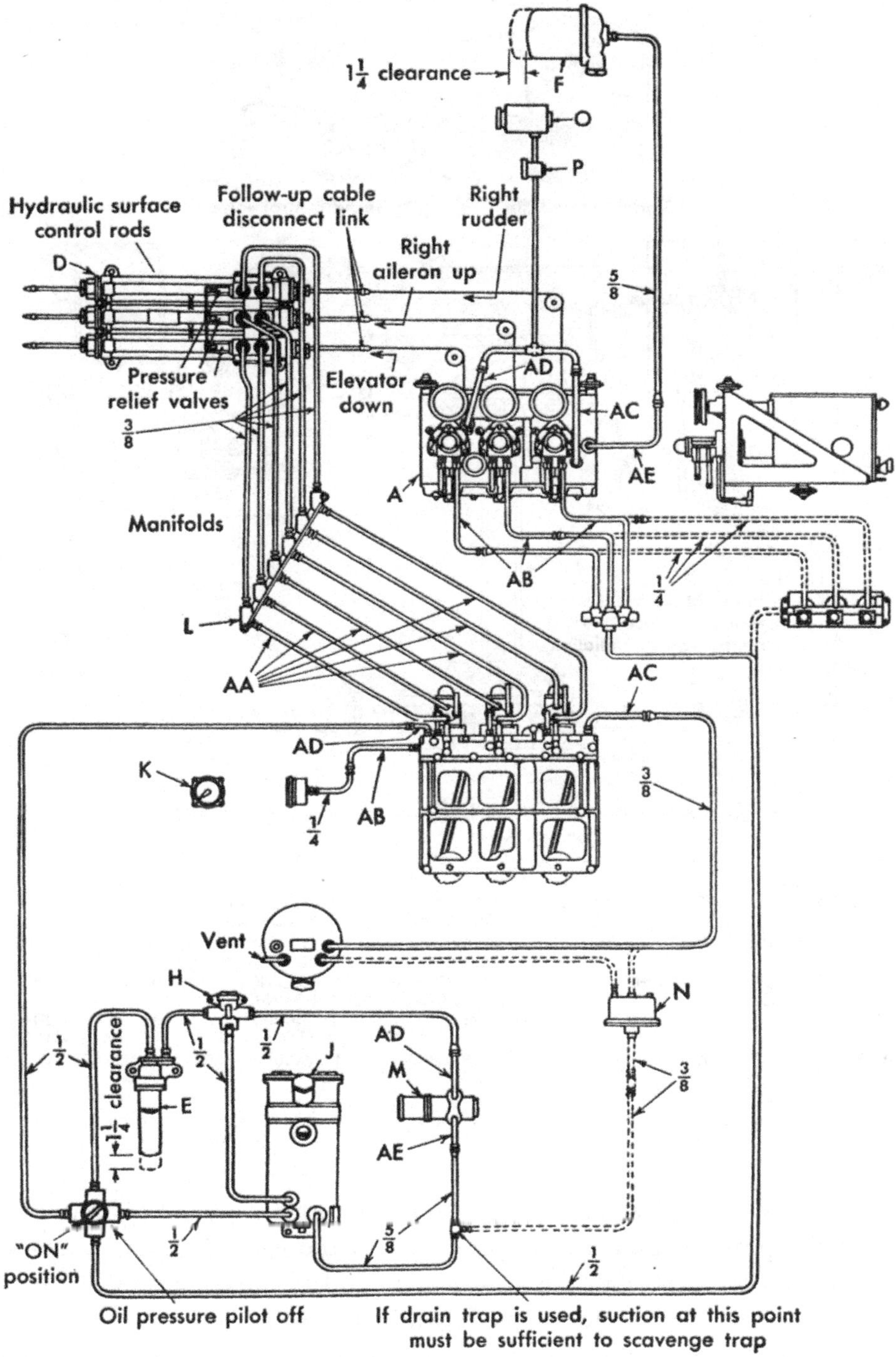

Fig. 301. A drawing of the installation of the automatic-pilot equipment. (Courtesy Jack & Heintz, Inc.)

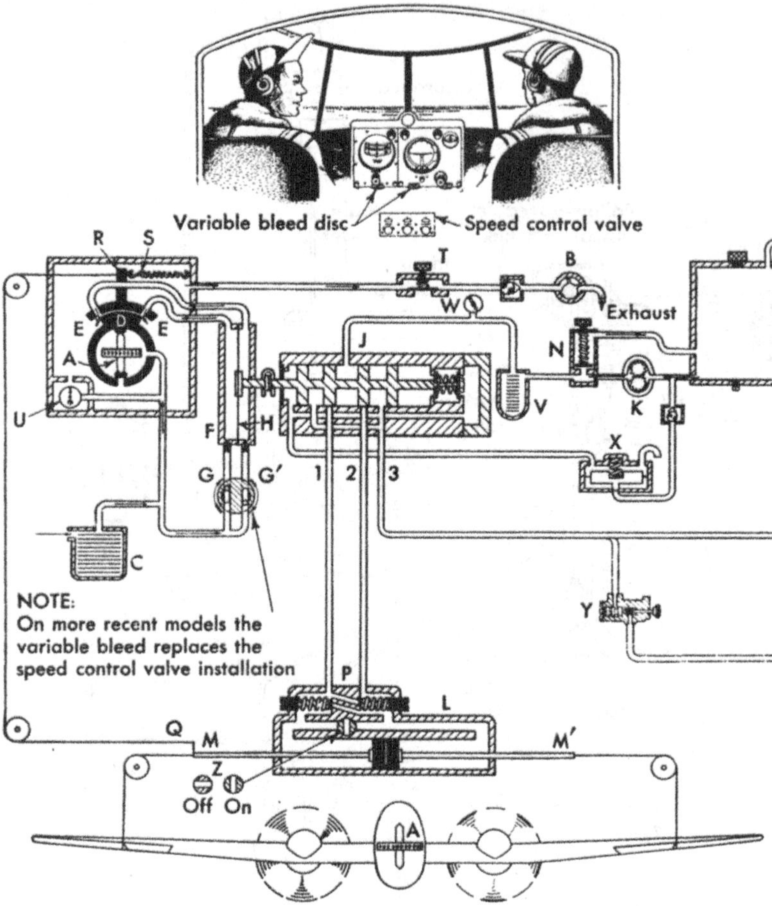

A GYRO ROTOR	F AIR RELAY	L HYDRAULIC SURFACE CONTROL	Q PISTON CONTROL CABLE	V OIL FILTER
B VACUUM PUMP	G-G' BLEEDS	M-M' PISTON ROD	R FOLLOW-UP LEVER	W OIL PRESSURE GAGE
C AIR FILTER	H DIAPHRAGM	N OIL PRESSURE REGULATOR	S BALANCE SPRING	X DRAIN TRAP
D SHROUD	J BALANCED OIL VALVE	O SUMP	T VACUUM REGULATOR	Y SPEED CONTROL VALVE
E-E' AIR VALVE PORTS	K HYDRAULIC PUMP	P OVERPOWER VALVE	U VACUUM GAGE	Z ON-OFF VALVE

Fig. 302. An automatic-pilot operational diagram, level flight. (Courtesy Jack & Heintz, Inc.)

The air signal system consists of the shroud, D, on the gimbal ring and the ports, E and E′, also mounted in the case. The air from the filter, C, passes through the air relay, F, and the ports, E and E′, past both edges of the shroud, D, as shown in Figure 302.

The air relay, F, has two inlet bleeds, G and G′. One bleed is on each side of the diaphragm which is connected to the spool of the balanced oil valve, J. A constant oil pressure is maintained in the balanced oil valve by the oil pump, K, and the oil-pressure regulator, N. Any movement of the spool of the balanced oil valve caused by a difference in the air pressure on either side of the diaphragm allows oil to flow to the hydraulic surface control, L. This oil moves the piston rod, M–M′, one way or the other. This piston rod is connected to the control cables, which operate the ailerons. In Figure 302, the airplane is flying level, and the automatic-pilot system is in neutral. The gyro axis is vertical. In this position, the shroud, D, allows equal amounts of air to be drawn through the ports, E and E′. The air pressure in the air relay, F, is equal on each side of the diaphragm, H. The balanced-oil-valve spool is in the center position, and no oil can flow through the hydraulic surface control. The pressure regulator, N, allows the oil to flow back to the sump, O.

In Figure 303, the airplane has been banked to such a position that the right wing is lower than the left. The gyro, because of its rigidity in space, maintains its axis in a position vertical to the earth's surface. The port, E, of the air signal system is closed by the shroud, D. Port E is open, and the vacuum on the left side of the diaphragm, H, is increased. This vacuum causes the diaphragm to move to the left. The movement of the diaphragm causes a corresponding movement to the left of the balanced-oil-valve spool. This movement of the oil-valve spool permits oil to flow through pipe No. 1 to the hydraulic surface control. The oil passes around the overpower valve, P, and enters the hydraulic surface-control cylinder. The oil pressure moves the piston to the right. This movement of the piston which is connected to the aileron control cables depresses the right aileron and raises the left aileron. This position of the aileron causes the airplane to rotate about its longitudinal axis to the left, bringing the airplane back to a level flight position. The oil from the other side of the surface-control cylinder returns to the balanced oil valve through pipe No. 2 and flows back to the sump through pipe No. 3.

The follow-up system, which is an important part of the hydraulic control system of the automatic pilot, is shown in Figure 304. The follow-up system is an arrangement by which the applied control is removed as the airplane is returning to its normal attitude. This is important to avoid overcontrol which would cause the airplane to rotate into a left-bank position. The follow-up system is arranged so that the control surface will be back in its neutral or centered position when the airplane is back in its normal flight attitude.

The air-valve ports, E and E′, Figure 302 are not fixed solidly to the gyro case. They can be moved in relation to it by means of the follow-up mechanism. A cable is connected to the hydraulic surface-control piston rod at Q. This cable is attached to the lever, R, on the follow-up

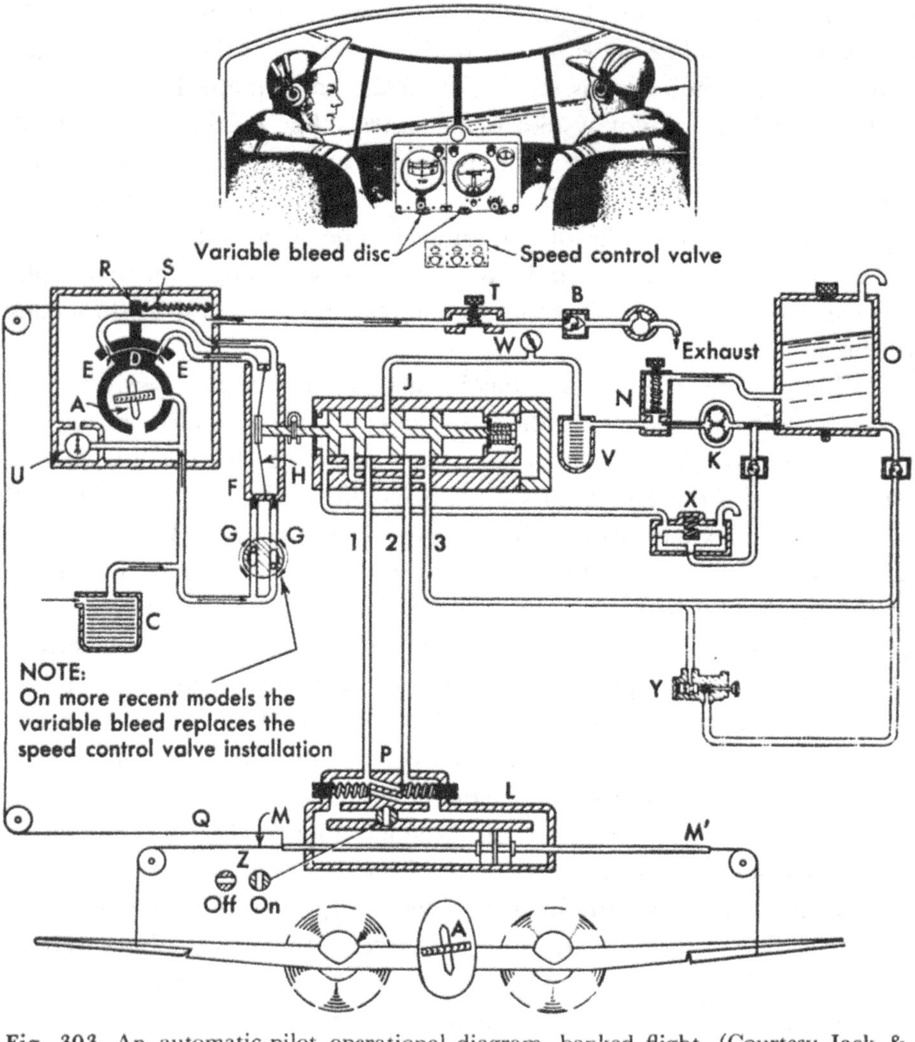

Fig. 303. An automatic-pilot operational diagram, banked flight. (Courtesy Jack & Heintz, Inc.)

assembly. When the servo piston, M–M′, moves to the right, the follow-up assembly is moved against the opposed balance spring, S. This moves the ports, E and E′, to the left. When these ports reach a neutral position, that is, when each of them is half open, the air relay and the balanced oil valves are centered. When the balanced oil valve is centered, the hydraulic-surface-control movement away from neutral is stopped. As the airplane moves back toward its normal attitude, the air ports, E and E′, which have driven ahead of the gyro case, pass beyond the neutral point and begin to cause hydraulic surface-control movement in the opposite direction. It must be understood that this is not opposite

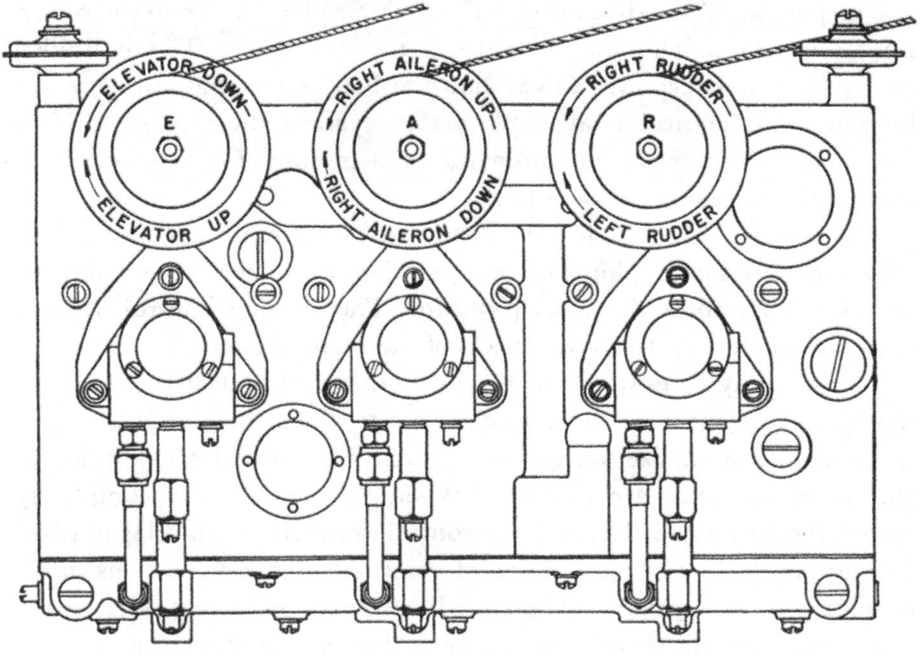

Fig. 304. An automatic-pilot follow-up pulley arrangement, right-hand installation. (Courtesy Jack & Heintz, Inc.)

control, but is simply removing the control originally applied. The mechanism is so arranged that the correct amount of control will be applied and removed at the proper rate as the airplane returns to its normal flight attitude.

The position of the air-valve ports can be governed by the human pilot by means of the manual control knobs. A remote-control system may be installed whereby this manual control may be operated by a remote-control operator, such as the flight engineer.

In Figures 302 and 303, the part, T, is a vacuum regulator which maintains the proper vacuum for the operation of the rotor and the air valves. This vacuum is maintained regardless of the speed of the vacuum pump. The vacuum in inches of mercury is indicated on the vacuum gauge, U.

The oil sump, O, carries the reserve oil. The part, V, is the oil filter. The pressure-relief valve, N, automatically regulates the oil pressure. The oil pressure is indicated on the oil gauge, W. The drain trap is shown by the letter, X. When installed, the drain manifold is below the level of the sump.

The purpose of the drain trap is to provide a check valve in the return lines to the sump. The sump is vented to the atmosphere. The hydraulic, surface-control-overpower valves shown at P are set so that ample power for automatic control is available yet they allow the human pilot to overcome the action of the automatic pilot in case of emergency, even though the system is in operation.

The by-pass valve, Z, is used to turn the automatic pilot on or off. When the automatic pilot is not in operation, this valve is open and the oil flows freely through the by-pass tube. The airplane control surfaces may then be moved freely by means of the manual controls.

The air-relay bleeds, G and G′, in Figures 302 and 303 govern the sensitivity and speed of the automatic pilot. When the bleeds are open to their maximum size, maximum sensitivity and speed of operation of the automatic pilot are obtained. When these bleeds are completely closed, the air relay will shut off the control to which the air relay applies. By varying the size of the opening of the air-relay bleeds, various stages of sensitivity and speed may be had. In some of the early models of the automatic pilot, the speed of operation of the controls was brought about by adjustment of the speed-control valve shown at Y in Figures 302 and 303.

Before attempting to use an automatic pilot, the operator should have a thorough understanding of its principles of operation and the relation between its component parts. Before taking off, the pilot should check the vacuum on run-up. The vacuum gauge should show from $3\frac{3}{4}$ to $4\frac{1}{4}$ in. of mercury.

The pilot should be sure that the bank-climb-gyro control unit is uncaged. He should set and uncage the gyro control unit. All controls should be centered. While the controls are in this position, the turn-gyro follow-up card should be set to match the turned-gyro card. The aileron follow-up index should be set to match the center graduation on the bank dial. The elevator follow-up index should be set to match the *elevator* alignment index.

On later models, the air bleeds should be opened to a point just below that at which oscillation occurs. On early models, the speed-control

Fig. 305. Operation of an automatic pilot. (1) Elevator manual control knob; (2) vacuum gauge; (3) bank index; (4) aileron follow-up index; (5) horizon bar; (6) horizon dial; (7) elevator follow-up index; (8) elevator alignment index; (9) "caged" signal; (10) miniature airplane; (11) caging knob; (12) miniature airplane adjusting knob; (13) caging knob; (14) turn gyro card; (15) turn gyro follow-up card; (16) aileron manual control knob; (17) rudder manual control knob; (18) speed valve adjusting knobs; (19) valve adjusting reference dials; (20) adjustable bleed dials. (Courtesy Jack & Heintz, Inc.)

valves should be opened to a maximum setting of 6, as shown on the dials. The automatic pilot should then be engaged, and the human pilot should check the operation of the controls by turning each control knob.

All oil pressures should be checked. The oil pressure should be within 10 lb. of the pressure recommended for that particular installation. The pilot should check the overpower valves by operating the controls manually against the automatic pilot. The automatic pilot should then be disengaged.

After taking off, the pilot should trim the airplane for hands-off level flight. On later models of the automatic pilot, the adjustable air bleeds should be set to 3, or the best setting used during the ground check. On early models, the speed-control valves should be wide open or at 6 if the best setting is not known.

The turn-gyro follow-up card should be set to match the turned-gyro card by turning the rudder knob. The elevator follow-up index is then set to match the elevator alignment index by turning the elevator knob.

The automatic pilot should be turned on by slowly moving the automatic pilot ON-OFF valve to the ON position. While holding the controls, the pilot can feel the automatic pilot take over.

The sensitivity dials are then set for the desired control response. The control response can be made very sluggish or extra sensitive. An extra-sensitive setting may cause fast, short oscillations of the control surfaces. To make course changes, rotate the rudder manual-control knob slowly and smoothly. If the turn to be made is through a considerable number of degrees, the bank should be set in with the aileron control knob. Correction for the desired fore-and-aft attitude should be made by rotating the elevator knob. The operating limit from the vertical, for use as a visual flight indicator when the pilot is operating the controls, is 50° for the bank-climb-gyro control unit, and 55° for the turn-gyro control unit. The operating limit from the vertical for use as an automatic pilot control instrument is 25° for both control units. For more detailed instructions in the operation of any type of automatic pilot, the pilot should refer to the automatic-pilot manufacturer's manual.

Minneapolis-Honeywell C-1 Autopilot. This automatic pilot is the electromechanical type. It is operated by electric currents and uses electrical motors instead of hydraulic actuating cylinders. Other types of automatic pilots depend upon air pressure and hydraulic pressure for their operation.

This automatic pilot, as with other automatic pilots, depends upon gyroscopic principles to detect changes in direction. Changes in direction include movements of the airplane about the vertical axis, the longitudinal axis, and the lateral axis. It is claimed that the automatic pilot is smoother and more precise than the average human pilot in making corrections to the controls in order to bring the airplane back to its proper flight attitude when it is displaced from it in any direction.

Fig. 306(a).

Fig. 306(b).

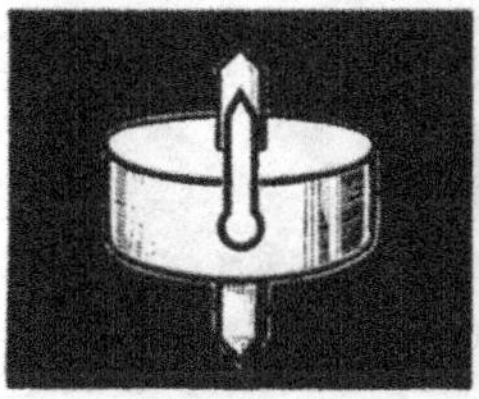

Fig. 306(c).

Fig. 306(d).

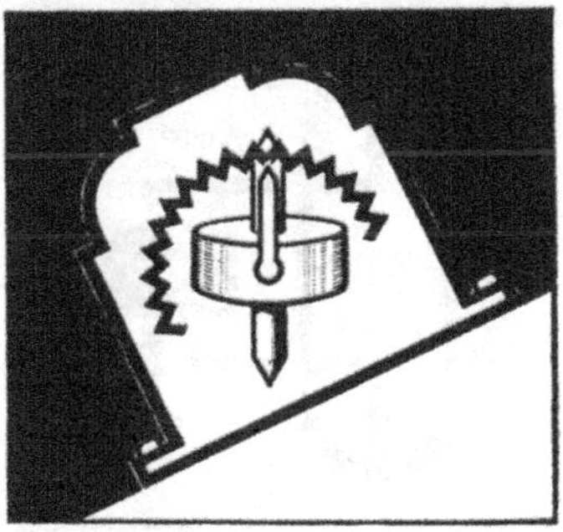

Fig. 306(e).

Fig. 306, a, b, c, d, e. How airplane deviations affect the gyro. (Courtesy Minneapolis Honeywell Regulator Company)

The system consists of a gyro-operated directional stabilizer, a banking pot and rudder pick-up, an amplifier and directional panel, a vertical-flight gyro, an elevator pick-up pot, an elevator servo unit, an aileron servo unit, a rudder servo unit, aileron pick-up pot, a skid pot, and the necessary controls.

The gyro case of each of the gyros is fastened to the airplane structure (Fig. 306a). The case tilts with the airplane, but the gyro maintains its position (Fig. 306b). The wiper is attached to the rotor (gyro) Cardan ring and moves over the pot winding which is attached to the case. A pot consists of a winding about a suitable core, making up a resistance coil over which a wiper unit moves. When the airplane tilts from straight and level flight, the proper gyro and its wiper stay in a fixed position (Figure 306c). The case moves about the rotor and carries the pot

Fig. 307(a).

Fig. 307(b).

Fig. 307(c).

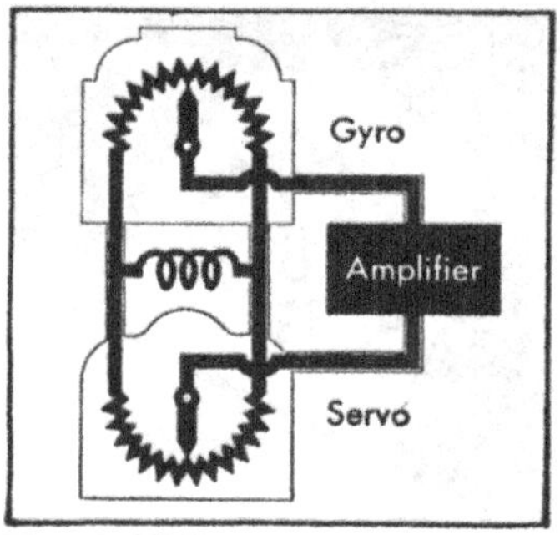

Fig. 307(d).

Fig. 307, a, b, c, d. Gyro pick-up pot and servo balance pot form a bridge circuit. (Courtesy Minneapolis Honeywell Regulator Company)

winding with it (Figure 306d). This movement causes the wiper to move across the pot winding (Figure 306e). A corresponding pot winding is attached to the servo case (Figure 307a). When the wiper of the rotor moves over its pot winding, it sets up an unbalanced electric current through the servo winding (Figure 307d) which affects the amplifier. The amplifier, connected by the two wipers, operates a servo unit in response to any unbalance in the bridge circuit.

As shown in Figure 308–1, when the airplane is in straight and level flight, the elevator pickup pot in the gyro is at its normal position, N, and the servo balance pot is centered. This holds the elevator control surfaces in the normal, level flight position. When the wipers are in these positions, the bridge circuits are in balance and no signal is transmitted to the amplifier.

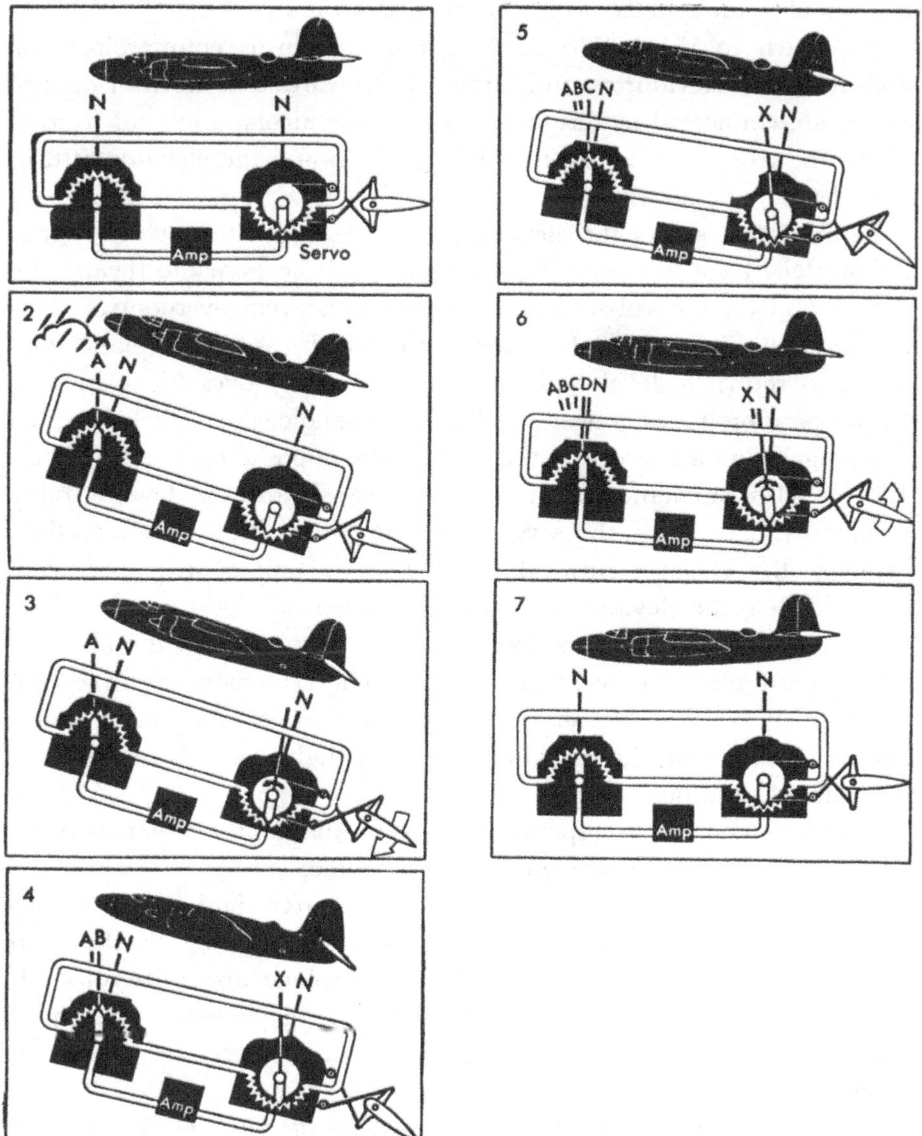

Fig. 308. How the automatic pilot corrects airplane deviations. (Courtesy Minneapolis Honeywell Regulator Company.)

In Figure 308–2, a sudden gust of air has raised the airplane's nose. This movement causes the gyro case and its attached, elevator pick-up pot to tilt with the airplane, thus causing a displacement of the pot wiper. The pot wiper is moved to position 1 and unbalances the bridge circuit. This sends a correcting signal to the elevator panel of the amplifier, causing the down-elevator relay and the amplifier to close and sending a down-elevator impulse to the elevator servo unit.

As shown in Figure 308–3, the servo unit turns counterclockwise and moves the elevator control surfaces downward. The elevator control cables are connected to this servo unit. As the airplane returns to level flight, the wipers return to their zero position and the elevator returns to its level flight position.

Figure 308–4 shows the elevator in the down position which brings the airplane back to a level flight attitude. It is necessary to reverse the direction of the control-surface movement to prevent overcontrol. This is shown in Figures 308–5, 308–6, and 308–7. As the airplane continues its return to level flight, the elevator pot moves to position 3. In this position it is no longer parallel to the balance pot. The bridge circuit is again unbalanced, but this time in the opposite direction (Figure 308–5). This unbalance of the amplifier causes the up-elevator relay to close, thus making the servo unit rotate in a clockwise direction, bringing the elevator upward. As the servo unit rotates clockwise, Figure 308–6, the elevator is returned to its normal flight position. The control-surface movement is slowly reduced in speed, and the airplane returns gradually to a level flight position. The movement of the control becomes slower and slower as the airplane approaches level flight. Figure 308–7 shows the airplane in normal flight with the elevator in its centered position.

The other controls are operated in exactly the same manner, correcting "roll" or "yaw" movements of the airplane.

The vertical-flight gyro acts as an automatic artificial horizon. Potentiometers in the vertical-flight gyro detect any tipping of the gyro housing caused by a displacement of the airplane about its roll-and-pitch axis. The electric signals produced by the movement of these pots cause the autopilot to operate the controls of the airplane and bring it back to a normal flight attitude.

Directional Stabilizer and Other Parts of the Automatic Pilot. The directional stabilizer detects any movement of the airplane about its vertical axis and sends out electrical impulses which, through the amplifier and servo unit, operate the rudder to bring the airplane back to its proper direction. This unit, like the vertical-flight unit, contains a gyroscope.

Because of precession, a gyroscope, which is spinning on a horizontal axis at 6 A.M., would be vertical at noon and would turn end over end in 24 hr. This precession is due to the movement of the earth about its axis.

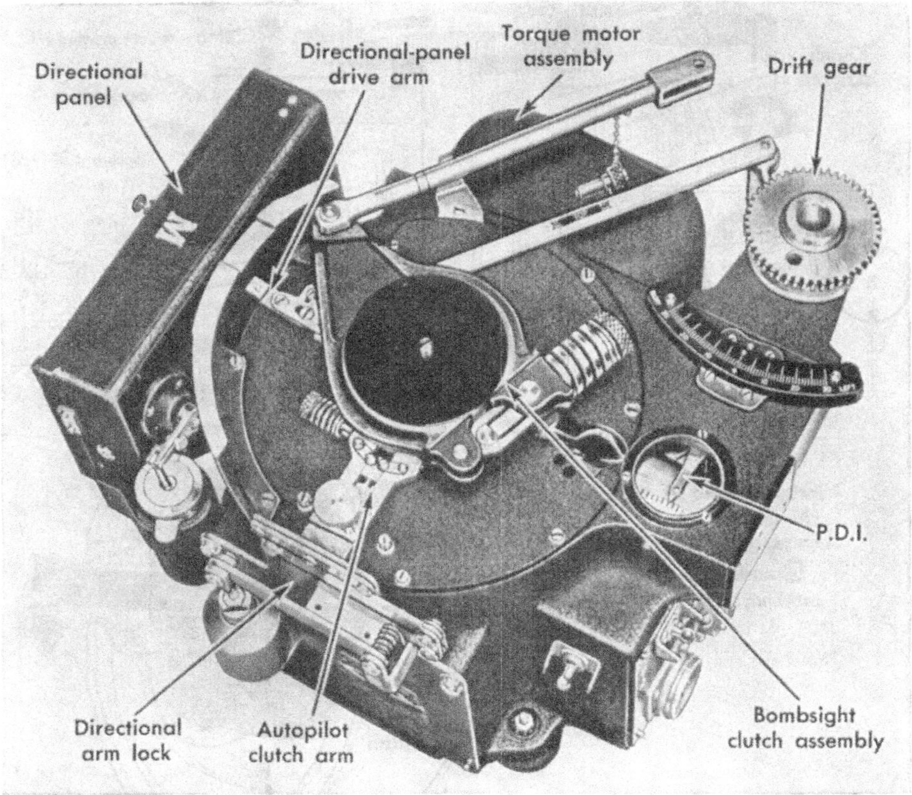

Fig. 309. An automatic pilot directional stabilizer. (Courtesy Minneapolis-Honeywell Regulator Company)

One function of the torque motor assembly is to maintain the axis of the gyroscope in its relative position to the floor of the airplane in straight and normal flight. The torque motor assembly corrects or counteracts any external forces which tend to precess the axis of the gyroscope away from its normal position.

Figure 310 is a schematic diagram to show the operation of the directional stabilizer. The clutch gears, D and E, are driven in opposite directions by the torque motor through the intermediate gears, B and C. When force is applied to the autopilot clutch drum which tends to turn the sector side of the Cardan toward the right, the resulting upward precession of the gyro brings the pick-up wiper attached to the rotor case into contact with the pick-up segment of the clutch contact sector. A tilt of 1° will make this contact. This contact completes a circuit through

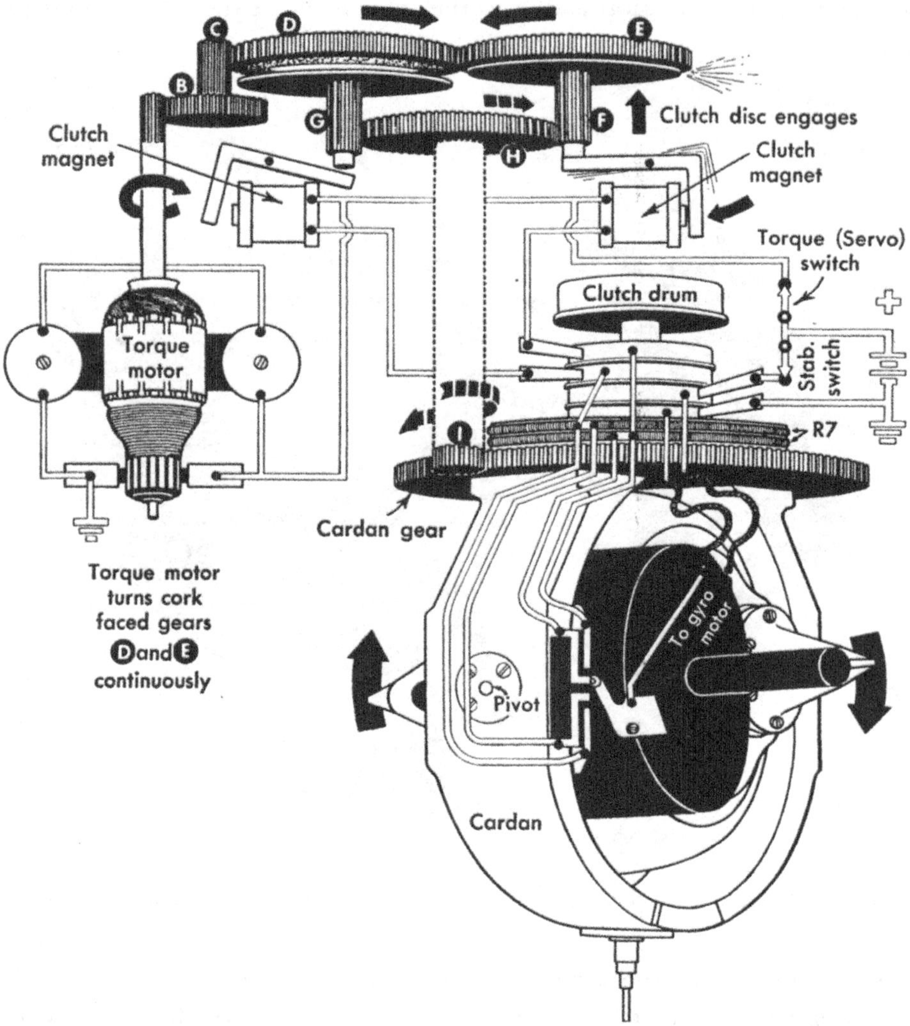

Fig. 310. A schematic diagram of a directional stabilizer. (Courtesy Minneapolis-Honeywell Regulator Company)

the lower resistance and the righthand clutch magnet. This magnet is energized and raises the corresponding clutch disk against a cork surface on the underside of the clutch gear, E. The frictional force applied to the clutch disk, F, by the clutch gear is transmitted through the intermediate gears, H and I, to the Cardan ring gear, as shown by the dotted arrows. This force tends to turn the Cardan and the rotor to the left, or in the opposite direction to the original force applied to the autopilot clutch drum. The directional armlock is closed while the airplane turns, the gyro Cardan and Cardan gear maintain their position, and the rest of the stabilizer, including the torque motor, gear train, and all other parts, moves around the Cardan gear to a new position.

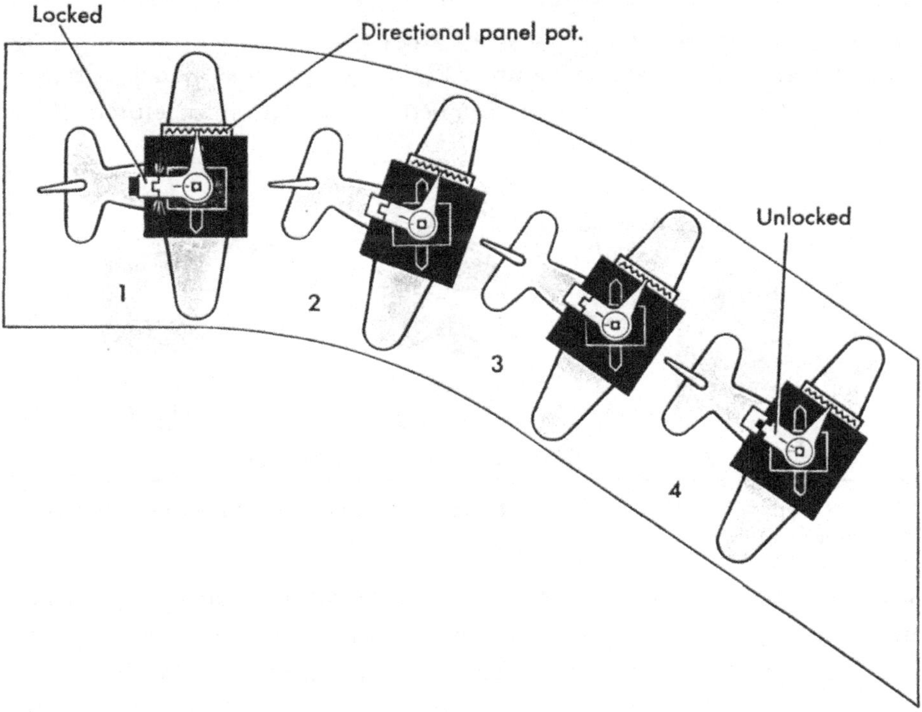

Fig. 311. Operation of the directional panel. (Courtesy Minneapolis-Honeywell Regulator Company)

The directional arm lock is a solenoid-operated clamp mounted on the rear side of the stabilizer, as shown in Figure 311. In the figure, this lock is represented by a square key and keyway at the left of the autopilot clutch. When the pilot desires to change the airplane's heading, he applies a signal calling for rudder and aileron movement by turning his "Turn Control" knob. The turn control includes a switch which automatically closes the directional arm lock (Figure 311–1) before any rudder signal can be introduced. The directional arm lock prevents movement of the wiper over the directional panel pots. Slippage is provided for in the automatic-pilot clutch which allows the gyro rotor to remain in its original position. This slippage continues until the pilot returns the control knob to its zero position. The directional arm lock is then released, and the autopilot will keep the airplane accurately on its new heading (Figure 311–4).

When the airplane turns, the frame of the directional panel with its attached pot windings moves with the frame of the airplane causing a displacement of the wipers with respect to the pots, as shown in Figure 311–1, –2, and –3 and Figure 312–1, –2, and –3. This displacement produces electric signals which cause the autopilot to correct the direction of flight. As shown in Figure 312, the directional panel is represented as a pot with its wiper fastened to the autopilot clutch. This

Fig. 312. Operation of the directional arm lock. (Courtesy Minneapolis-Honeywell Regulator Company)

wiper is centered when the airplane is flying straight on a given heading, as in Figure 312–1. When abrupt air currents change the airplane heading as in Figure 312–2, the stabilizer case and attached pot windings turn with the airplane, while the wiper is held in position by the rotor. This displacement of the pot applies corrective voltage signals, and the autopilot system operates the rudder and ailerons in such a manner as to return the airplane to its original heading, as shown in Figure 312–3.

Figure 313 is an internal view of the directional panel. This panel consists of a cast-aluminum housing containing a sliding block on which are mounted the rudder pickup pot wiper and the dual banking pot wipers. The pot windings are mounted rigidly to the inside of the back of the housing as shown in the figure. The slide is moved from side to side by the directional-panel drive arm which extends through the slot at the back of the housing into a hole drilled in the back of the slide. The directional-panel drive arm is an extension of the autopilot clutch on the directional stabilizer, and any change in the airplane's direction moves the slide and its attached pot wipers across the pots attached to the directional panel.

Above the dual banking pots is a spring leaf-switch which is operated by cams attached to the directional-panel slide. The operating cams are so spaced that, when the slide is moved ⅛ in. or more in either direction from center position, the switch will close a circuit which energizes the erecting cutout magnet in the vertical-flight gyro. Below the rudder pickup pot is a resistor card which is a part of the dual banking-pot circuit. A dash pot is fastened to one end of the directional panel.

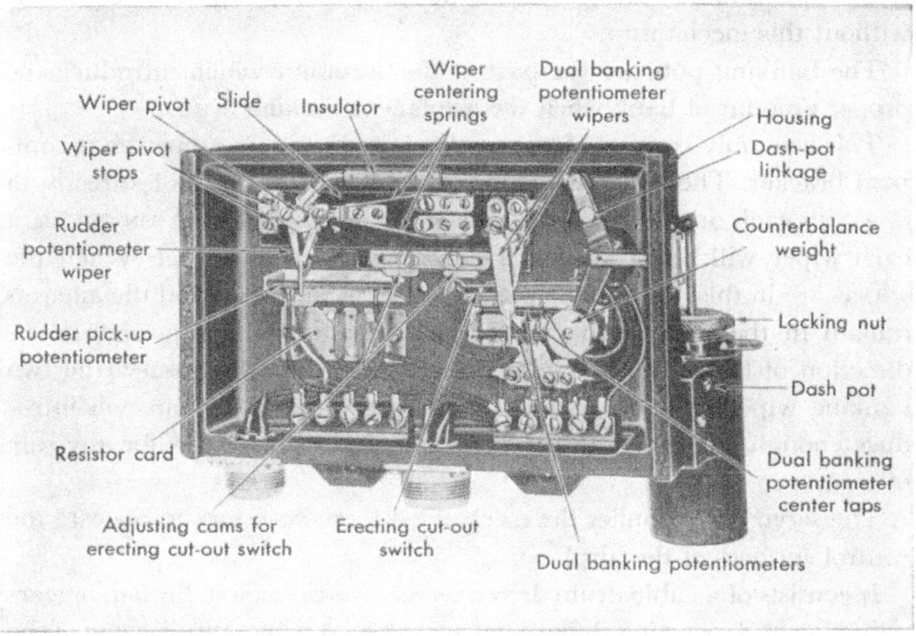

Fig. 313. Internal view of the directional panel. (Courtesy Minneapolis-Honeywell Regulator Company)

The rudder pot wiper is mounted to the slide on a pivot. It has its upper end linked to the oil-filled dash pot by means of a horizontal bar and two crank levers connected by a shaft which passes through the back of the case. When the top of the rudder pot wiper is restrained by the dash pot and linkage, any movement of the slide will cause the wiper to rotate on its pivot with a greater movement of its lower end along the pot. This rotation is opposed by a pair of spring leaves which tend to center the wiper. When the slide movement stops, the springs recenter the wiper and bring it back to a vertical position. The amount of rotation is limited by two stops attached to the slide.

When an airplane's directional deviation is smooth and gradual, the effect of the dash pot will not be enough to deflect the springs. The wiper therefore will not pivot, and its displacement will be proportional to the deviation of the airplane. When the deviation is sudden, the restraining effect of the dash pot will be greater and will overpower the springs. This causes rotation of the wiper and results in extra movement of the wiper proportional to the abruptness of the deviation. This assures that sudden deviation will instantly result in a stronger correcting signal to the amplifier. This strong signal will cause greater rudder correction to be applied more quickly to arrest the deviation than would be possible without this mechanism.

The banking pots are the part of the autopilot which introduce the proper amount of bank when the airplane is making a turn.

This assembly consists of two center-tapped pots mounted on a common bracket. The two pot wipers are mounted on the slide directly in line with each other. When the slide is at the middle point (ON COURSE), each wiper will be at the exact electrical center of its pot. When the wipers are in this position, no voltage signal is picked up and the ailerons remain in the trim position. When the airplane is displaced from its direction of flight and a rudder correction becomes necessary, the two banking wipers will move across their pots. This movement will introduce enough aileron control to bank the airplane correctly for any rate of turn.

The servo unit supplies the mechanical force necessary to operate the control surfaces of the airplane.

It consists of a cable drum driven by an electric motor through a gear-reduction and reversing-differential assembly. A schematic drawing of the power transmission system is shown in Figure 314. Three identical servo units are required in each airplane. One unit operates the ailerons to correct roll; another unit operates the rudder to correct directional deviations; and a third unit operates the elevators to correct pitch deviations. Each unit is connected to the control surfaces by $\frac{1}{8}$-in. flexible steel cables. These cables are attached to the airplane's main control cables.

The servo unit is designed to respond instantly with its full rated power to operate the airplane's control surfaces in either direction. The unit has enough power to operate the control surfaces at any speed. It also has sufficient braking power to hold the control surfaces firmly in any position between control movements.

An electrical balancing system regulates the amount of control in proportion to the airplane's change in direction.

Safety limit switches prevent the servo unit from driving the control surfaces to their mechanical limits.

The servo unit (Figure 314) is powered by a shunt-wound, $1/20$-hp., 26-volt, d-c motor, which operates on current supplied directly from the airplane's batteries. The motors in all three servo units run continuously in the one direction while the autopilot is in operation.

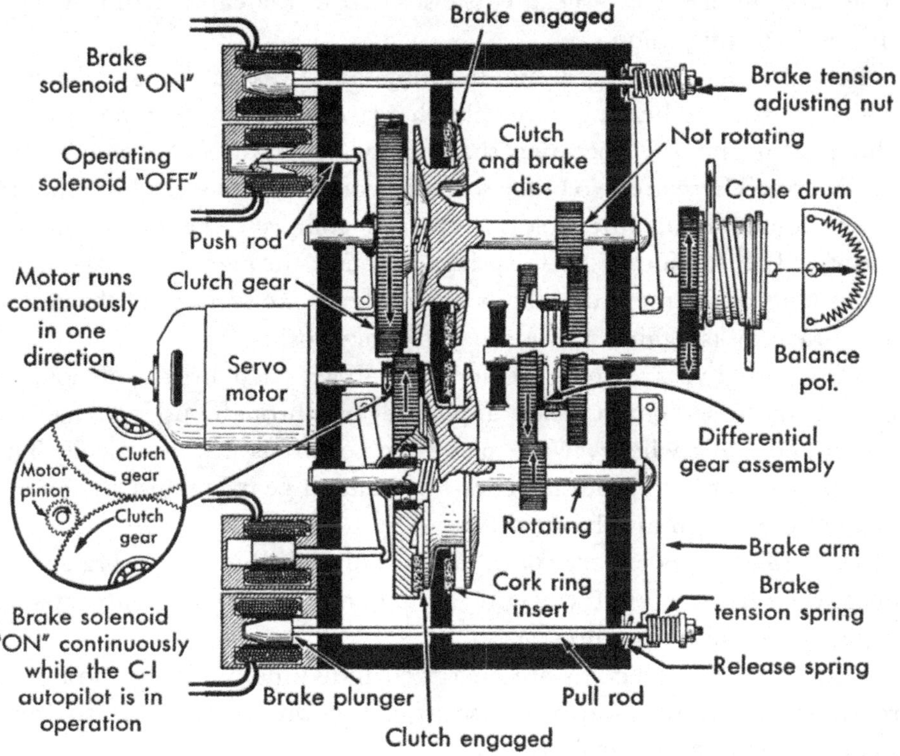

Fig. 314. Schematic drawing of the power transmission system. (Courtesy Minneapolis-Honeywell Regulator Company)

The power of the motor is transmitted to the control cables by a power transmission system made up of two gear trains, two brake solenoids, two operating solenoids, four frictional surfaces, a differential assembly, and the cable-drum drive assembly. The motor pinion meshes with one of two, large clutch gears which, in turn, drives the other clutch gear. The two clutch gears, each of which has a cork insert on one side, rotate continuously in opposite directions while the autopilot is in operation.

Attached to each operating shaft is a clutch and brake disk. Each disk has two, flat friction surfaces. One surface faces the cork insert on the adjoining clutch gear. The other faces a similar cork insert in the brake ring fastened to the servo-unit frame.

Attached to each operating shaft is a small, operating-shaft gear which meshes with one of the differential gears. The two differential gears rotate freely on the differential shaft. One gear is on each side of the differential crosshead which operates the two crosshead beveled pinions. The rotation of the crosshead is transmitted to the cable drum through the cable-drum pinion.

Between each clutch gear and its clutch-and-brake disk is a coil spring which presses the bearing in the clutch gear in one direction against the clutch arm and the operating shaft in the other direction against the brake arm. There are two brake solenoids, one on each side of the motor. These solenoids operate pole rods which compress the brake tension springs. The tension of these springs, applied through the brake arms to the ends of the operating shafts, forces the brake surfaces of the clutch and brake disks against the cork-faced brake springs.

There are two solenoids located one above each brake solenoid at the rear of the servo unit. These apply pressure through push rods against the clutch arms whenever the operating solenoids are energized. The clutch arms force the rotating cork-lined clutch gears against the clutch faces of the clutch and brake disks.

The brake solenoids are connected in parallel. Both are energized at once when the servo-unit engaging switch on the autopilot control panel is placed in the ON position.

The operating solenoids are energized individually by the closing of relays in the amplifier in response to correction signals received from the corresponding control bridge.

In series with each operating solenoid is a cam-operated limit switch which prevents the servo unit from driving the control surfaces to their mechanical stops. A balance pot controls the amount of cable-drum rotation resulting from a given deviation of the airplane. The direction of the drum rotation is determined by which of the two operating solenoids is energized.

The amplifier is the unit which controls the operation of the servo units. The amplifier receives the signals from the control bridge brought about by the movement of the wipers over their pots in the gyro units. The amplifier operates relay switches which energize the operating solenoids in the servo unit. The deviation of the wiper over its pot in one direction energizes one operating solenoid, causing the control to be moved in one direction. A movement of the wiper in the opposite direction causes the amplifier to energize the other operating solenoid, which causes a movement of the control in the opposite direction.

The autopilot control panel is equipped with knobs by means of which the pilot can introduce signals to the rudder and aileron bridge circuits which will maneuver the airplane in properly banked turns while the airplane is flying under autopilot control. Before engaging each of the servo units, the pilot must mechanically trim the airplane for straight and level flight. This will place the balance pot wiper in a position on its pot which corresponds to the mechanical trim position of its control surface. The pilot then adjusts each centering knob to line up the electrical balance point of each bridge circuit with this mechanical trim position of the balance pot wiper. After the engaging switches have been turned on, the autopilot system, by always returning the balance pot to its trim position, keeps the airplane in straight and level flight.

Directly above each centering control knob is a pair of telltale or indicating lights. Each pair of lights is connected in parallel with the operating solenoids of its corresponding servo unit. One light or the other will glow when the autopilot amplifier is calling for servo-unit rotation. This happens whether or not the servo unit is engaged. When the pilot is engaging the autopilot, the illumination of either telltale light indicates that the balance point of the bridge circuit is not lined up with the trim position of the balance pot wiper. Further adjustment of the corresponding center control is required whenever one or the other of the lights is lighted.

XXII NAVIGATION INSTRUMENTS

Magnetic Compass. The magnetic compass is probably the first instrument used by man to determine direction. Before he invented the compass, man guided himself by the sun or other heavenly bodies. The fixed position of the North Star was discovered early in man's history, and various other stars were used as direction finders at night. The magnetic compass in its simplest form is a magnet suspended either by a thread or on a pivot so that it is free to swing. The cardinal points of the compass are north, east, south, and west, indicated respectively by the capital letters, N, E, S, and W.

In the magnetic aircraft compass, instead of a single magnet, a bundle of magnets is usually used. Mounted on the magnet is a circular card carrying not only the cardinal points, but a complete circumference usually marked off in 5° intervals. The compass needle and the card are enclosed in a cell filled with a liquid, such as alcohol, to dampen out small movements and cause the instrument to rotate smoothly and without violent oscillation. If the compass were free to swing in the open air, the needle would oscillate before coming to rest in any direction. A white line called the lubber line is mounted vertically on the transparent window in the front of the compass, or it may be a wire mounted back of the opening. The card is read with reference to the lubber line which indicates the proper reading. In other words, the card remains stationary and the lubber line is carried past the card by the turning of the airplane.

The earth itself acts like a great magnet, being surrounded by the earth's magnetic field. The North Magnetic Pole is not located at the geographic North Pole, but is at present in the neighborhood of 71° north latitude and 96° west longitude. The South Magnetic Pole, likewise, is not located at the geographic South Pole, but at approximately 72° south latitude and 150° east longitude. The magnetic poles are not definite points on the surface of the earth, but are the approximate centers of comparatively large magnetic areas which are constantly but very slowly changing their position. The earth is surrounded with an immense magnetic field, the lines of which extend between the North and South Magnetic Poles. Any free-moving magnet in the form of a bar will rotate until its long axis is parallel to the magnetic lines of the earth at the point where the magnet is located. At all points on the earth, except near the magnetic equator, the north end of the needle tends to dip in the Northern Hemisphere and the south end tends to dip in the Southern Hemisphere.

Fig. 315. A magnetic aircraft compass. (Courtesy Kollsman Instrument Division, Square D Company)

Fig. 316. A magnetic aircraft compass. (Courtesy Eclipse-Pioneer Division, Bendix Aviation Corporation)

Compasses to be used in the Northern Hemisphere usually have a small weight in the form of a wire wrapped around the compass needle on the south-seeking half of the needle to balance against this dipping. In the Southern Hemisphere, the weight is on the opposite end of the needle. The magnetic lines of force are not parallel with the earth's surface except at the magnetic equator of the earth. Because the North and South Magnetic Poles are not located at the same place on the earth's surface as the geographic North and South Poles, magnetic compasses at most places on the earth's surface do not point directly toward the North Pole. This error in compass reading is called the variation.

Maps containing lines of equal magnetic variation may be had. These lines, called isogonic lines, show the variation of the compass at any point on the line. A line from the North Magnetic Pole to the South Magnetic Pole which passes approximately through the eastern part of

Lake Superior and into the ocean near Savannah, Georgia, is the line of zero variation. This line is called the agonic line. This line continues on around the earth back to the North Magnetic Pole.

In most aircraft, the compass is subject to another error called deviation. This error is caused by the magnetic fields set up in the airplane itself or by magnetic substances used in the structure of the aircraft which affect the magnets of the compass. These effects are partially

Fig. 317. A cutaway view of a magnetic compass. (1) and (2) Compass magnets; (3) float; (4) pivot; (5) pivot jewel; (6) compass card; (7) compass bowl; (8) expansion chamber; (9) flexible diaphragm; (10) compass lens; (11) lubber's line. (Courtesy Eclipse-Pioneer Division, Bendix Aviation Corporation)

overcome by placing compensating magnets near the compass. By adjusting these magnets with screws, most of the deviation may usually be eliminated.

The simple magnetic compass consists of magnets which are usually made of cobalt steel. Most airplane compasses contain two of these magnets or compass needles. These needles are attached to the bottom side of a float, one on each side of a compass jewel which rests on a pivot. The compass card carrying the degree markings and the cardinal points, as well as other points shown by figures, is attached to the float. This entire assembly is placed in the compass bowl which is filled with liquid. This liquid imparts a certain amount of buoyancy to relieve the pivot and the jewel of a part of the weight of the compass assembly. An expansion chamber, or a flexible diaphragm, takes care of the expansion and contraction of the liquid.

Directional Gyro. A directional gyro is a gyroscopic instrument which supplies a deadbeat directional reference, but which does not have direction-seeking ability like that of a compass. It consists of a rotor mounted in a gimbal that is free to rotate in a horizontal plane. The bearings of the gimbal mounting are vertical. The rotor spins on a horizontal axis and is operated either by a stream of air or by an electric current, depending upon the type of instrument. The air-operated directional

Fig. 318. An electrically driven directional gyro. (Courtesy Sperry Gyroscope Company, Inc.)

gyro is connected by means of tubing to a vacuum air supply. The vacuum may be developed either by an engine-driven vacuum pump or by a Venturi tube. The air stream acts against buckets in the rims of the rotor or gyro wheel. The wheel is enclosed within a case which protects it from air currents. Recent types are equipped with an erector system which uses the exhaust air from the rotor shroud to keep the gyro horizontal.

The gyro rotor has a speed of approximately 17,000 r.p.m. When the rotor is spinning, the principle of gyroscopic rigidity maintains the axis of the gyroscope in a fixed direction. The airplane turns around the rotor, gimbal, and vertical-ring assembly within the case. Attached to the vertical ring is a compass card which is read against a lubber line in the window in the front of the instrument. A caging knob is provided by which the instrument may be set on any desired heading or caged in one position to prevent injury during maneuvers which exceed its angular operating limits. When the caging knob is pushed in, the rotor and gimbal-ring assembly is held level while the card is turned to the desired heading. The knob is then pulled out to allow the instrument to function as a gyroscope. The directional gyro will continue to

Fig. 319. A cutaway view of an electrically driven directional gyro: (1) Gyro; (2) gimbal ring; (3) vertical ring; (4) caging mechanism. (Courtesy Sperry Gyroscope Company, Inc.)

indicate the heading within very close tolerances for periods of 15 to 20 min., or longer, even in rough air. It can then be checked with the magnetic compass while the aircraft is in straight flight, and reset if necessary.

This instrument operates on a suction of approximately 4 in. of mercury. Approximately 1.5 cu. ft. of air per minute at sea level pass through the instrument. A Venturi tube will furnish a suction of 4 in. of mercury when mounted in the airplane's slipstream. A single Venturi will only furnish enough suction to operate one gyro horizon or one directional gyro.

The dial markings may be either fluorescent or fluorescent radioactive. An air filter is installed to prevent particles of dust or foreign matter from entering the instrument case.

The electric directional gyro is similar to the air-operated instrument with the exception that the gyroscope is rotated by an electric current. An electrically driven directional gyro is designed for high-altitude flying and gives satisfactory operation at altitudes as great as 40,000 ft. and at temperatures as low as − 58° F. No air flows through this instrument. The gyro is held in a level position by means of an erection system which consists of a leveling switch and a torque motor.

Gyrosyn Compass. The Gyrosyn compass combines the functions of both a magnetic compass and a directional gyro. This instrument is fundamentally a directional gyro with a magnetic sense — a directional gyro

Fig. 320. Gyrosyn compass with repeater and flux valve. (Courtesy Sperry Gyroscope Company, Inc.)

which is synchronized with the earth's magnetic meridian by means of a flux valve. It is a gyro-stabilized compass. The Gyrosyn compass provides a stable directional indication under all conditions of rough air. It is north-seeking in the same sense that a magnetic compass is. It is also without northerly-turning error, oscillation, or swinging; it does not drift off the meridian nor does it require resetting.

The indications on this instrument may be transmitted to a number of compass repeaters in other parts of the airplane. The transmission system is self-synchronous, and the repeaters never require resetting. The readings on the repeating compasses are always the same as those on the Gyrosyn compass itself. The Gyrosyn-compass system consists of the directional indicating instruments mounted on the instrument panel, the flux valve, and as many as six repeater compasses which may be placed at various points in the airplane. The stabilizing gyro element is basically the same as the electrically driven directional gyro.

The flux valve is the direction-finding part of the Gyrosyn compass. The flux valve detects the direction of the lines of force of the earth's magnetic field and transmits this information electrically to a precessing device within the Gyrosyn. This flux valve is small, light in weight, and sealed in an airtight case. It is usually installed at the tip of a wing, at the top of the vertical fin, or in some other part of the aircraft which is located as far as possible from any magnetic disturbances.

The flux valve has no rotating parts. It is mounted as a pendulum free to swing in any direction, within a sealed case so that it normally remains in a horizontal position, regardless of the position of the airplane. It is necessary that the instrument remain horizontal because it obtains its directional indications from the horizontal component of the earth's magnetic field.

In rough air, the flux valve may swing out of the horizontal position or may oscillate under all except smooth air conditions. This causes fluctuating signals which, if used directly, would be useless for directional indications. However, the gyro in the indicating instrument, consisting of a 20-oz. rotor electrically driven at 23,500 r.p.m., reacts at a very slow rate and effectively dampens out all short-period fluctuations. The dampening-out of these short-period fluctuations gives a continuous, accurate reading even in very rough air.

The Gyrosyn receives the signals from the flux valve by means of an electric cable. The Gyrosyn consists primarily of the following elements: a gyroscope which spins about a horizontal axis and provides a stable directional reference; a precession or "slaving" device that applies a torque or twisting force to the inner gimbal ring, which causes the gyro to precess about its vertical axis; an amplifier that is installed to control the slaving device; signal-transformer Selsyn that relays to the amplifier the displacement signal or angular difference between the gyroscope's axis and the direction of the lines of force of the earth's magnetic field; and a Selsyn transmitter that is used for transmitting the headings of the Gyrosyn to the repeater compasses. A small annunciator for indicating the direction of slaving is used to set the Gyrosyn quickly to the magnetic heading for the first synchronization.

The gyro itself is the rotor of a three-phase induction motor. The stator and the rotor are enclosed in the gyro housing. The entire gyro housing is free to swing on horizontal bearings in a vertical gimbal ring. The vertical gimbal ring swings about a vertical axis on bearings in the upper and lower parts of the case. Like all gyro-operated instruments, the gyro element remains stationary, and the airplane turns the case of the instrument about it.

A pair of gears transmit the relative angular rotation from the gyro element to a pointer on the face of the instrument. The dial is mounted on the front of the instrument with the north and south markings at the top and bottom of the face. The east is at the right of the face and the west at the left. As the pointer moves, it indicates both the angle of turn and the change in heading of the airplane.

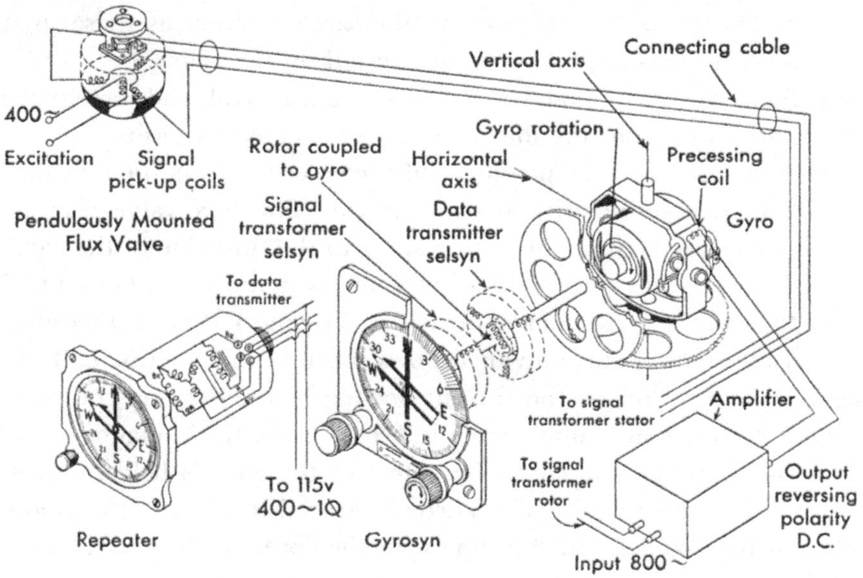

Fig. 321. A schematic drawing to show construction and operation of a gyrosyn compass. (Courtesy Sperry Gyroscope Company, Inc.)

The direction sensed by the flux valve is transmitted electrically to a signal transformer that is mechanically coupled to the gyro. Any difference between the indicated and the magnetic heading is detected by a voltage generated in the rotor of the signal-transformer Selsyn. This voltage is amplified and passed through precession coils which apply a twisting force to the gyro to bring it in line with the magnetic heading. A leveling switch and torque motor are used as a leveling device. The leveling switch and torque motor counteract any tendency of the gyro to leave the horizontal position relative to the gyro case.

The repeaters duplicate the position of the pointer on the Gyrosyn. Each repeater contains three stator coils and a rotor coil that is mechanically connected by a shaft to the repeater pointer. The three coils of the

Selsyn transmitter in the Gyrosyn itself are connected to the corresponding coils in the repeater. The Selsyn rotors are energized by one phase of the 400-cycle supply. The rotor of the transmitter induces a voltage in each of its three coils. These voltages are reproduced in the repeater. If the pointer of the repeater is not in the same position as the pointer of the indicator, a twisting force will be exerted on the rotor as a result of the interaction of the stator and rotor fluxes. This results in a twisting force being applied to the repeater rotor which brings it to the same position as the transmitter rotor. When the magnetic fields set up about the rotor and the stator coils are in the same position as those in the transmitter, no twisting force will be exerted on the rotor.

The flux valve is composed of a core, an exciting coil, and three pickup coils. Lead wires from the four coils are connected to terminals in the top terminal plate. The pendulumlike element is mounted inside a hemispherical case which is sealed airtight. The flux valve acts as a detector, sensing its direction with respect to the direction of the earth's magnetic field. The core of the flux valve is acted upon both by the lines of force of the earth's magnetic field and by the 400-cycle three-phase 115-volt alternating current which flows through the exciting coil. The magnetic field set up by the current flowing in the exciting coil affects the core material and thus serves to "feel" the earth's magnetic field as it passes through the core. The effect of the earth's field induces a current in the pickup coils. This current varies directly with the amount of the earth's magnetic field in line with the core. As the core is turned in the magnetic field of the earth, the current flowing in each pickup coil varies with reference to the direction of the lines of force in the earth's magnetic field.

To operate the Gyrosyn compass, the electric supply is turned on, and the instrument is set to the airplane's approximate heading by turning the set knob at the lower right of the dial. The magnetic course indicator is then set to the magnetic course to be flown by turning the "Set Course" knob at the lower left of the dial. Before making a turn, the course indicator can be set to the desired new heading. The airplane may then be turned until the direction-indicating pointer is again lined up with the course indicator. The repeaters also have course indicators which are used in exactly the same manner as the Gyrosyn compass itself. No initial setting is necessary for the repeaters as they are self-synchronous.

To fly any course, it is only necessary to set the desired magnetic course to be flown and fly in such a direction that the Gyrosyn indicator is between and parallel to the course-indicating pointers.

This instrument operates on a 115-v., three-phase, 400-cycle a-c supply. The power consumption is 25 w. with a power factor of 0.55. A phase adapter may be used which will convert a 115-v., single-phase, 400-cycle power to a 115-v., three-phase, 400-cycle power, thus making possible the operation of three-phase aircraft instruments from a single-phase power supply.

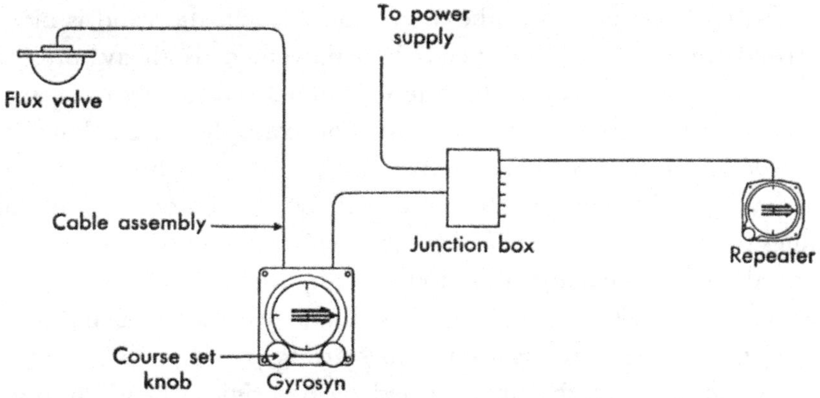

Fig. 322. A diagram to show installation of a gyrosyn compass. (Courtesy Sperry Gyroscope Company, Inc.)

In case the aircraft's power supply is direct current, such as that supplied by a 12-v. or 24-v. battery, a small inverter may be used to change the supply to 115-v., three-phase, 400-cycle alternating current.

Aviation Clocks. An aviation clock is standard equipment along with the navigation instruments. These clocks are designed and constructed to meet the strenuous requirements of aircraft use. Two general types of eight-day clocks are available: the standard aviation clock and the elapsed-time clock. Both of these types have the sweep-second hand required by the Civil Aeronautics Authority when such clocks are used for instrument flying. Most aviation clocks are of heavy construction and built to resist vibration and shock. They are usually rustproof, which is important in damp climates. Accuracy is maintained through all temperature variation by means of special balance wheels and hairsprings. These clocks do not gain or lose time at any stage of winding. Most aviation clocks are not affected by magnetism, either permanent or induced. Wherever possible, nonmagnetic materials are used.

Most aviation clocks are really three clocks in one. These clocks have two small dials on the main dial. The main dial shows standard time, while one small dial shows the elapsed time when using the sweep-second hand in connection with the stop-watch feature. The other small dial shows the elapsed time in minutes for distance computation in navigation problems. Most aviation clocks have luminous markings on the dial which also react to fluorescent lighting or ultraviolet rays.

Drift Meter. A drift meter is an instrument used to measure the angle at which an airplane is being carried off its regular course by the wind. It is also used to figure ground speed. An airplane flying through the air is carried by the wind in the direction in which the wind is blowing at a speed equal to the wind speed. This movement is always present in any aircraft unless the air is absolutely calm and is entirely independent of the forward air speed of the aircraft. For example, an airplane flying with its longitudinal axis constantly along an east-west line with a wind from the north blowing at 30 m.p.h. will be, at the end of one hour, 30 miles south of the east-west line along which it started. If the forward air speed of the airplane was 100 m.p.h., the airplane would have advanced from east to west 100 miles, but it would be 30 miles south of a point 100 miles directly west of its starting point.

The angle between the lines formed by its original heading and the line representing the path it actually followed is called the "drift angle." In order that the airplane follow the true east-and-west line, it would have had to be headed into the wind at an angle equal to the drift angle. With this heading, the airplane would be flying northward at the same rate the wind was blowing it southward. These two movements would neutralize each other and make it possible for the airplane to follow the true east-west line.

The ground speed is the speed at which the airplane actually passes over the surface of the earth. The air speed of the airplane, with constant throttle setting and other things being equal, remains at 100 m.p.h. whether the airplane is flying directly into the wind, across the wind, or down wind. The ground speed varies with the direction of the wind in relation to the airplane's heading. If an airplane with an air speed of 100 m.p.h. is flying directly into a 30-mile wind, the ground speed is 70 m.p.h. If the airplane is flying down wind with an indicated air speed of 100 m.p.h. and the tail wind is equal to 30 m.p.h., the ground speed is 130 m.p.h. As the airplane flies at other angles to the wind direction, the ground speed varies between 70 and 130 m.p.h. The only time that the air speed and ground speed are equal is when the airplane is flying in absolutely calm air.

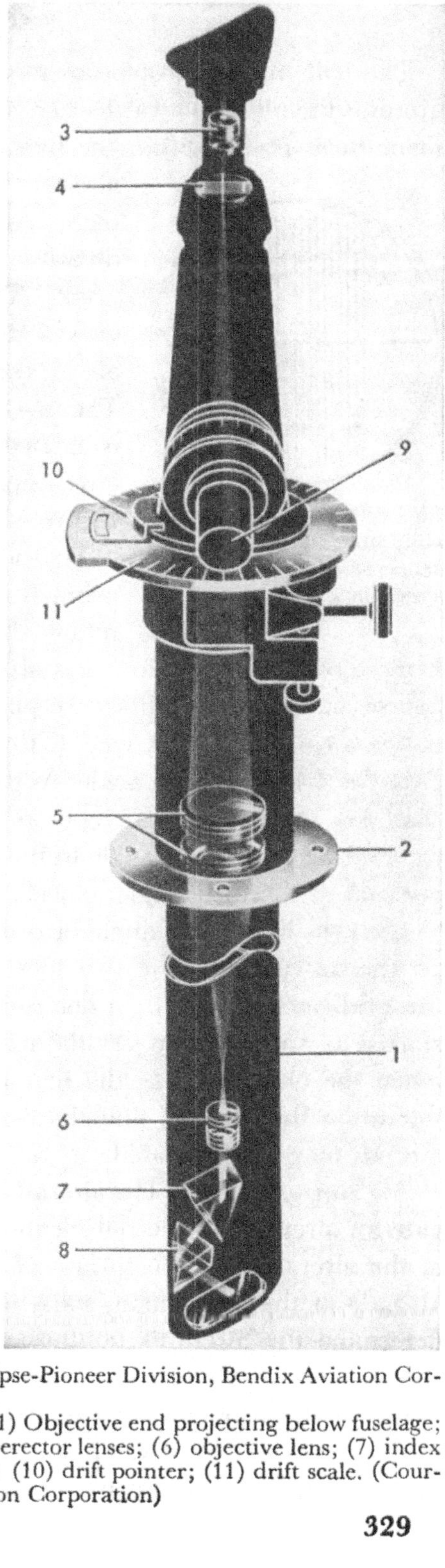

Fig. 323 (*left*). A drift meter. (Courtesy Eclipse-Pioneer Division, Bendix Aviation Corporation)
Fig. 324. A phantom view of a drift meter. (1) Objective end projecting below fuselage; (2) flange; (3) eyepiece lens; (4) reticle; (5) erector lenses; (6) objective lens; (7) index prism; (8) reflecting prism; (9) sight handle; (10) drift pointer; (11) drift scale. (Courtesy Eclipse-Pioneer Division, Bendix Aviation Corporation)

The drift meter must be so mounted that the operator can see the ground directly downward when looking into the eyepiece. The instrument turns freely within the flange which holds it to the body of the aircraft. The optical system consists of an eyepiece lens, a reticle, two erector lenses, an objective lens, an index prism, and a reflecting prism. The rays of light from the objects viewed through this instrument pass through all of the optical elements to the eyepiece. The image is viewed against the reticle which is etched with fore-and-aft lines, a center cross-line, and two speed lines, as shown in Figure 325.

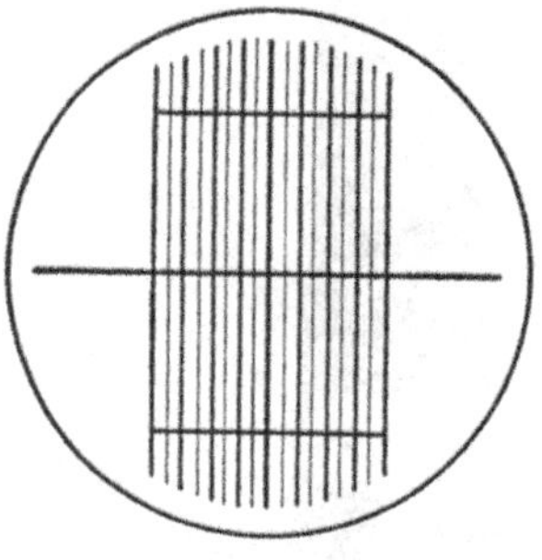

Fig. 325. A reticle etched with fore-and-aft lines, a center cross-line, and two speed lines.

To measure the angle of drift, the index prism is adjusted with the sight handle which rotates the instrument until objects move from front to back along the fore-and-aft lines. If the airplane is drifting off its course and these grid lines are placed parallel to the longitudinal axis of the airplane, objects viewed through the instrument on the ground cross the grid line at an angle. As the instrument is rotated in the proper direction, these objects travel more nearly parallel to the grid lines. With the grid lines set at an angle to the longitudinal axis of the airplane that is equal to the drift angle, objects viewed on the ground move parallel to the grid lines. The angle of drift is read on the drift scale mounted on the instrument. The two speed lines, which are at right angles to the grid lines, are used in determining ground speed. A stop watch is started as an object crosses the leading ground-speed line and stopped when the object crosses the rear ground-speed line. Taking into consideration the altitude and the elapsed time, the ground speed of the aircraft may be calculated.

The Aircraft Sextant. The aircraft sextant is an instrument used to navigate an aircraft by celestial means. It is used to determine the position of the aircraft over the surface of the earth by measuring the angular altitude of the sun, moon, stars, or planets. A sextant, alone, will not determine the aircraft's position; it only measures the angle between the celestial body viewed and the horizontal plane of the earth's surface immediately below the airplane. In addition to the sextant, a navigator requires an accurate watch, an up-to-date air almanac, and proper tables. With this equipment, he may calculate very closely the position

Fig. 326. An aircraft sextant. (Courtesy Eclipse-Pioneer Division, Bendix Aviation Corporation)

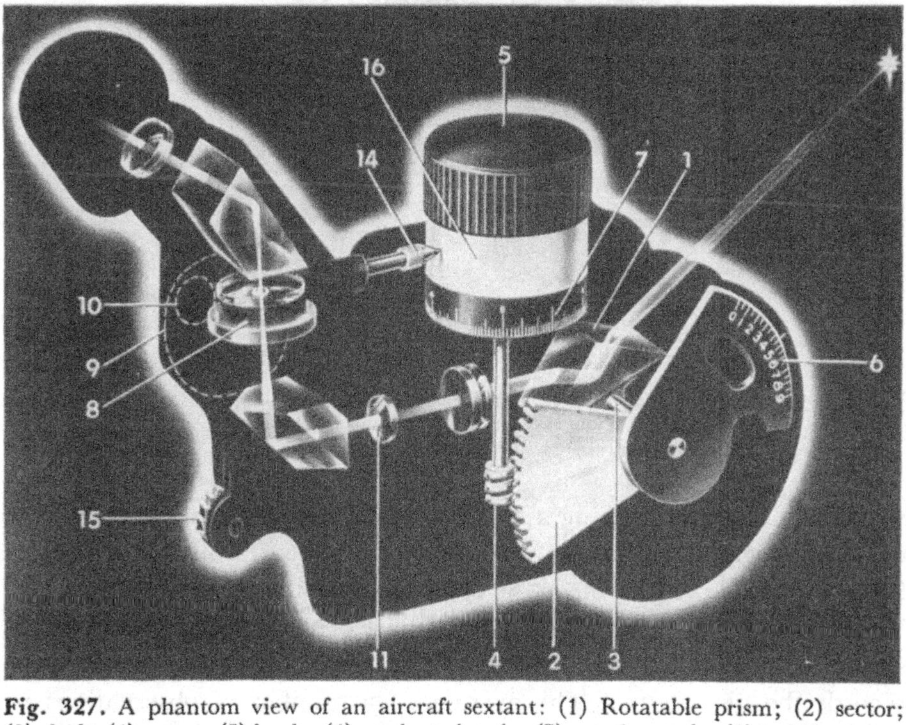

Fig. 327. A phantom view of an aircraft sextant: (1) Rotatable prism; (2) sector; (3) shaft; (4) worm; (5) knob; (6) graduated scale; (7) rotating scale; (8) bubble chamber; (9) diaphragm chamber; (10) knurled knob; (11) astigmatizer; (14) trigger-operated pencil; (15) ratchet; (16) marking surface. (Courtesy Eclipse-Pioneer Division, Bendix Aviation Corporation.)

of the aircraft over the surface of the earth. Accurate navigation depends upon the skill with which the sextant is used.

The body of the sextant contains a prism which may be rotated in such a manner that the object viewed can be made to coincide with a bubble which is seen in the field of view. This bubble, when centered in the field of view, represents the true horizon, or, actually, an artificial horizon which would coincide with the true horizon. After the bubble is formed by turning a knob on the sextant clockwise, a celestial body is sighted and, by turning another knob, its image is brought alongside the bubble. The image of the heavenly body and the bubble are lined

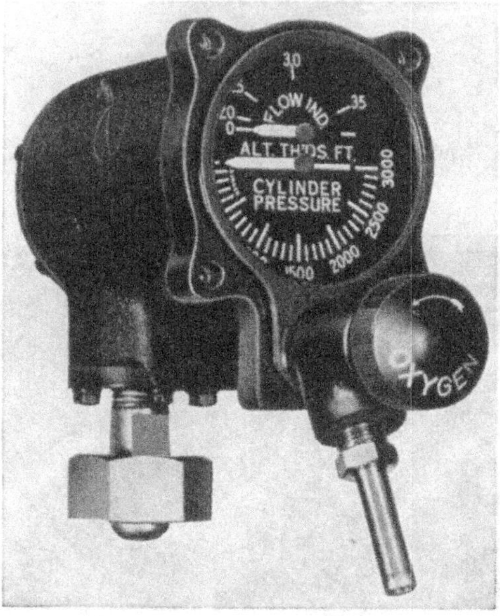

Fig. 328. An oxygen regulator. (Courtesy Eclipse-Pioneer Division, Bendix Aviation Corporation)

up horizontally. By pressing a switch handle, the image of the heavenly body is elongated until it looks as though it passed through the bubble from side to side. Then, the angle of elevation of the heavenly body may be read on the proper dial which is a part of the sextant. The sextant shown in Figure 327 has a trigger arrangement and a pencil which records the observation. The average of a number of observations is usually taken to eliminate errors.

Oxygen Regulator. The oxygen regulator has been classed with aircraft instruments since it is necessary for high-altitude flight. This instrument regulates the flow of oxygen from the supply tank to the mouthpiece or mask on the pilot's face. The pressure in the oxygen tanks may be as great as 3000 p.s.i. This high pressure must be reduced to a low pressure suitable for breathing. The dial of the regulator has two indicating needles: the smaller needle shows the flow of oxygen through the regulator; the larger needle indicates the pressure of the oxygen in the tank. When in use, the pointer is set by opening or closing the needle valve marked "Oxygen" in Figure 328 until the pointer stops at the graduation which reads the altitude in thousands of feet at which the aircraft is to be flown. With the pointer set at 30, the regulator automatically

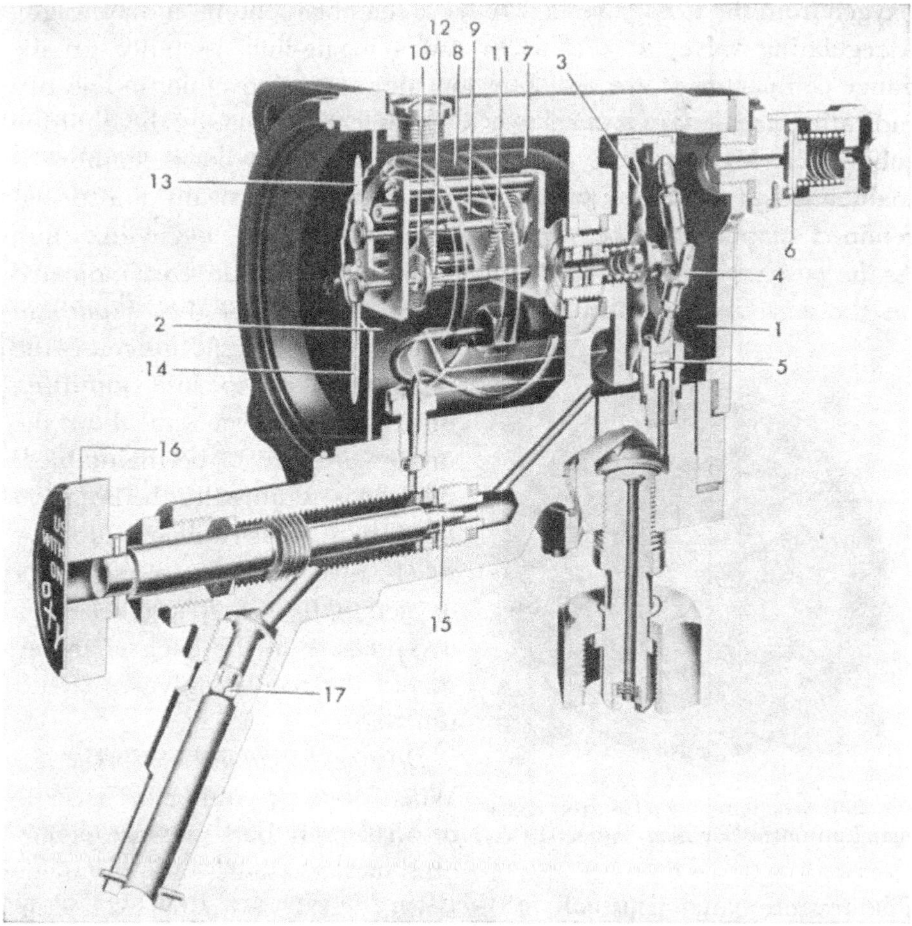

Fig. 329. A cutaway view of an oxygen regulator: (1) Back chamber; (2) front gauge compartment; (3) diaphragm; (4) toggle-link assembly; (5) regulating valve; (6) relief valve; (7) and (8) Bourdon tubes; (9) and (10) sectors; (11) and (12) pinions; (13) flow pointer; (14) tank-pressure pointer; (15) needle valve; (16) knob; (17) calibrated orifice. (Courtesy Eclipse-Pioneer Division, Bendix Aviation Corporation)

allows the correct amount of oxygen to flow to meet the requirements of the normal pilot at an altitude of 30,000 ft.

Because of the decreased pressure at high altitudes, the blood does not receive enough oxygen to supply the needs of the body. At approximately 15,000 ft., the average individual experiences tiredness, headache, and increased heartbeat. At altitudes above 20,000 ft., unless additional oxygen is breathed, the average individual becomes unconscious.

The oxygen regulator is divided into two chambers: the front chamber is the gauge chamber; the back chamber receives the high-pressure oxygen from the storage tank. The back chamber contains a diaphragm, a regulating valve, a relief valve and a toggle-link assembly. In the gauge compartment are two Bourdon tubes and two pinions. The two indicating needles are rotated over their respective scales by the Bourdon tubes. The high-pressure oxygen which enters the back chamber is maintained at a reduced pressure in the chamber by means of a spring-retained diaphragm, the regulating valve, and the toggle-link unit. As the pressure in the gauge chamber builds up to the pressure indicated on the dial by the altitude needle, the movement of the diaphragm attached to the toggle link causes the regulating valve to close, admitting only enough oxygen to allow the proper pressure to be maintained. Oxygen is admitted to the Bourdon tube by a small capillary tube. A relief valve permits the escape of oxygen to the atmosphere, relieving overpressure in the back chamber should the regulating valve fail to function.

Fig. 330. A suction gauge. (Courtesy Kollsman Instrument Division, Square D Co.)

De-icing Pressure and Suction Gauges. When de-icing equipment is used on an aircraft, both pressure gauges and suction gauges are installed. The pressure gauge is usually of the Bourdon type and indicates the air pressure in the de-icing boots. These gauges indicate pressures up to approximately 10 lb. per sq. in. To hold the boots firmly in place when not in use, suction is applied which is measured by a suction gauge, usually of the Bourdon type.

XXIII ENGINE INSTRUMENTS

The earliest aircraft engines were operated entirely without instruments. Many aircraft, after World War I, were equipped with only a tachometer, an oil-temperature gauge, and a water-temperature gauge.

Fig. 331 A typical instrument panel for a twin-motored aircraft showing the automatic pilot installed. (Courtesy Sperry Gyroscope Company.)

These engines were of comparatively low horsepower, and their low compression ratios did not develop the high temperatures of later engines. Most engines were operated at comparatively low altitudes, and supercharging was in its infancy. The throttle and the ignition switch were practically the only engine controls with which the pilot was concerned, although a few engines had a spark advance and a carburetor choke. Modern engine instruments include the tachometer, oil-pressure gauge, fuel-pressure gauge, manifold-pressure gauge, suction gauges, and temperature indicators for oil, air, the carburetor, and the cylinder heads.

Pressure Gauges. Pressure gauges may operate on the principle of the Bourdon tube, the aneroid cell, or by means of a diaphragm.

The Bourdon tube consists of a sealed metal tube which has been pressed into an elliptical cross section shown in Figure 340 and bent into the form of an arc. One end of the tube is fastened securely to the instrument case, while the other end is left free to move. An opening in the fastened end of the tube allows liquid or gas to enter the tube under pressure. Pressure applied within the tube tends to straighten the tube and thus causes the free end to move. The free end of the tube is fastened by means of suitable linkage to an indicating pointer which is rotated over the face of the dial by the movement of the free end of the Bourdon tube.

The aneroid cell is a flattened, metal, disklike cell. The sides of this cell are usually corrugated as shown in Figure 335, and the air has been exhausted from it. When this cell is mounted under spring tension, it becomes extremely sensitive to pressure changes. The cell is connected by means of the proper linkage to an indicating pointer designed to rotate over the face of a gauge.

Fig. 332. An oil-pressure gauge indicator. (Courtesy Electric Auto-Lite Company)

The diaphragm type of pressure gauge has a sealed-in diaphragm which is deflected by pressure against it.

One of the most commonly used instruments which operates on the principle of the Bourdon tube is the oil-pressure gauge. All aircraft engines are lubricated by oil under pressure furnished to the bearings by oil pumps. These pump the oil either from the wet sump within the engine or from a remote oil supply. Each engine lubricating system is designed for a predetermined oil pressure. An oil-pressure-relief valve and surge chamber are commonly installed to prevent excess pressure or sudden changes in pressure in the oil lines. The oil-pressure gauge is usually a simple gauge of the Bourdon-tube type and is usually calibrated for pressures from 0 to approximately 200 lb. of pressure. The surge chamber, which is a closed chamber in which air is trapped, prevents violent oscillation of the needle because of the action of the pump. Heavy oils, when cooled, may show excess pressure. Gauge readings should not be depended upon until the oil has reached the correct operating temperature. On engines which operate under comparatively low oil pressures, the surge chamber is not as important as it is on engines which operate under high oil pressures. A comparatively low oil pressure might be approximately 60 lb. per sq. in., while high oil pressures are in excess of 100 lb. per sq. in.

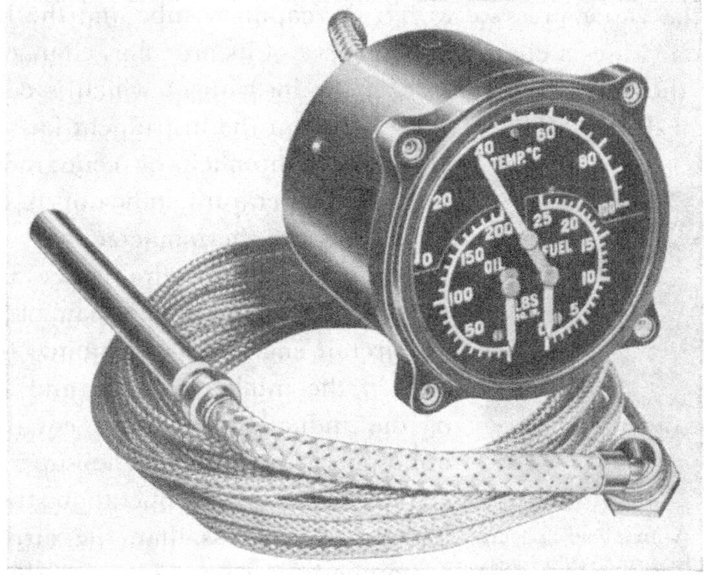

Fig. 333. An engine gauge unit indicating oil pressure, fuel pressure, and temperature. (Courtesy Electric Auto-Lite Company)

Most fuel-pressure gauges operate on the Bourdon-tube principle and are calibrated from 0 up to approximately 25 lb. per sq. in. When fuel-pressure gauges are used with pressure discharge carburetors, they are usually calibrated up to 25 lb. per sq. in., while gauges used on other types of carburetors have ranges from 0 to 10 lb. per sq. in.

An engine-gauge unit is simply a unit in which three instruments are combined. This instrument shows fuel pressure, oil pressure, and oil temperature. The oil-pressure and fuel-pressure instruments are the ordinary Bourdon-tube type with the fuel pressure graduated from 0 to 10 lb. per sq. in. and the oil pressure graduated from 0 to 200 lb.

per sq. in. The oil temperature is graduated from 0 to 100° C. Each of the units operates entirely separately from the others. The thermometer unit is the vapor-pressure type. The fuel-pressure and oil-pressure, Bourdon-tube gauges are operated by the actual pressure of the fuel and oil on the Bourdon tube.

The vapor-pressure type of Bourdon-tube instrument consists of the Bourdon tube which is connected by a capillary tube that may be as much as 30 ft. in length with a cell containing a highly volatile liquid. The cell and the Bourdon tube are hermetically sealed to opposite ends of the capillary tube, and the cell is placed at the point where the temperature is to be measured. Any change in the temperature of the cell changes the vapor pressure within the capillary tube and the Bourdon tube which causes a change in the curve of its arc. This change is indicated by the needle on the face of the instrument which is calibrated in terms of degrees. The degrees shown on the instrument face may be either in Fahrenheit or centigrade. This type of temperature indicator is called a vapor-pressure thermometer.

Fig. 334. A manifold-pressure gauge indicator. (Courtesy Eclipse-Pioneer Division, Bendix Aviation Corporation)

The manifold-pressure gauge indicates the pressure in the intake manifold. When an aircraft engine is not running, the pressure in the intake manifold and all parts of the induction system is equal to the atmospheric pressure. In a nonsupercharged engine, when it is in operation, the manifold pressure is less than the surrounding atmospheric pressure, because of the suction caused by the air-fuel mixture being drawn into the cylinder. The particular function of the supercharger is to increase the pressure within the manifold until it equals or exceeds normal sea-level pressure. The manifold-pressure gauge is usually graduated from 10 to 50, the figures indicating inches of mercury pressure.

The design of the combustion chamber in the engine determines the compression ratio which is most effective for the operation of the engine. The atmosphere grows less dense with altitude so that a nonsupercharged engine can get less air into the cylinders at greater altitudes. As the density of the air decreases, the cylinder is at a lower pressure and does not contain as much air by weight. Supercharging builds up the pressure by forcing more air into the induction system. By proper supercharging, the manifold pressure may be maintained at 30 in. of mercury even at high altitudes. Since the horsepower of the engine depends upon the amount of oxygen burned with the gasoline in the combustion chamber, it will develop full horsepower only when the manifold pressure is maintained at 30 in. Without a manifold-pressure gauge, the pilot cannot know the exact pressure within the manifold. A supercharged

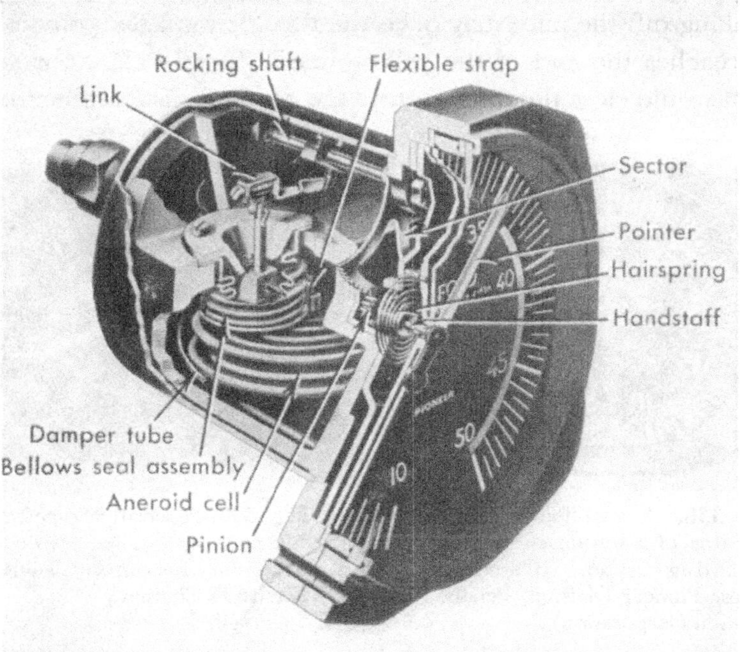

Fig. 335. A cutaway view of a manifold-pressure gauge showing its construction. (Courtesy Eclipse-Pioneer Division, Bendix Aviation Corporation)

engine which would develop its maximum horsepower at 10,000 or 12,000 ft. at full throttle would, if the supercharger were left in operation, be seriously damaged by the increased compression near sea level. With the supercharger in operation, the pilot, by watching the indication of the manifold-pressure gauge, reduces the manifold pressure simply by closing the throttle until the needle indicates 30 or the desired pressure.

Engines may be operated at higher than 30 in. of mercury for short periods of time, such as during take-off and climb, and the pilot must know how far the throttle may be opened with safety. For example, on the take-off, the engine manufacturer may recommend as much as

35 or 40 in. of manifold pressure. It is only by means of the manifold-pressure gauge that the pilot can tell just how far to open the throttle to obtain this pressure. Usually, the normal operating range is indicated on the instrument by a green band running from 30 to 36. A high-boost operating range is indicated by a yellow band extending from 36 to 42 on the instrument. Above 42 is a red band which indicates dangerously high pressures. The range covered by these bands varies with different engines in accordance with the manufacturer's specification.

In taking off, the pilot may open the throttle until the pointer on the gauge reaches the end of the yellow band. Shortly after the take-off, the pilot would close the throttle until the needle indicates the maximum

Fig. 336. A manifold-pressure indicator of an Autosyn remote indicating type. (Courtesy Eclipse-Pioneer Division, Bendix Aviation Corporation)

Fig. 337. A sensitive type manifold-pressure gauge. (Courtesy Kollsman Instrument Division, Square D Company)

manifold pressure for climbing. As he approaches his cruising altitude, he gradually throttles back until the needle indicates the correct cruising pressure. The manifold-pressure gauge operates by means of an aneroid cell or a double metal diaphragm. The instrument is equipped with a two-cell diaphragm. The linkage system starts with a flexible shaft which connects the diaphragm to the bellows-seal assembly. A link ties the bellows to a rocking shaft which is connected with a pinion and a hand staff and pointer. A hairspring is installed to keep all parts of the linkage snugly in position. A sealed chamber, in which the diaphragms are placed, is connected with the manifold by an airtight tube. If the pressure in the manifold exceeds 30 in. of mercury, the diaphragms start to collapse and this motion is transmitted through the linkage to the indicator needle which indicates pressures above 30 in. of mercury. If the pressure within the manifold becomes less than 30 in. of mercury, the diaphragms expand and the needle indicates pressures below 30.

It is necessary that the pilot be warned immediately of any decrease in pressure in the oil or fuel system. It is not possible for the pilot to watch all of the gauges continuously, and the oil or fuel gauge indication may drop suddenly and remain in this position a considerable length of time before the pilot notices it, unless some definite warning is given. Pressure-warning units are designed either to operate warning lights or to sound signals or do both to warn the pilot or flight engineer of this condition. This relieves the person operating an aircraft of the strain of continually watching instruments during flight. These instruments are designed to attract the attention of the operator as soon as there is any pressure drop below a minimum, safe operating value.

Fig. 338. An oil-pressure-warning unit. (Courtesy Eclipse-Pioneer Division, Bendix Aviation Corporation)

Fig. 339. A cutaway view of an oil-pressure-warning unit showing its construction. (Courtesy Eclipse-Pioneer-Division, Bendix Aviation Corporation)

Warning units are installed in the modern airplane to indicate a number of different conditions. The pressure-warning units, of course, indicate low or high pressures.

A fuel-pressure-warning unit consists of a fuel chamber, a pressure-sensitive diaphragm, a bellows, and two contacts. As the pressure drops below a predetermined value, two contacts are closed, completing an electric circuit which lights a warning light on the instrument panel. The bellows prevents the gasoline in the fuel chamber from reaching the electrical contacts should the diaphragm, by any chance, spring a leak. The fuel-pressure unit is operated by the fuel pressure which enters the instrument from the fuel line. The fuel enters the fuel chamber, one wall of which is a diaphragm. The diaphragm expands under the pressure of the liquid and, by pressing the electrical contacts apart, opens

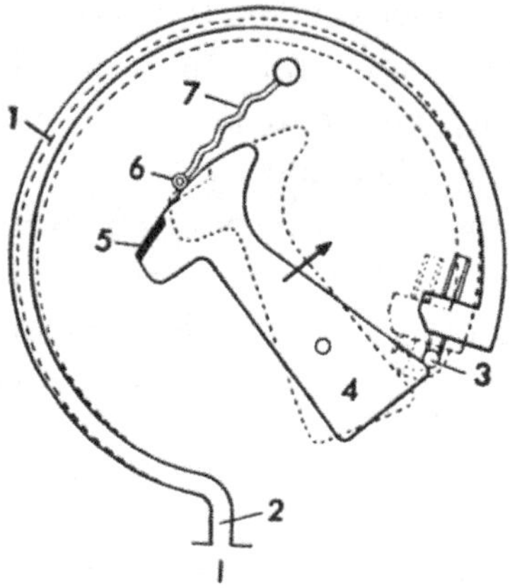

Fig. 340. A diagram showing the operation of a Bourdon tube. (1) the Bourdon tube; (2) Bourdon tube connector; (3) adjusting pin attached to end of Bourdon tube; (4) commutator assembly; (5) contact surface; (6) contact part; (7) wiping spring which closes circuit. (Courtesy Eclipse-Pioneer Division, Bendix Aviation Corporation)

the electric circuit. If the fuel pressure falls, the diaphragm contracts. Any movement of the diaphragm is followed by the bellows. The lever, to which is attached the multiple contact, moves with the bellows. As

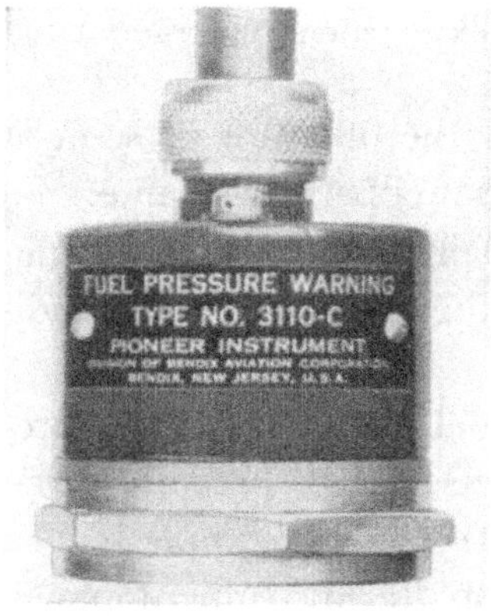

Fig. 341. A fuel-pressure-warning unit. (Courtesy Eclipse-Pioneer Division, Bendix Aviation Corporation)

Fig. 342. A cutaway view of a fuel-pressure-warning unit, showing its construction. (Courtesy Eclipse-Pioneer Division, Bendix Aviation Corporation)

the pressure reaches the predetermined safe minimum, the contact closes and causes a lamp on the instrument panel to light, a horn to sound, or both. It is possible for this type of gauge to give warning of

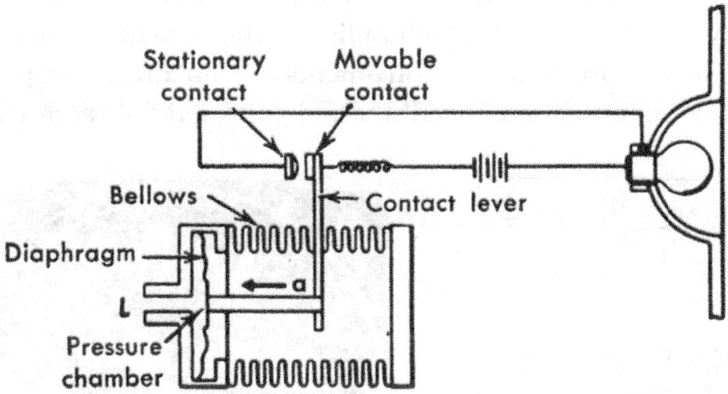

Fig. 343. A drawing showing the operation of a fuel-pressure warning unit.

excessive pressure by use of a double contact. One contact closes with low pressure, and the other closes with high pressure, the contact point being carried by the expanding and contracting bellows.

The oil-pressure-warning unit is a Bourdon tube which is normally extended by the oil pressure which holds the electrical contacts apart. As the pressure falls, the Bourdon tube contracts and closes the contacts, which turns on a light on the instrument panel or sounds a warning. A switch opens the electric circuit to prevent the light from burning when the engine is not in operation.

Fig. 344. An oil-pressure indicator. (Courtesy AC Spark Plug Division, General Motors Corporation)

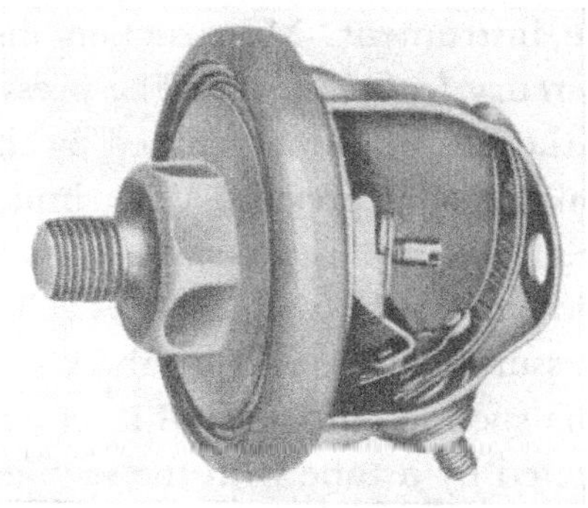

Fig. 345. An oil gauge engine unit which operates on a variable electrical resistance principle. (Courtesy AC Spark Plug Division, General Motors Corporation)

Many aircraft instruments, including the bank-and-turn indicator, contain a gyroscope which is driven by a stream of air produced by suction. This suction may be developed either by a vacuum pump, a Venturi tube, or the suction in the induction system of the engine. It is necessary to have a gauge which indicates the amount of suction being developed to be sure that the instruments are operating properly. The suction gauge consists of a metal aneroid cell or diaphragm connected

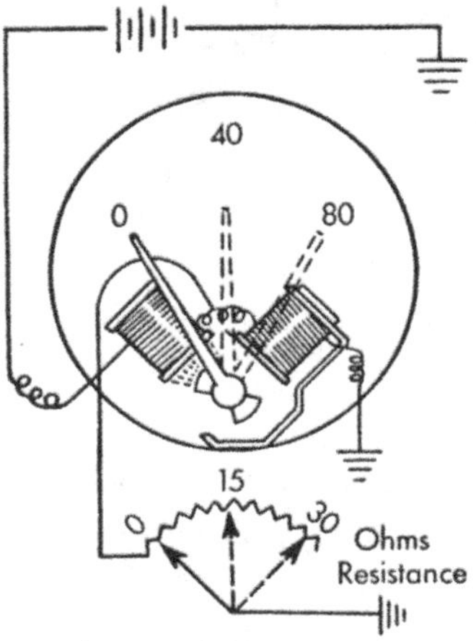

Fig. 346. A phantom view of an oil gauge indicator used in connection with a variable resistance type engine unit. (Courtesy AC Spark Plug Division, General Motors Corporation)

Fig. 347. A suction gauge which indicates suction in inches of mercury. (Courtesy Eclipse-Pioneer Division, Bendix Aviation Corporation)

with the proper linkage to move the indicator needle over the face of the instrument. Most suction instruments are graduated in inches of mercury from 0 to 10. The pressure-sensitive diaphragm cell measures variations in suction caused by the expansion of the cell under reduced conditions of pressure. It is similar to an altimeter working under artificial changes in pressure. The altimeter cell expands as it passes from a low to a high altitude. The cell in the suction gauge expands when the pressure is decreased by the vacuum pump or other sources of suction. The suction-gauging cell is, of course, mounted in an airtight case connected by a tube with the suction-producing mechanism.

Tachometers. Tachometers are instruments which indicate engine r.p.m. They are operated by centrifugal force, electric or magnetic forces, or by a timing arrangement such as that used in a watch.

It is necessary to have a tachometer for each engine. The tachometer informs the pilot whether or not the engine is functioning properly. For instance, before taking off, the engine is run up a short time at full power to determine whether or not it is developing its required r.p.m. The tachometer is also necessary to determine whether or not the engine is being operated within its proper speed limits. The tachometer illustrated in Figure 349 is calibrated from 500 to 2500 r.p.m. Other instruments may be calibrated within different ranges. When used in con-

Fig. 348. A cutaway view of a suction gauge showing its construction. Note aneroid cell. (Courtesy Eclipse-Pioneer Division, Bendix Aviation Corporation)

Fig. 349. A centrifugal tachometer. (Courtesy Eclipse-Pioneer Division, Bendix Aviation Corporation)

junction with a controllable-pitch or constant-speed propeller on a supercharged engine, the tachometer reading is combined with that of the manifold-pressure gauge and the propeller pitch to arrive at the power developed by the engine.

The centrifugal tachometer consists of a flyweight assembly connected to a linkage system which moves the pointer over the face of the instrument to indicate the engine r.p.m. The flyweight mechanism consists of three weights which are connected by links to a top and a bottom collar. This tachometer operates by means of these whirling weights. The rotating part is called a "governor." The three weights are pivoted at each end, allowing them to move outward when acted upon by centrifugal force. As the shaft spins, these weights move outward. The more rapidly the shaft spins, the farther the weights move. As the weights move outward, the bottom collar slides upward and squeezes the governor spring together. The distance which the bottom collar moves as it compresses the spring measures the engine speed. This motion is transmitted to a pointer indicating the r.p.m. A hairspring maintains tension on the linkage, forcing the indicator to follow all movements of the linkage. As the engine speed decreases, the centrifugal force becomes less and the governor spring pulls the weights back toward the shaft. This action allows the governor spring to lengthen and, as the bottom collar moves downward under the pressure of the spring, lower r.p.m. are indicated.

Fig. 350. A cutaway view of a centrifugal tachometer showing its construction. (Courtesy Eclipse-Pioneer Division, Bendix Aviation Corporation)

The electric tachometer is used in the same manner as the centrifugal tachometer. While the centrifugal tachometer is usually used on single-motor installations where the flexible shaft is comparatively short, the electric tachometer is used on multiengine installations, and the shaft is replaced by electric wires. It is necessary to have one tachometer for each engine and, in large multiengine aircraft, the length and weight of the flexible cable necessary for a centrifugal tachometer becomes excessive. On most four-engined aircraft there is one tachometer for each engine in the pilot's cockpit and another at the flight engineer's station, so that the aircraft has eight tachometers. Such installations are much simpler with electric wires than with long flexible shafts.

The electric tachometer consists of two widely separated units. One unit is an alternator which is mounted on the engine itself and driven by the engine. The other unit is the tachometer indicator which is mounted on the instrument panel. The panel unit contains a synchronous motor which receives its power and speed-controlling frequency from the alternator by means of wire connections. The synchronous motor driven by the alternator, in turn, controls the tachometer indicating mechanism. The alternator consists of a stator and a rotor. This unit is connected to the tachometer drive shaft of the airplane engine. As the

Fig. 351. Two electric tachometer indicators. (Courtesy Eclipse-Pioneer Division, Bendix Aviation Corporation)

engine rotates, the alternator drive shaft rotates with it and thus spins the rotor. The rotor, revolving inside the stator coil, generates an alternating current, the frequency of which varies directly with the r.p.m. of the engine. Electric wires conduct this current to the synchronous motor mounted in the instrument panel. This motor, which operates the r.p.m. indicator, matches the speed of the alternator on the engine. The self-synchronous motor also contains a stator coil and a rotor. A small permanent magnet and its return path rotate at the same speed as the rotor. The function of the whole system is to spin the magnet assembly at a speed equal to that of the tachometer drive.

The rest of the mechanism contained in the tachometer converts the motion of the magnet to a pointer indication of the engine r.p.m. The indicator mechanism consists of a drag cup which is mounted on the shaft connected to the pointer. A hairspring is mounted between the pointer and the drag cup. The drag-cup wall is between the magnet and its return path. There is no mechanical connection between the drag cup and the other two units. Lines of force, set up by the field of the magnet, loop from the magnet to its return path and back to the magnet. These lines of force leave the magnet at its north pole and re-enter the magnet at its south pole, thus completing a magnetic circuit. As the magnet spins, these lines of force rotating with it sweep through

Fig. 352. A cutaway view of an electric tachometer showing its construction. (Courtesy Eclipse-Pioneer Division, Bendix Aviation Corporation)

the metal wall of the drag cup. A magnetic field moving through metal sets up eddy currents in the metal. These induced eddy currents in the drag cup produce a rotative force causing the drag cup to follow the magnet around. The purpose of the hairspring is to resist the rotation of the drag cup. These two forces are balanced in such a way as to provide a pointer deflection which indicates the r.p.m. of the engine. The faster the engine rotates, the faster the magnet spins. As the magnet spins faster and faster, the eddy currents set up in the drag cup become greater, and the drag cup twists with a greater force against the hairspring. This causes the hairspring to be twisted into a tighter spiral permitting the pointer on the dial to move to indicate the higher r.p.m. of the engine.

A chronometric tachometer is an instrument used to indicate engine r.p.m. The instrument may be used in the same manner as other tachometers except that it is usually limited to single-engine airplanes because it is driven by a shaft. Standard types of chronometric tachometers usually have a range of from 0 to 3500 r.p.m. This type of tachometer is designed to rotate at one half the crankshaft speed and is provided with a reversing mechanism to prevent damage due to engine kickback and allow the drive unit to be driven in either direction.

There are four principal parts of the chronometric tachometer: (1) the driving mechanism, (2) the counting mechanism, (3) the watch mechanism, and (4) the synchronizing cams which time the counting arrangement. The watch mechanism which times the action of the synchronizing cams and the counting mechanism which counts the revolutions during one-second periods are connected with a pointer gear which shows the results on the dial. The chronometric tachometer actually totals the revolutions which occur during each alternate second. These revolutions are automatically measured by a special watch escapement. A chain of gears, driven by the shaft from the engine, operates the escapement cam and counting system. The escapement cam is operated by a friction drive which causes a gear to become meshed with the counting gear intermittently for one-second intervals. During the one second that these gears are in mesh, the large gear rotates a distance proportional to the speed of the drive system. This movement is transmitted to the pointer gear which rotates the indicator hand through the same distance as the counting gear. At the end of the second, the counting gear is disconnected and returned by a spring to its starting position. The indicator hand is held stationary during this interval. If, during the next second, the engine has run faster than before, the counter gear pushes the indicator pointer farther around the dial. If the engine has run more slowly than it did during the preceding second, the pointer is released and a spring drops it back to the position of the counter gear at the end of the second. The pointer moves by jerks, and the indicated reading at any instant is the speed of the engine during the previous second. This cycle of operation continues as long as the drive shaft continues to rotate. After the engine is stopped, a ticking may be heard which is caused by the escapement continuing to run due to the tension of the main spring until the spring loses its stored-up energy. This ticking may last as long as 30 sec.

Self-Synchronous Instruments. Self-synchronous instruments are usually of the remote-indicating type. The transmitting part of the instrument transmits the desired information to a remotely located instrument panel. These instruments may indicate such conditions as fuel pressure, oil pressure, manifold pressure, engine r.p.m., landing-gear position, tail-wheel position, flap position, temperatures, liquid flow, or fuel levels. By means of selector switches, one instrument on the panel may be used in connection with several transmitters. For example, the temperature of four engines may be indicated on a single instrument by using a selector switch in one of four positions. Some indicators may be equipped with more than one indicating needle; one needle indicates a condition in the right engine, while the other needle indicates the corresponding condition in the left engine.

There are several, satisfactory, self-synchronizing remote-indicating systems in common use. The trade names of three systems are the Selsyn, the Telegon, and the Autosyn. The operation of these various systems is similar, but they are designed to operate on different electric currents. The Selsyn systems are designed to operate from a standard 12- or 24-volt storage battery; the Telegon instruments are designed to operate on a 110-volt alternating current; while the Autosyn instruments are designed to operate on a 26-volt, 400-cycle current. Some of these instruments that require a higher voltage or a different cycle of alternating current may, by means of an adapter, use currents from batteries or from a source having a different phase from that desired.

Fig. 353. An electric indicator tachometer of the sensitive type. (Courtesy Kollsman Instrument Division, Square D Company)

The Autosyn system is a system of remote indication. This system makes it possible to transmit to the pointer of an instrument on the instrument panel the actions of a distantly placed measuring instrument or mechanism. The system is composed of two units, a transmitter and a receiver or indicator. Electric wiring joins these two basic points of the Autosyn remote-indicating system. Both transmitter and indicator units contain an Autosyn. These instruments, while they may be used on small aircraft, are more generally used on large multiengine aircraft where considerable distance separates the point where the measurement is to be made and the instrument panel. The transmitter contains a measuring device and an Autosyn. An Autosyn indicator consists of an Autosyn plus a pointer. The pointer is attached to the Autosyn rotor shaft which rotates it about the instrument dial by means of the shaft. The Autosyn represents an adaptation of the synchronous motor principle. The self-synchronous motor principle consists of two separated motors which operate in exact timing with each other. The rotor of one motor turns exactly the same distance as the rotor of the other motor. In the Autosyn, the rotors neither spin nor produce power. The electrical design of the Autosyn is such that the rotors of two connected Autosyns match each other's position when energized by means of an electric current. The rotor of one Autosyn moves only the distance necessary to match any movement of the rotor of the first Autosyn.

Transmitter and indicator Autosyns are much alike, both in construction and in electrical characteristics. Each has a rotor and a stator. When a rotor is energized by an alternating electric current, a transformer action causes three distinct voltages to be induced in the secondary stator windings. The values of the voltage vary with the position of the rotor in relation to the stator. Any change in the position of the rotor causes a new and completely different combination of the three voltages which are induced. When two Autosyns are connected, the rotors of both Autosyns occupy the same position in relation to their stators. Both sets of induced voltages are then equal and opposite, and no current flows in the interconnected stator leads. Under this condition both rotors remain stationary. When the two rotors do not have the same position in relation to their stators, the voltages in the two stators are not alike. When the voltages are not alike, rotation of the second rotor occurs and continues until both rotors are in identical positions in their electric fields. When they are in the same relative position, the induced voltages are in balance and the rotation stops. When a measuring element is used to turn a rotor of a transmitter Autosyn to a certain position, the rotor of the indicator Autosyn takes the same position in its field. The indicator Autosyn remains fixed until a change takes place in the transmitter. The Autosyn system moves the indicating pointer of an aircraft instrument by direct action. The measuring part of the transmitter unit is connected to any device which will cause the rotor to move, indicating the position of the mechanism. The measuring device may be a Bourdon tube, a pressure diaphragm, a flyweight governor, or any type of prime mover. Instead of moving the pointer by means of a linkage system, the gauge element is moved by the indicating rotor. The movement of the transmitter rotor causes a movement of the Autosyn indicator rotor which, in turn, moves the pointer. Two or more indica-

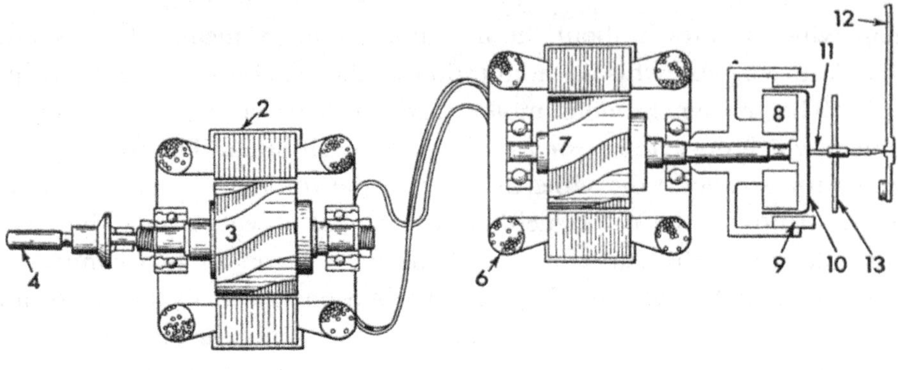

Fig. 354. A diagram showing the operation of an electric tachometer: (1) The alternator unit; (2) stator; (3) rotor; (4) alternator drive shaft; (5) a synchronous motor; (6) stator coil; (7) synchronous motor rotor; (8) permanent magnet; (9) magnet's return path; (10) drag cup; (11) staff; (12) pointer; (13) hairspring. (Courtesy Eclipse-Pioneer Division, Bendix Aviation Corporation)

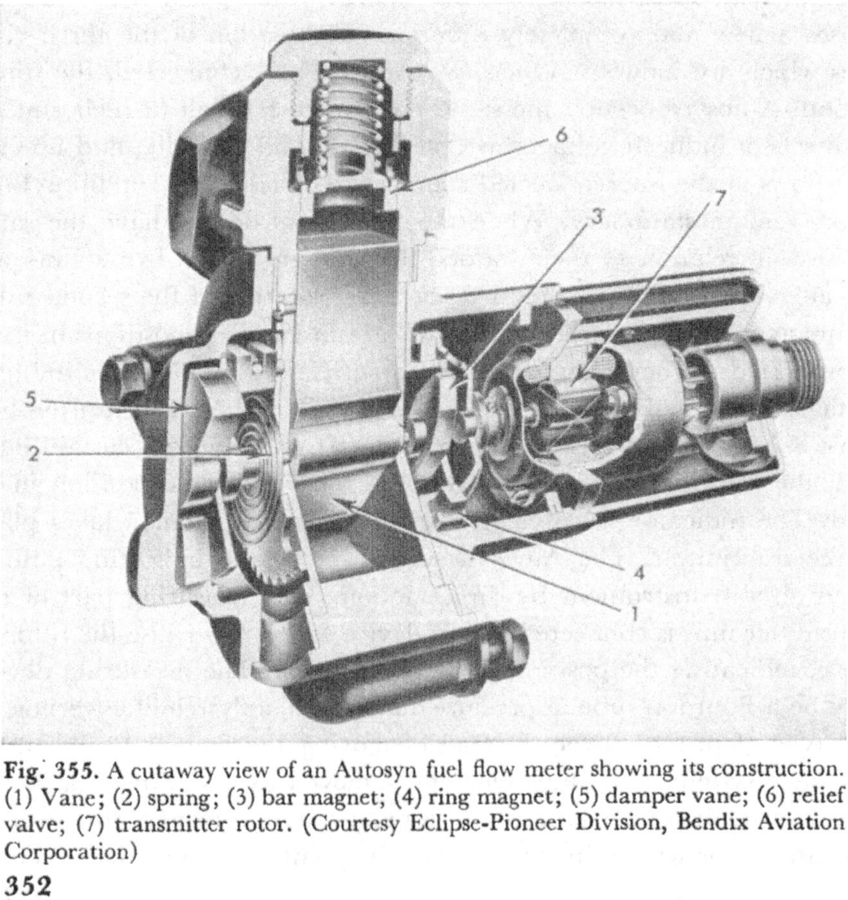

Fig. 355. A cutaway view of an Autosyn fuel flow meter showing its construction. (1) Vane; (2) spring; (3) bar magnet; (4) ring magnet; (5) damper vane; (6) relief valve; (7) transmitter rotor. (Courtesy Eclipse-Pioneer Division, Bendix Aviation Corporation)

tions may be shown on a single dial. The Autosyn system may be used to show oil pressure, manifold pressure, oil temperature, tachometer readings, fuel pressure, fuel flow, fuel quantity, position of landing wheels, position of flaps, or position of tail wheels.

Fig. 356. A fuel flow indicator for a twin-engine aircraft which operates on the Autosyn principle. (Courtesy Eclipse-Pioneer Division, Bendix Aviation Corporation)

The Autosyn flow meter consists of two units. One is the transmitter, and the other is the detector or indicator. The two units may be widely separated and are connected by electric wires. This instrument indicates at all times the rate at which gasoline flows from the tanks to the engine. Some instruments indicate how many gallons per hour are flowing, but most instruments indicate the number of pounds per hour of gasoline that the engine is using.

Flow meters are primarily used to get the most economical performance of the engines. An overrich mixture is indicated by a higher rate

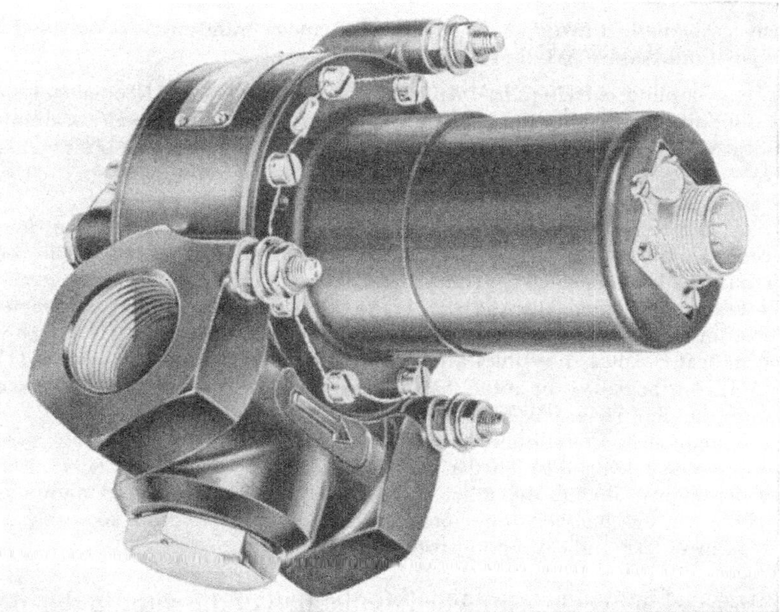

Fig. 357. An Autosyn fuel flow meter. (Courtesy Eclipse-Pioneer Division, Bendix Aviation Corporation)

of flow than should be used to develop the indicated horsepower. This is corrected by adjusting the mixture controls. The transmitter unit is mounted between the fuel tank and the carburetor on the engine side of the fuel pump. The gasoline flowing to the engine passes through the metering chamber in the transmitter. In the metering chamber of the transmitter is a movable vane. This vane extends into the stream of gasoline flowing through the transmitter. The more rapid the flow of gas, the greater the distance the vane is moved out of the neutral position. The vane moves the rotor in the transmitter. This movement of the rotor causes the rotor in the indicating unit to place itself in the same relative position to its stator as the rotor in the transmitter is to its stator.

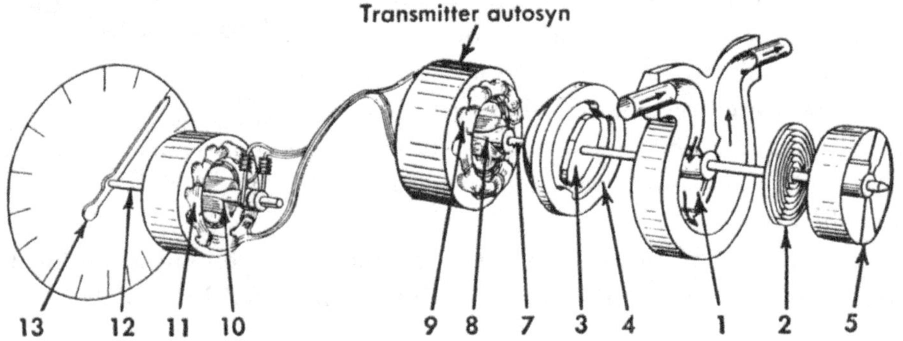

Fig. 358. A schematic drawing of the Autosyn flow meter transmitter. (Courtesy Eclipse-Pioneer Division, Bendix Aviation Corporation)

OPERATION. Gasoline entering the inlet port strikes the vane (1). Impact of the fuel plus pressure drop across the vane, causes the vane to move against the restraining force of a calibrated spring (2). The two forces are so balanced that any static position assumed by the vane represents a measure of the rate at which fuel is passing through the metering chamber.

The bar magnet (3), attached to the vane shaft, moves with the vane. The ring magnet (4) repeats motions made by the bar magnet, since the two are magnetically coupled. The ring magnet is attached to the rotor shaft (7) of the transmitter Autosyn; therefore, motions of the ring magnet cause the rotor (8) to change position relative to the stator (9).

In accordance with the Autosyn principle of operation, the rotor (1) of the panel-mounted indicator Autosyn assumes an identical position relative to its stator (11). The pointer (13) is attached to the rotor shaft (12) of the indicator Autosyn. Motion of this rotor causes the pointer to deflect on a calibrated dial.

The damper vane (5) cushions wide, misleading fluctuations sometimes caused by large air bubbles passing through the metering chamber. Should the flow exceed the capacity of the instrument, the relief valve automatically opens and allows fuel to by-pass the metering chamber, and closes again when the flow drops to within the instrument's range. (10) Indicating unit rotor.

The indicating needle is attached to the shaft of the rotor in the indicating unit. As in all Autosyn units, the rotor in the indicator matches the position of the rotor in the transmitter. As the pivoted vane in the metering chamber moves back and forth, moving the rotor in the transmitter, the rotor in the indicating unit matches this movement. The farther the pivoted vane moves, the farther the indicating needle moves over the face of the indicator dial.

Temperature Indicators or Thermometers. Temperature is indicated by means of a thermometer. Any device for indicating temperature may be classed as a thermometer.

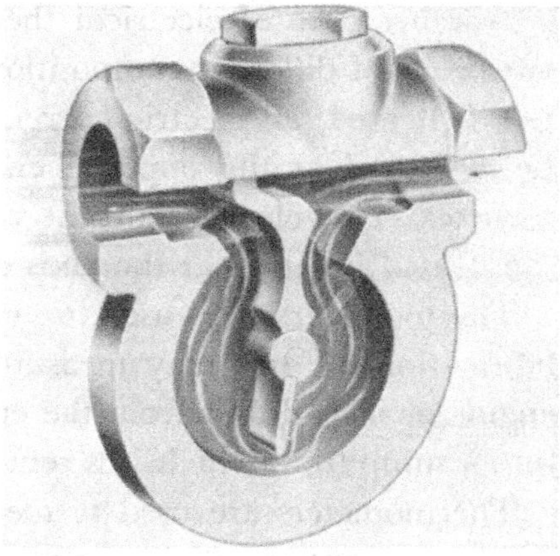

Fig. 359. The vane of an Autosyn fuel flow meter. (Courtesy Eclipse-Pioneer Division, Bendix Aviation Corporation)

The ordinary thermometer depends upon the expansion of a liquid contained in a bulb which is connected to a capillary tube. The capillary tube is graduated to indicate temperature degrees. Changes in temperature of the liquid in the bulb cause the liquid to expand and contract, extending or retracting the column of liquid in the capillary tube.

Vapor-pressure thermometers depend upon changes in vapor pressure within a sealed system, due to changes in temperature.

Fig. 360. A vapor pressure thermometer. (Courtesy Electric Auto-Lite Company)

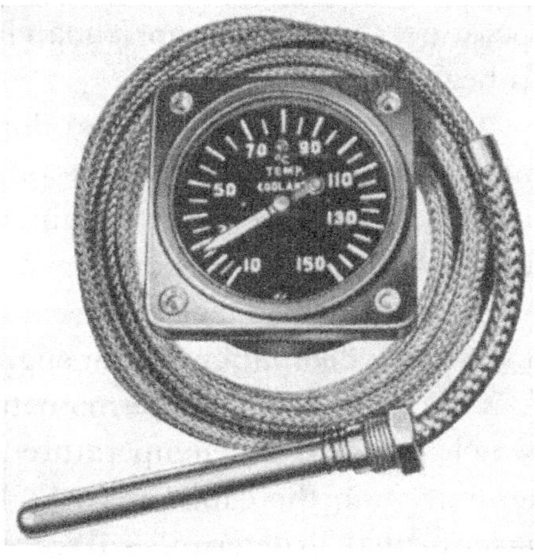

Fig. 361. A vapor pressure coolant thermometer. (Courtesy Electric Auto-Lite Company)

Electrical thermometers usually depend upon the variation in electrical conductivity of a conductor due to changes in temperature. The electrical conductivity varies inversely with the temperature. Most conductors carry less current at high temperatures than they carry at low temperatures.

Another type of electrical thermometer is the thermocouple. When two wires of different composition are joined at one end and this junction is heated, an electric current is generated in the wires which may be measured at the opposite ends of the wires by means of a galvanometer. This electric current varies directly with the temperature of the heated junction at the ends of the wires.

Thermometers are used to determine the temperature of engine lubricating oil and may measure the oil temperature as it enters the engine or as it drains from the engine, either into a crankcase sump or into a sump from which it is removed for return to the storage tank.

Thermometers are used to measure the temperature of the air both inside and outside of the cockpits and cabins and in the carburetor. The measurement in the carburetor is usually made in the carburetor throat just before the air is mixed with the fuel. The temperature of the liquid in liquid-cooled engines is taken to determine the operating temperature under various flight conditions. Before taking off, the temperature of the cooling liquid, if the engine is liquid cooled, and the temperature of the oil are always checked. The cooling liquid should reach a temperature of approximately 70° C. or 160° F. The oil temperature should be approximately 30° C. or 90° F. before the take-off is begun.

Thermometers enable the pilot to adjust the oil radiator shutters or other controls to maintain the proper oil temperature. Carburetor thermometers are used to determine when the temperature is such that ice might form in the carburetor. When carburetor heat is used, the temperature must be checked because hot air admitted to the carburetor may cause detonation in the engine's cylinders.

The term, free-air thermometer, is used to indicate thermometers which measure the temperature of the atmosphere either outside the aircraft or in the cabin or cockpits. The outside temperature is important, in that it determines largely the conditions under which ice may form on the various parts of the airplane. When thermometers are used to indicate ice temperatures, they are sometimes called "ice-warning indicators." Thermometers used for this purpose are accurately calibrated and may be used to check other thermometers used in the aircraft. Most thermometers are calibrated either to the Fahrenheit scale or to the centigrade scale. On the Fahrenheit scale, 32° are equal to 0° on the centigrade scale, and this indicates the freezing point of water; 212° on the Fahrenheit scale correspond to 100° on the centigrade scale, and this is the temperature at which water begins to boil under normal conditions of pressure. Oil thermometers are usually calibrated from 0° to 100° C., which correspond to 32° and 212° F.

The thermometers used to indicate the temperature of the liquid coolant are usually calibrated from 0° to 200° C., which correspond to 32° F. and 390° F. Free-air thermometers are usually calibrated to indicate from approximately − 40° C. to + 50° C., which correspond to approximately − 40° F. and + 155° F. Since it is necessary to have the temperature indicated on the instrument board for the convenience of the pilot, it is often necessary to transmit to the instrument board the temperature at a point considerably remote.

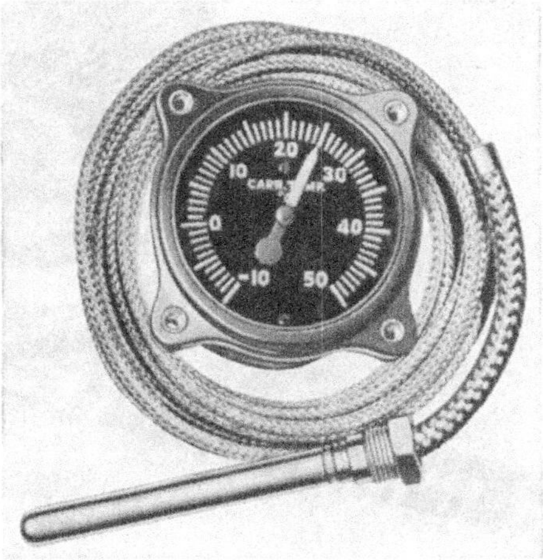

Fig. 362. A vapor-pressure-type carburetor thermometer. (Courtesy Electric Auto-Lite Company)

The vapor-pressure thermometer is a remote-indicating type. This thermometer consists of the indicating instrument mounted on the instrument board, a long capillary tube, and a metal bulb located at the point where the temperature is to be measured. The bulb is usually a hollow brass cylinder with an outside diameter of approximately ½ in. and a length which may vary from approximately 2 in. to approximately 4 in. The indicator, itself, consists of a Bourdon tube connected by a linkage system to the indicator needle. The capillary tube is fastened to the stationary end of the Bourdon tube with an airtight joint. The other end of the capillary tube forms an airtight joint with the bulb which contains a highly volatile liquid such as methyl chloride.

As the temperature of the liquid in the bulb changes, the vapor pressure in the entire system changes and affects the Bourdon tube. Vapor pressure does not vary directly with temperature, and a series of stops must be arranged to regulate the movement of the Bourdon tube. The capillary tube is usually of copper protected by a braided metal covering. When carefully calibrated, this type of thermometer gives quite accurate temperature indications. The vapor-pressure thermometer may be used to measure the temperature of free air, cabin air, oil, coolants, carburetor air, or carburetor mixtures. The capillary tube may be any length up to more than 20 ft.

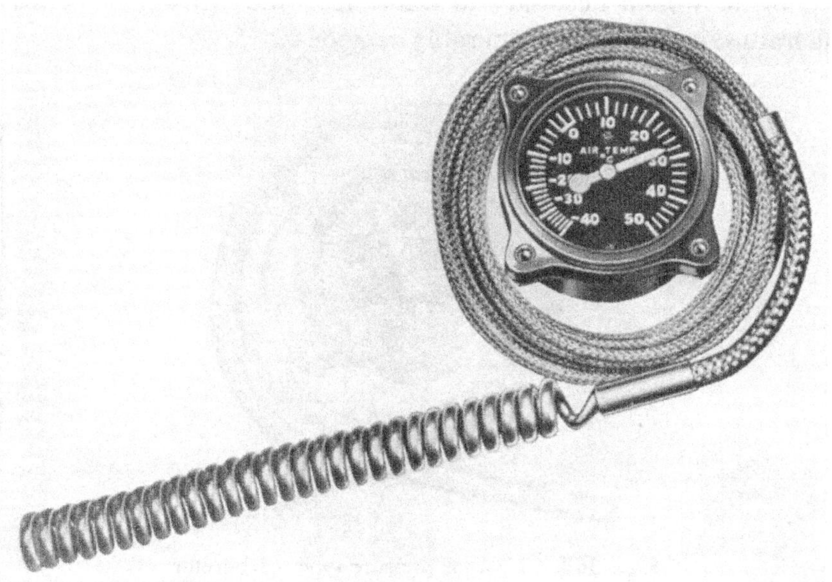

Fig. 363. A free air electric thermometer. (Courtesy Electric Auto-Lite Company)

These thermometers consist of the temperature-sensitive element, or bulb, and wires connecting the bulb with the instrument which is installed on the instrument panel. The indicating ranges on these thermometers are similar to those on the pressure type. These thermometers may be used to measure any of the temperatures desired in the same manner as the pressure type.

Electrical thermometers are of two general types: one type makes use of the Wheatstone bridge, and the other type works similarly to a regular voltmeter, having a permanent magnet and a moving coil. The first type with the Wheatstone bridge contains three resistance coils with a resistance of 100 ohms each. These coils are of manganin wire, and they form three arms of the Wheatstone bridge. The sensitive element or bulb is a coil of pure nickel wire wound on an anode-treated aluminum tube. This coil has exactly 100 ohms' resistance at a temperature of 0° C. or 32° F., depending on the scale used. This coil is inserted into a tube of Monel metal and is soldered into place. The two ends of the nickel coil are soldered to a Bakelite plug which has molded into it two silver-plated, brass female inserts. These inserts are arranged to receive two split pins made up of silver-plated brass. These split pins are connected with the wires leading to the indicator part of the instrument on the instrument panel. This sensitive bulb forms the fourth arm of the Wheatstone bridge. Metals, as a rule, are better conductors of

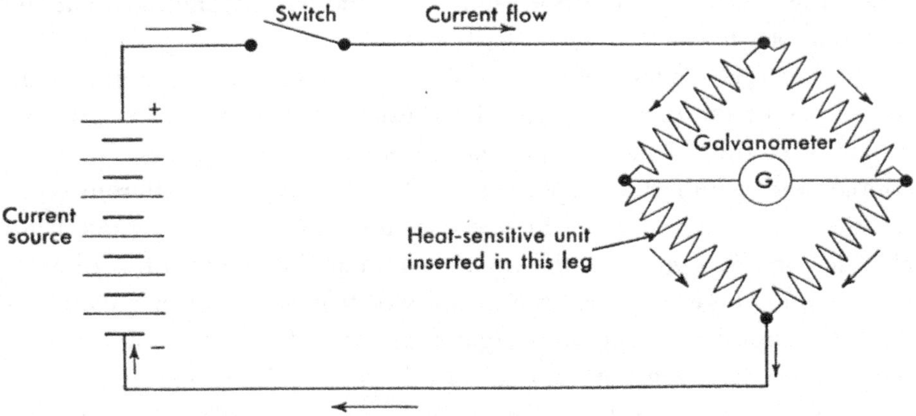

Fig. 364. A diagram showing the Wheatstone bridge.

electricity when cold than when hot. The indicating instrument, which is similar to a voltmeter, is connected across the Wheatstone bridge. When the temperature is at 0° C., the sensitive element and one of the 100-ohm coils are balanced against the other two 100-ohm coils, and no current flows through the circuit in which the instrument is inserted. With any change in the temperature of the sensitive element, this balance is upset. If the temperature rises, the resistance increases in the sensitive element and part of the current flowing through the bridge arrangement passes through the indicating instrument. This causes a deflection of the indicating needle toward the right, and this deflection indicates accurately the temperature of the bulb. If the temperature falls below 0° C., the resistance in the bulb is decreased and current flows through the instrument circuit in the opposite direction, deflecting the needle to the left. The instrument itself is electrically shielded to protect it against the magnetic effects of other instruments in the electrical system of the aircraft.

The other type of electrical thermometer contains a movable coil and a permanent magnet and is called the "ratio type" of thermometer. This thermometer is an electrical arrangement made up of a circuit having two parallel branches. One branch has a fixed resistance in series with a coil of known resistance. The other branch has the sensitive element or bulb in series with a second coil of known resistance. As the current varies in the circuit, due to changes in the resistance of the sensitive element because of changes in temperature, the current flows through the instrument in one direction for temperatures above zero, and in the other direction for temperatures below zero, deflecting the indicating needle either to the right or to the left depending upon the direction of current flow.

Another type of temperature indicator or thermometer is based on the theory of the thermocouple. This instrument consists of a sensitive indicator of the voltmeter type, the wires connecting with the thermocouple, and the thermocouple itself. When two wires of different composition are joined together at one end and this junction is heated, an electric current is set up in the wires which may be detected by connecting the other ends of the wires across a sensitive voltmeter or galvanometer. The number of the volts generated depends on the difference in temperature between the hot ends of the wires and the cold ends of the wires. The ends of the heated wires are sometimes called the hot junction, and the opposite ends are called the cold junction. One wire or lead of the thermocouple is usually of iron, and the other of an alloy of copper and nickel called constantin. These two metals are selected because they develop a definite electromotive force per degree change in temperature. The two wires in close contact are brazed to the surface of a solid copper gasket which is designed to replace a standard spark-plug gasket. This gasket is inserted in place of a spark-plug gasket and it indicates accurately the temperature of the cylinder head at that point. A thermocouple may have any desired form of ending and may be inserted under a crankcase stud nut or clamped to any part of the aircraft engine to indicate the temperature. The cold junction must have a compensating device to cancel out the effect of temperature changes of the instrument itself. A bimetal spiral spring is attached to one of the controlled springs of the instrument for this purpose. The indicator needle is moved, not only by the voltage of the thermocouple, but also by the temperature surrounding the instrument itself. When a thermocouple is disconnected, the instrument indicates the temperature of the surrounding air. An arrangement is also included to compensate for the temperature of the moving coil.

Air-Fuel-Mixture Indicators (Exhaust-Gas Analyzer). The air-fuel-mixture indicator or exhaust-gas analyzer is an instrument which indicates the ratio between air and fuel in the mixture being used by the engine. The air-fuel mixture is determined by the composition of the exhaust gases. When air and fuel are burned together, a number of by-products are given off in the exhaust gases. The exhaust gas from the engine contains water vapor, carbon dioxide, carbon monoxide, oxygen, hydrogen, nitrogen, and a small percentage of other gases. The ratio between the air and fuel burned in the engine cylinder affects the amount of each of these gases in the exhaust gas. Without a fuel-mixture indicator the pilot usually determines, as nearly as possible, the correct mixture by moving the mixture control toward the rich position until the r.p.m. of the engine begin to fall off. The control is then moved toward the lean position until the r.p.m. of the engine again begin to decrease. He then picks out the position between these two points where the engine develops maximum r.p.m. With the fuel-mixture indicator, he can determine, by reading the indication on the gauge of the instrument, when the mixture is in the proper ratio. The instrument consists of a Wheatstone-bridge arrangement, an indicator of the sensitive voltmeter type, a current supply, and a ballast tube. The ballast tube is merely a resistance coil sealed in a hydrogen-filled glass tube to protect it from temperature, pressure, and altitude effects. The ballast tube maintains the current supply at 4.1 v. Two arms of the Wheatstone bridge are arranged so that they are exposed to the exhaust gases from the engine. The other two arms are sealed in a cell filled with air which is kept saturated with moisture by means of a removable wick. The principle of operation is based upon the difference in the heat conduction of hydrogen and carbon dioxide. Hydrogen conducts heat about six times as readily as does air, while carbon dioxide has a heat conductivity of approximately one half that of air. Hydrogen, therefore, conducts heat about twelve times as readily as does carbon dioxide.

Enough current is allowed to flow through the Wheatstone bridge to maintain the resistors at a temperature of about 260° F. A continuous sample of gas from the exhaust manifold is carried through the cell containing two arms of the Wheatstone bridge. The rate at which the heat is carried away from these two arms indicates the composition of the exhaust gases. The indicator needle is so set that, with the proper mix-

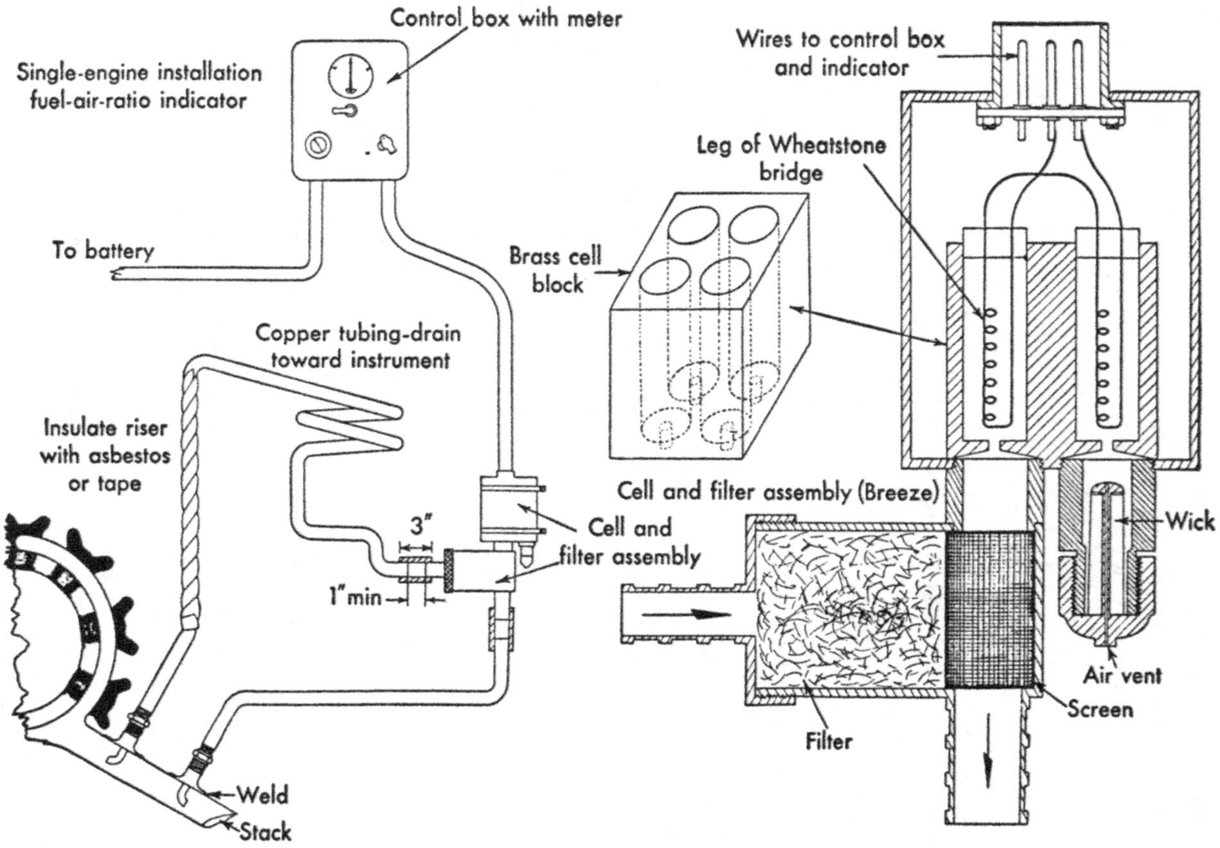

Fig. 365. A diagram showing the construction and installation of an air-fuel mixture indicator, or exhaust gas analyzer.

ture, it indicates zero or remains at the center of the dial. A rich mixture, which is one having more fuel than is necessary for the correct air-fuel ratio, produces a larger proportion of hydrogen, and the carbon dioxide content is reduced. Hydrogen carries away the heat from the coils more rapidly than carbon dioxide. This rapid carrying-away of heat lowers the temperature of the coils, and the needle on the indicator swings toward a rich mixture. The richer the mixture, the farther the needle will swing because the richer mixtures increase the content of hydrogen in the exhaust gas. As the mixture becomes lean, the amount of carbon dioxide increases, and the amount of hydrogen decreases. This combination of gases removes less heat from the coils, and the needle on the indicator swings to the lean position. As the mixture is made more lean, the amount of hydrogen decreases so that less heat is carried away from the coil and the indicator swings farther to the lean side.

There is one condition which the pilot must keep in mind and that is, when detonation in the cylinder occurs, the amount of hydrogen in the exhaust gas is greatly increased. Detonation occurs with a lean mixture, and the needle, under these conditions, will indicate an overrich mixture. As the mixture is leaned still further, more detonation occurs and the needle indicates a still richer mixture. If, at any time, as the mixture is leaned, the needle suddenly indicates rich, detonation is indicated, and the mixture lever should be moved toward the rich position.

Engine Synchronizers. The engine synchronizer is an instrument which informs the pilot whether the propellers are rotating at exactly the same speed. The pilot may, by operating the throttles or propeller controls, cause the propellers to rotate at exactly the same speed as indicated by the instrument. When two or more propellers are operating at different speeds, vibration and throbbing noises may be set up. This instrument consists of a very sensitive indicator of the voltmeter type and is adjusted to measure a difference in voltages between two similar electric tachometers. The indicator instrument is usually calibrated to indicate a variation in engine r.p.m. of approximately 40 for each engine. The instrument is connected into the circuit with the tachometer indicators and generators. There is usually a three-position control switch having an OFF position, a tachometer position, and a synchronizer position. The tachometer indicators cannot be used when the synchronizer is in operation. This instrument will show a difference in engine revolutions of approximately 2 per min. The engine throttles should be so adjusted that the pointer remains at zero.

Another type of engine synchronizing instrument is called the engine synchroscope. This instrument consists of an indicator of the voltmeter type. A condenser unit is installed to smooth out the pointer movement. Coils of different resistances may have to be installed depending upon the type of engine magneto used. The synchroscope, with the proper resistance coils, is connected between the primary circuits of two magnetos. One magneto is used for each engine. When connected to the two magneto switches, two alternating voltages, the frequencies of which

Fig. 366. An engine synchroscope for a twin-engine aircraft. (Courtesy Kollsman Instrument Division, Square D Company)

Fig. 367. An engine synchroscope for a four-engine aircraft. (Courtesy Kollsman Instrument Division, Square D Company)

depend on the engine r.p.m., are passed to the instrument. When two currents having different frequencies are applied, a series of beats or reoccurring excessive waves appear at regular intervals. If the difference in r.p.m. of the two magnetos is great, these beats cause a wide swing of the synchroscope pointer. If the magnetos are rotating at exactly the same speed, the steady alternating current causes no beats and the pointer remains stationary. The pilot adjusts the throttles to get the slowest swing possible on the needle, as it is not always possible to get the magnetos to rotate at exactly the same speed. The greater the difference in the speed of the magnetos, the more rapid the oscillations of the synchroscope indicator needle.

Fuel-Level Indicators. Fuel-level gauges vary from the simple sight-reading gauge, consisting of a float which operates an indicator in a glass tube projecting from the bottom of the fuel tank, to very complex remote-indicating devices. The simplest form is a graduated sight glass mounted directly on the fuel tank. In this glass tube is an indicator connected with the float in the tank which shows to the pilot by its position in the glass tube the fuel level in the tank.

Another type of indicator consists of hydraulically operated bellows. This system consists of a closed hydraulic system having four bellows connected by tubing. One pair of bellows is connected with a float lever in the fuel tank. The other pair of bellows is connected with a movable arm which indicates on a scale the fuel level or the contents of the fuel

Fig. 368. A combination tachometer and synchroscope for a twin-engine aircraft. (Courtesy Kollsman Instrument Division, Square D Company)

tank. As the float rises in the tank, it releases the pressure on one bellows and compresses the other. This causes a movement of the two bellows connected with the indicator, moving the indicator over the dial.

When the tank is located conveniently, a rotating needle may be arranged under a glass dial which is rotated directly through a gear train attached to a float in the tank. Remote-indicating instruments, however, usually consist of a float arrangement in the tank which moves a contact over a resistance coil, causing a variation in the flow of electric current which is measured by a voltmeter type of indicator on the instrument panel. This type is commonly used in automobiles.

Another kind of fuel indicator operates on the hydraulic pressure developed in the tank by the fuel. The greater the depth of the fuel, the greater the pressure on the instrument, which causes a deflection of a bellows or a diaphragm which is transmitted to the instrument board by means of one of the remote-indicating devices, such as a Selsyn. The float causes the rotor in the transmitter to turn, causing a corresponding turn of the rotor in the indicator.

Torque Meter or Horsepower Indicator. A torque meter measures the torque reaction on the stationary gear by the propeller reduction gear. While this instrument measures the torque reaction, it is used to indicate engine brake horsepower. The effective power of the engine is used to turn the propeller. A measure of the force exerted by the engine in

Fig. 369. Nose of a large aircraft engine showing the torquemeter or horsepower indicator. (Courtesy Wright Aeronautical Corporation)

turning the propeller is a close indication of the engine's power output. This instrument is mounted on the front section of the crankcase.

The instrument consists of a torque arm which has a ball race on its inside diameter. The torque arm is assembled to a support by means of 86 balls inserted between the races, permitting relative motion between the torque arm and the support and acting as an antifriction device. The torque arm at its outer end acts on a balance valve. This valve is actuated in one direction by the torque reaction of the stationary gear and is balanced in the other direction by the oil pressure on the valve

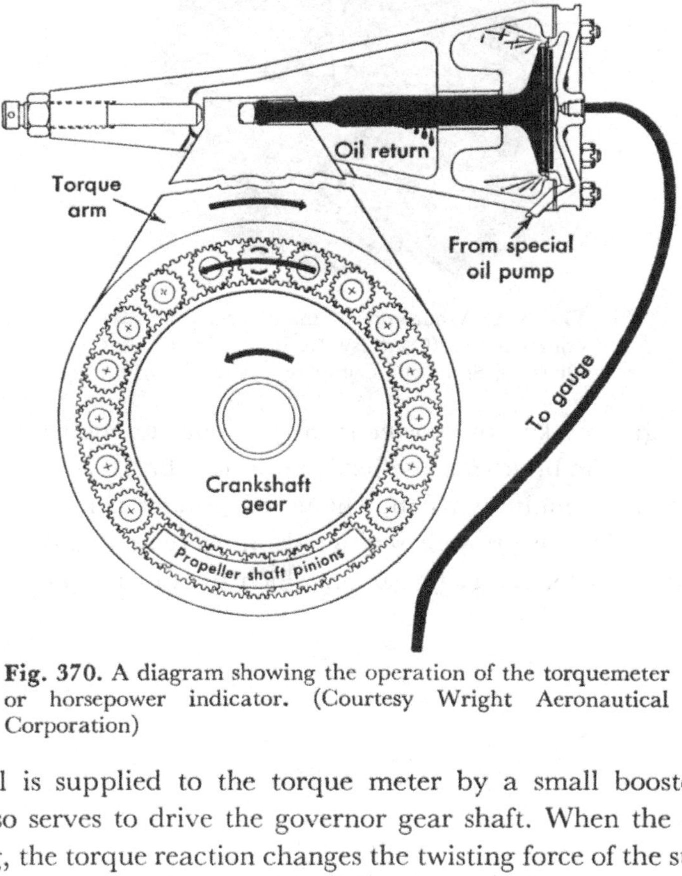

Fig. 370. A diagram showing the operation of the torquemeter or horsepower indicator. (Courtesy Wright Aeronautical Corporation)

head. Oil is supplied to the torque meter by a small booster pump which also serves to drive the governor gear shaft. When the engine is operating, the torque reaction changes the twisting force of the stationary gear to a slight outward movement of the valve. This results in an increase in the oil pressure on the valve head. The oil pressure increases until it is just sufficient to overcome the torque reaction and thus causes a slight inward movement of the valve. This inward movement of the valve opens metering slots in the sleeve in which the valve moves. The oil escaping through these openings is led to the crankcase. The oil pressure on the head of the valve then decreases until the torque reaction is not sufficient to overcome the oil pressure. When another outward movement of the valve is produced, a balance of oil flow is quickly reached. An oil-pressure-gauge line is connected into the valve housing and is led to a special oil-pressure gauge calibrated to read "Brake-mean-effective pressure."

Fig. 371. A horsepower indicator for an aircraft engine. (Courtesy Kollsman Instrument Division, Square D Company)

Since engine brake horsepower is proportional to engine revolutions per minute and to brake-mean-effective pressure, brake horsepower may be calculated by multiplying the engine revolutions per minute by the brake-mean-effective pressure registered on the gauge and divided by a constant value for the engine upon which the instrument is installed.

XXIV THE TURBO-SUPERCHARGER ELECTRONIC CONTROL SYSTEM

The electronic control system for turbo-superchargers is designed to maintain constant carburetor inlet pressure by automatically regulating the position of the exhaust waste gate. The system may be installed to operate in connection with a single engine or may be installed in multi-engine airplanes. In multi-engine airplanes the installation consists of one turbo-boost selector and one main junction box plus one set of the other units required for each engine.

The parts of the system necessary for each engine consist of a waste-gate motor, an induction system, a "pressuretrol," an amplifier, a turbo-governor, a turbo-boost selector, a nacelle junction box, and a main

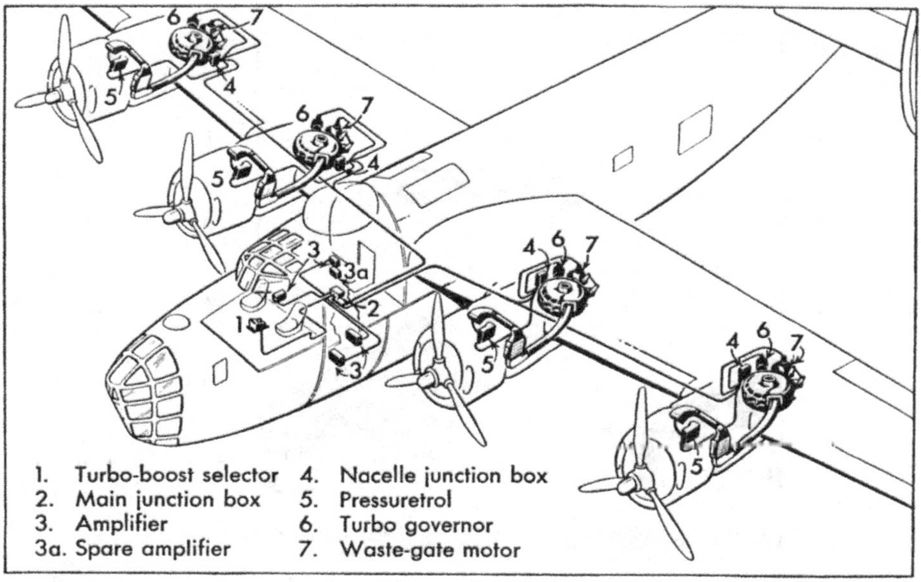

Fig. 372. The central system for turbo-superchargers installed in a four-engine airplane. (Courtesy Minneapolis-Honeywell Regulator Company)

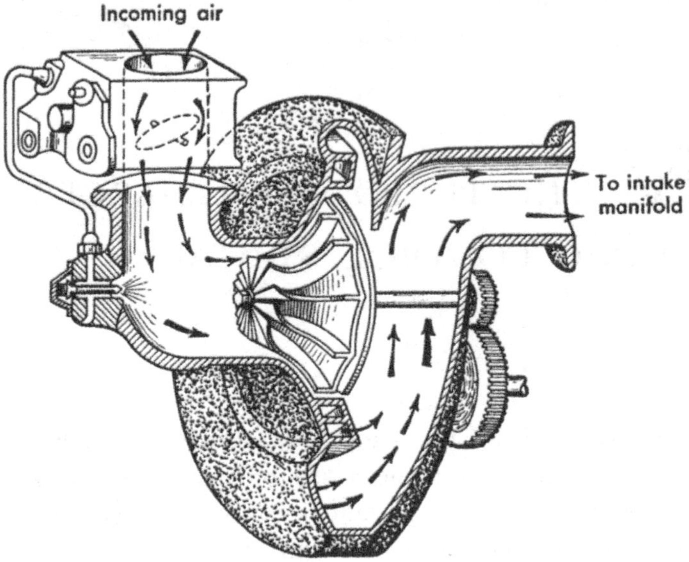

Fig. 373. A supercharger. (Courtesy Minneapolis-Honeywell Regulator Company)

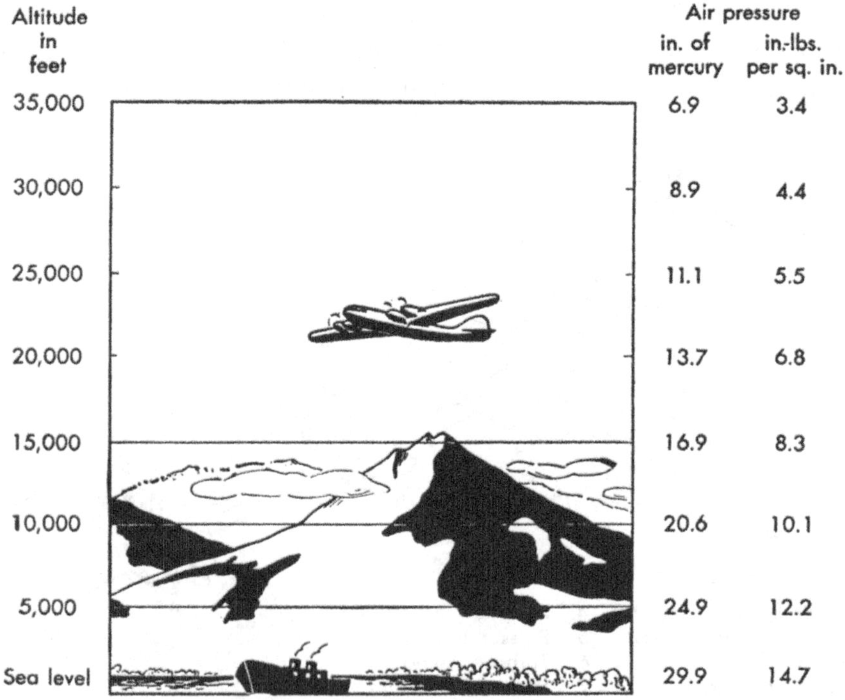

Fig. 374. Air pressure changes with altitude. (Courtesy Minneapolis-Honeywell Regulator Company)

junction box. The installation in a four-engine aircraft is shown in Figure 372.

The purpose of the turbo-supercharger is primarily to maintain engine horsepower. The horsepower output of an engine depends almost entirely upon the total weight of the air-and-fuel mixture entering the cylinders each second. This depends upon the manifold pressure. To obtain greater power from a given engine, it is often desirable to maintain the manifold pressure above atmospheric pressure by means of a supercharger (Figure 373). The decrease in atmospheric pressure at higher altitudes is shown in Figure 374. It is only by maintaining constant manifold pressure that the full rated horsepower of the engine is developed at high altitudes.

Figure 375 is a drawing of a turbo-supercharger. The turbo-supercharger is located in the airplane at such a point that it makes use of the exhaust gases from the engine. The speed of the supercharger, which consists of a gas-driven turbine that is directly connected to the air-compressor element consisting of an impeller unit and a diffuser unit, is regulated by opening and closing the waste gate (Figure 376). When the waste gate is open, the exhaust gases from the engine escape without passing through the turbine. When the waste gate is closed, the gases are forced to act against the turbine wheel and spin the compressor. The amount of turbo-boost may be made to suit the requirement of the engine at any altitude by regulation of the waste gate.

Figure 377 shows the various parts of the automatic turbo-supercharger regulator. The source of electric power for this unit is a 400-cycle inverter. Usually two inverters are installed, but only one is used at a time. The inverter supplies 115-v., 400-cycle alternating current to operate the turbo-control system.

The turbo-boost selector is installed within reach of the pilot. It contains four small potentiometers which require adjustment only to compensate for small differences in engine or turbo-supercharger performance. This unit is shown in Figure 379. Once the calibrators are set, the pilot can control the turbo-boost on all four engines by turning the large, central control knob. Figure 380 shows part of the instrument installation on a four-motor aircraft. The turbo-boost selector unit is mounted just to the right of the pilot's control column.

The pressuretrol (Figure 382) is a sensing element which measures electrically the pressure of the air supplied by the turbo-supercharger to the carburetor. This unit controls the automatic operation of the system to maintain whatever pressure the pilot has selected.

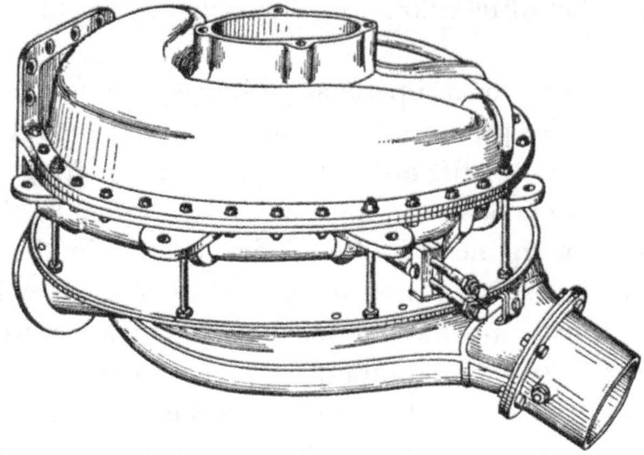

Fig. 375. A turbo-supercharger. (Courtesy Minneapolis-Honeywell Regulator Company)

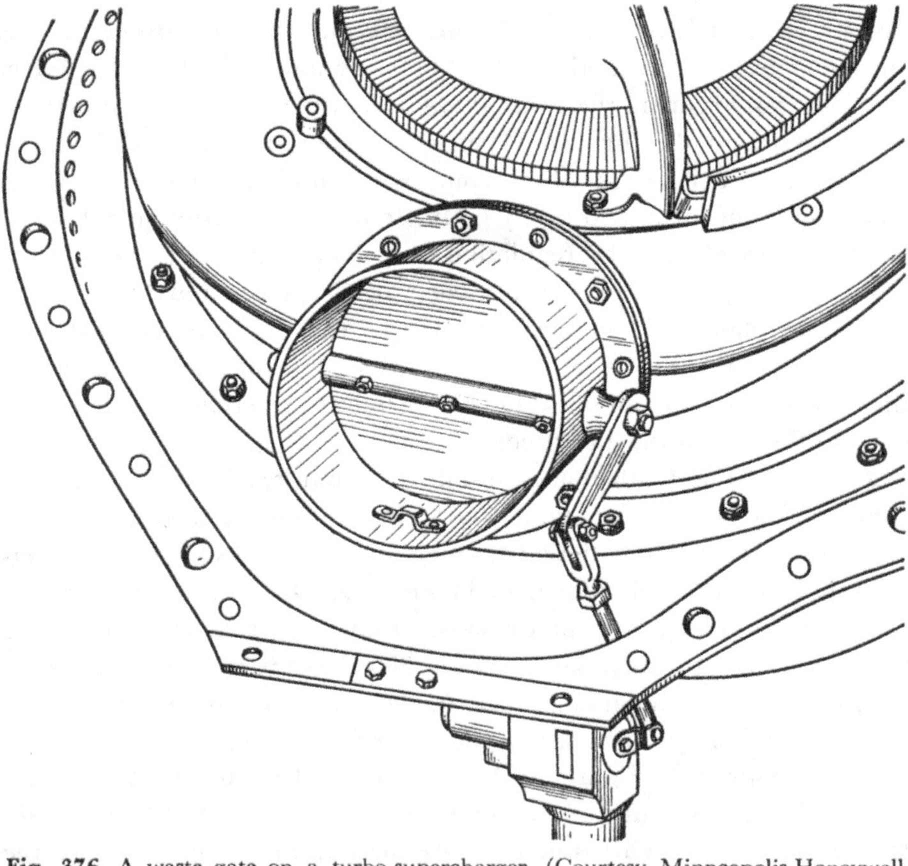

Fig. 376. A waste gate on a turbo-supercharger. (Courtesy Minneapolis-Honeywell Regulator Company)

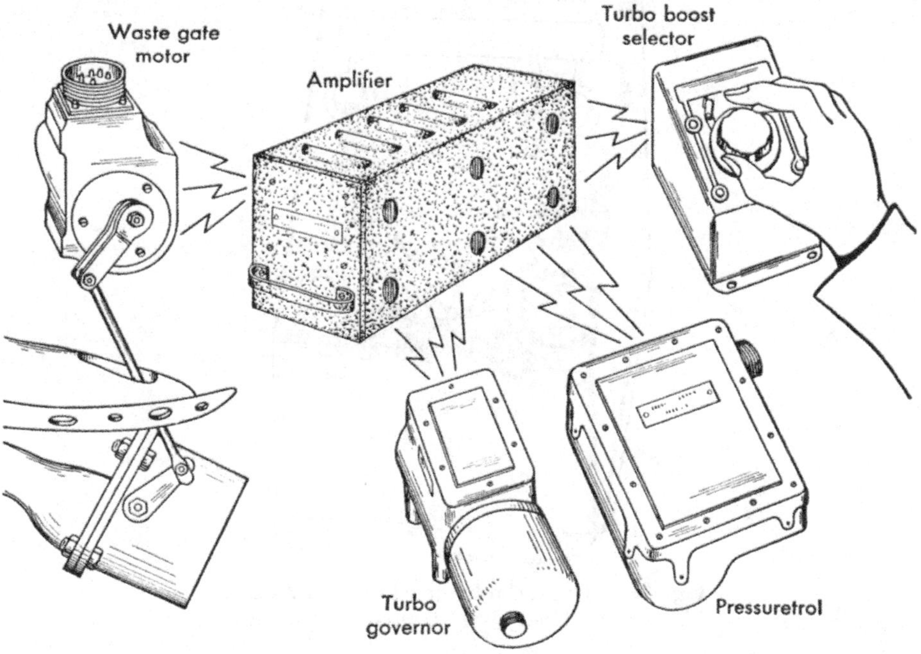

Fig. 377. The turbo-supercharger control system. (Courtesy Minneapolis-Honeywell Regulator Company)

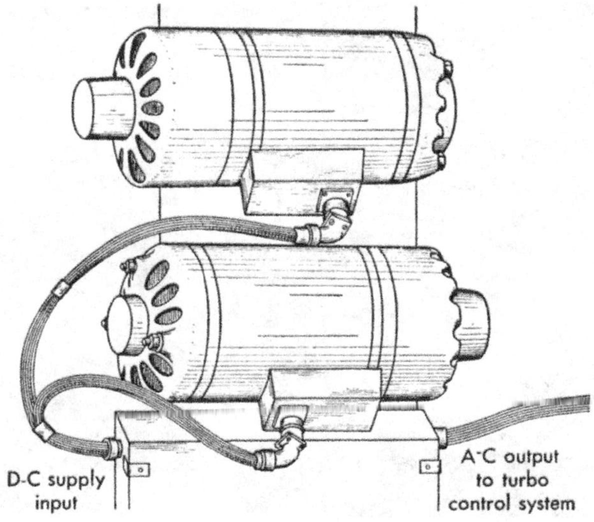

Fig. 378. The airplane's inverters. (Courtesy Minneapolis-Honeywell Regulator Company)

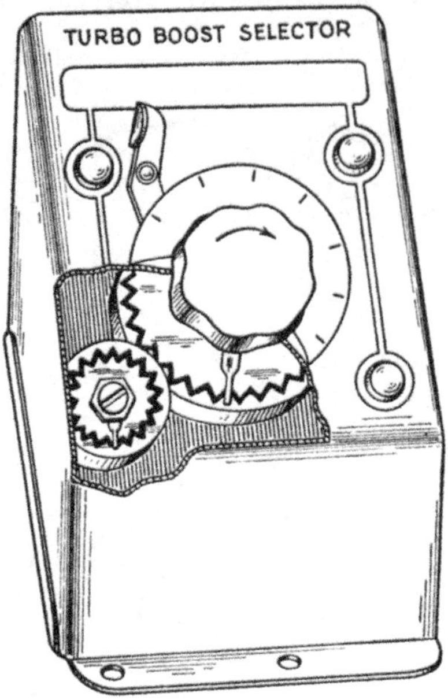

Fig. 379. The turbo-boost selector. (Courtesy Minneapolis-Honeywell Regulator Company)

Fig. 380. Part of the instrument installation on a four-engine aircraft. (Courtesy Minneapolis-Honeywell Regulator Company)

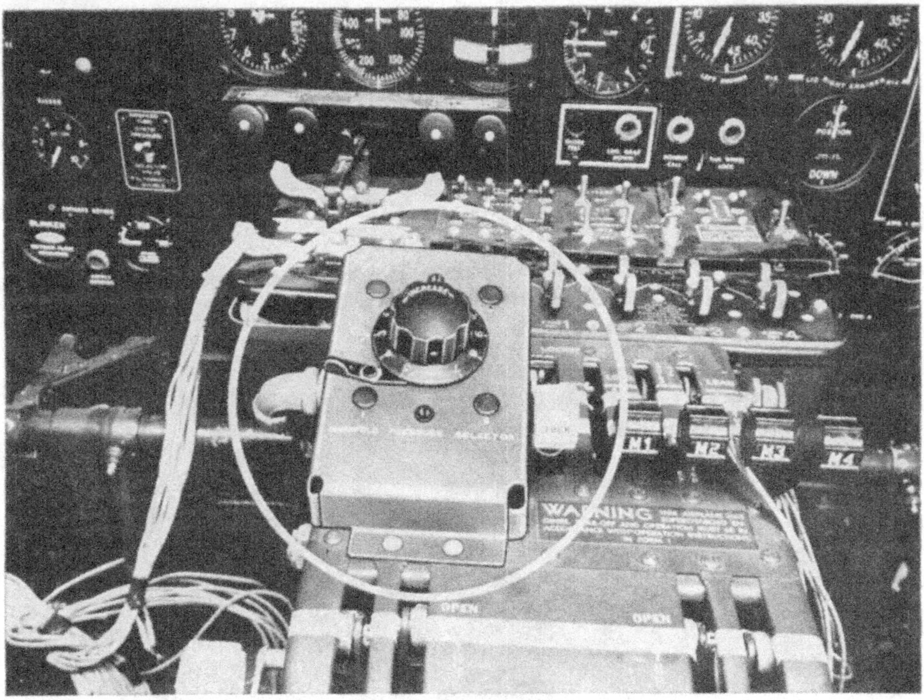

Fig. 381. A manifold pressure selector. (Courtesy Minneapolis-Honeywell Regulator Company)

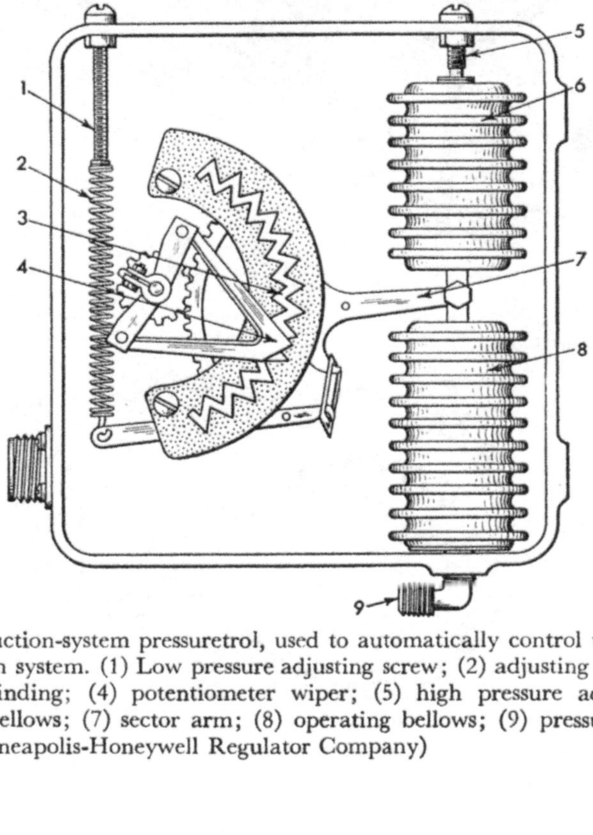

Fig. 382. Induction-system pressuretrol, used to automatically control the air pressure in the induction system. (1) Low pressure adjusting screw; (2) adjusting spring; (3) potentiometer winding; (4) potentiometer wiper; (5) high pressure adjusting screw; (6) reference bellows; (7) sector arm; (8) operating bellows; (9) pressure connection. (Courtesy Minneapolis-Honeywell Regulator Company)

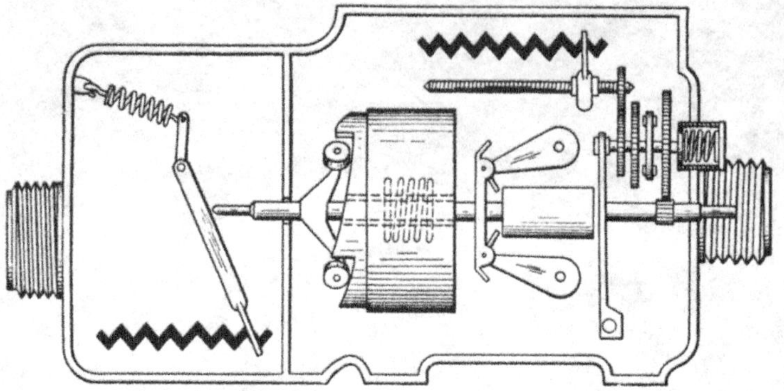

Fig. 383. A turbo-governor. (Courtesy Minneapolis-Honeywell Regulator Company)

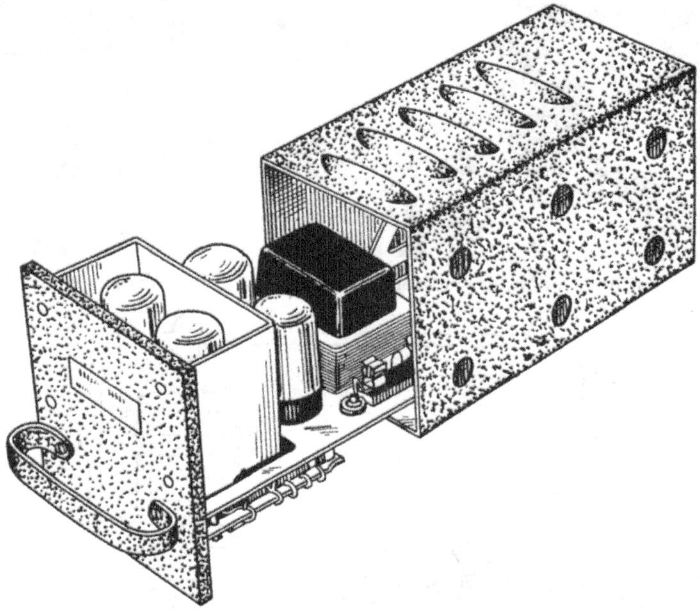

Fig. 384. A turbo amplifier. (Courtesy Minneapolis-Honeywell Regulator Company)

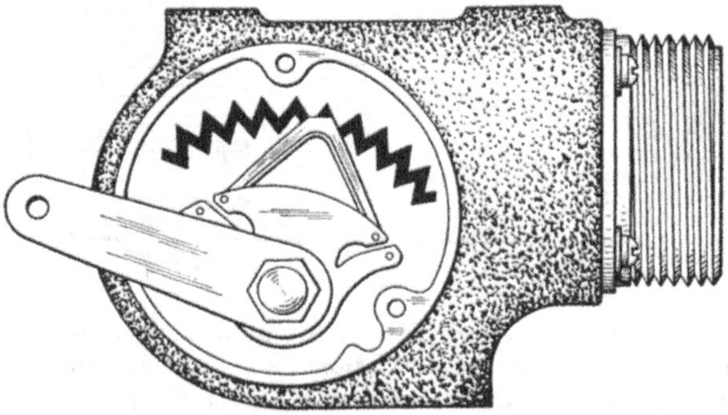

Fig. 385. The waste-gate motor. (Courtesy Minneapolis-Honeywell Regulator Company)

The turbo-governor (Figure 383) is a dual safety device driven by a flexible control shaft geared to the supercharger. The overspeed control prevents the turbo from exceeding its safe operating speed.

The amplifying unit (Figure 384) relays the signals which control the waste-gate motor. One signal will cause the waste gate to be closed; the other signal will open the waste gate. The waste-gate motor (Figure 385) operates the waste gate in response to control signals received from the amplifier.

The turbo-boost selector (Figure 386) is really the manifold-pressure selector. To decrease or increase the manifold pressure, it is necessary only for the pilot or copilot to turn the knob on this unit to the desired

Fig. 386. A manifold boost selector. (Courtesy Minneapolis-Honeywell Regulator Company)

reading. Figure 387 is a cutaway view of the turbo-boost selector. The dial is graduated from 1 to 10. Usually 8 is the maximum take-off power. Emergency power may be obtained by releasing a catch allowing the dial to be rotated to 9 or 10. This range is used for emergency power only. The unit consists of a main selector potentiometer assembly. This is the large cylindrical winding and a wiper which contacts this winding on its inner surface. This wiper is moved by means of the knob on the outside of the case. Calibrator potentiometers are each connected to the control system of a corresponding engine.

The pressuretrol unit (Figure 388) is directly acted upon by pressure variations at the carburetor inlet. This unit consists of a voltage-dividing potentiometer operated by a pair of bellows connected to the induction

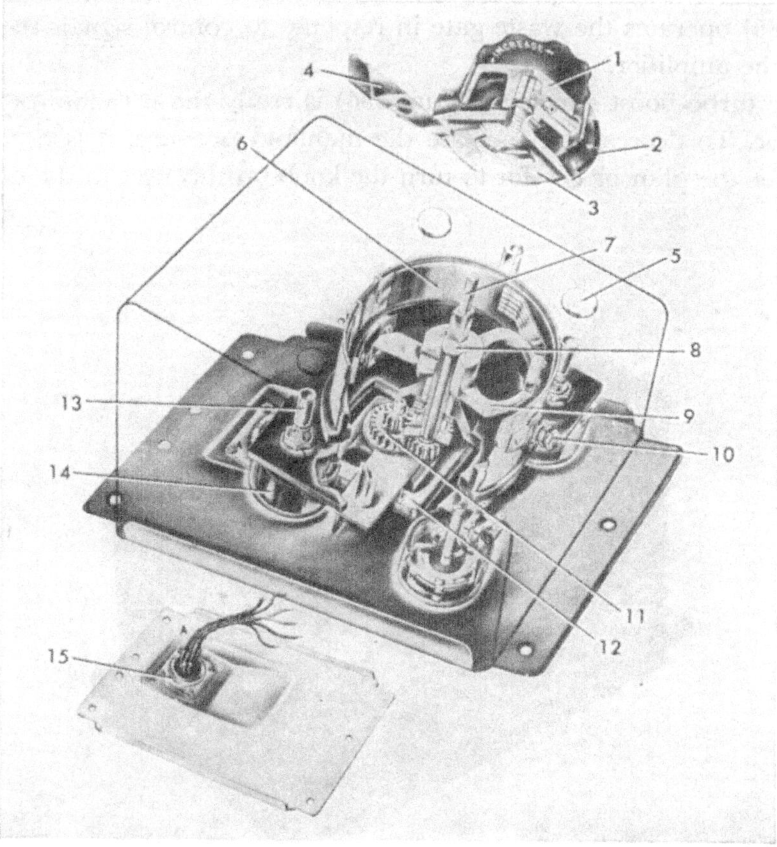

Fig. 387. A cutaway view of a turbo-boost selector. (1) Knob; (2) dial; (3) dial-stop plate; (4) latch; (5) snap plug; (6) selector potentiometer winding; (7) Vernier-drive shaft; (8) wiper shaft; (9) selector potentiometer wiper; (10) potentiometer center top; (11) Vernier gears; (12) transformer; (13) guide sleeve; (14) calibrator potentiometer; (15) connector. (Courtesy Minneapolis-Honeywell Regulator Company)

system near the carburetor inlet. One bellows contains an internal balancing spring, while the other is acted upon by carburetor inlet pressure. The bellows are connected to the wiper arrangement in the potentiometer. Since both bellows are subjected to atmospheric pressure from the outside, changes in atmospheric pressure do not cause variation in the bellows. In the spring-loaded bellows is a partial vacuum, and the spring is used to balance the bellows setup. This unit is similar to an aneroid barometer.

Fig. 388. A pressuretrol, external view. (Courtesy Minneapolis-Honeywell Regulator Company)

The turbo-governor (Figure 389) is an overspeed unit which automatically limits the turbine wheel in the supercharger to a safe top speed. Figure 390 shows a phantom view of the turbo-governor.

Overspeed is regulated by a pair of flyweights and a clutch. When the turbo overspeeds, the flyweights operate a clutch lever which results in an electrical signal which causes the waste gate to open and slow down the turbo-supercharger. As the turbo-supercharger slows down,

Fig. 389. The turbo-governor. (Courtesy Minneapolis-Honeywell Regulator Company)

the flyweight governor releases the clutch resulting in a downward movement of the potentiometer wiper. This movement brings the supercharger back to its normal speed.

The amplifier shown in Figure 391 supplies power to operate the waste gate in response to signals received from the control unit. It contains a fuse for the system, a power transformer, resistors and condensers, and four vacuum tubes. Figure 392 shows the internal wiring of the amplifier. The amplifier receives electric signals from the control units of the system and amplifies these signals to cause operation of the

Fig. 390. A phantom view of the turbo-governor. (1) Housing cover plate; (2) clutch gear; (3) counter-clockwise drive gear; (4) clutch disk; (5) clockwise drive gear; (6) clutch-disk shaft; (7) clutch spring; (8) threaded connector; (9) splined governor shaft; (10) clutch-lever adjusting screw; (11) thrust collar; (12) mounting bracket; (13) clutch-lever pivot; (14) clutch lever; (15) flyball weight; (16) governor spring; (17) spring barrel; (18) inertia rotor; (19) torque spring; (20) locking plate; (21) cam plate; (22) guide; (23) cam-roller assembly; (24) potentiometer wiper; (25) push rod; (26) cross-head; (27) wiper arm; (28) frame assembly; (29) potentiometer winding; (30) connector; (31) accelerator cover; (32) return spring; (33) fulcrum adjustment; (34) wiper-arm pivot; (35) terminal block; (36) nut on governor shaft; (37) bearing plate; (38) governor shaft; (39) governor housing; (40) declutching extension arm; (41) potentiometer winding; (42) collector bar; (43) lead screw; (44) potentiometer wiper; (45) lead-screw nut; (46) lead-screw drive gear. (Courtesy Minneapolis-Honeywell Regulator Co.)

waste-gate motor by energizing one of the waste-gate-motor windings. The current provided by the amplifier to operate the motor will be either in phase with the line current or 180° out of phase. The phase of this output current, which is the same as the phase of the incoming signal, controls the direction of rotation of the waste-gate motor.

The waste-gate motor (Figure 393) is the operating unit which moves

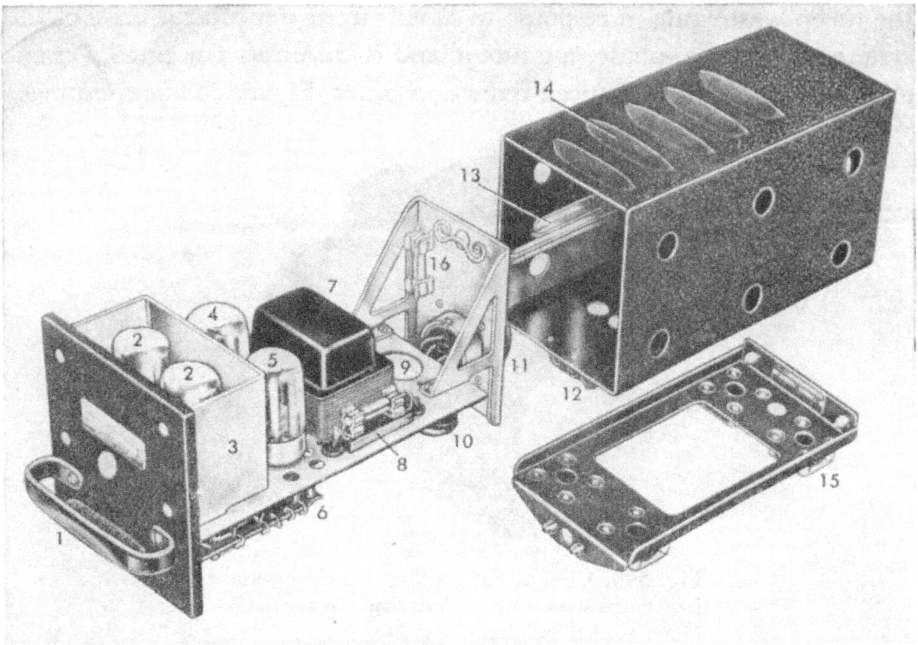

Fig. 391. A turbo-control system amplifier. (1) handle; (2) 705 tubes; (3) Heat Shield; (4) 7 x 4 tube; (5) 7 F 7 tube; (6) resistor card; (7) Power transformer; (8) Fuse; (9) Amplifier phase condenser; (10) shielded grid lead; (11) AN connector; (12) Dzus fastener; (13) guide rails; (14) Louvers; (15) Lord shock mounts; (16) spare fuse. (Courtesy Minneapolis-Honeywell Regulator Company)

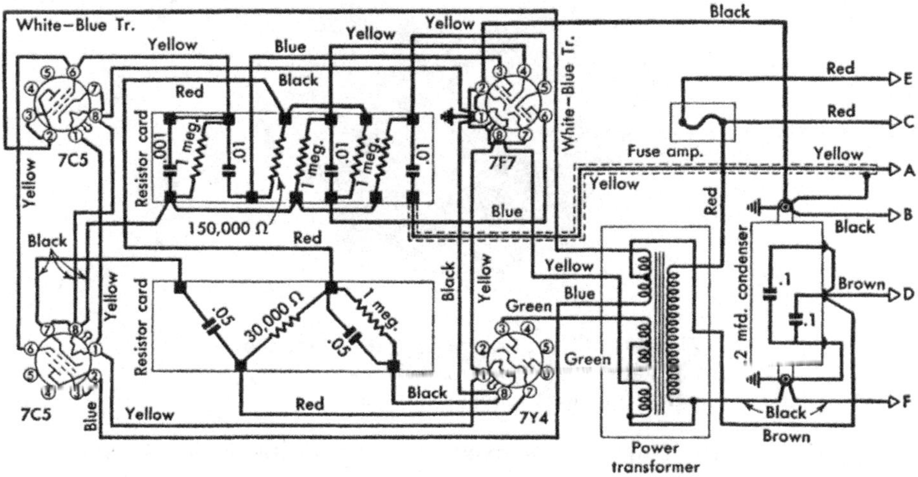

Fig. 392. The internal wiring of the amplifier. (Courtesy Minneapolis-Honeywell Regulator Company)

the turbo-waste gate in response to signals from the control units of the system. It is a two-phase, a-c motor and it transmits power to a crank arm through a train of speed-reduction gears. Figure 395 is a cutaway

Fig. 393. View of the waste-gate motor showing crank-arm end. (Courtesy Minneapolis-Honeywell Regulator Company)

view of the waste-gate motor. The crank arm of the motor is connected to the waste gate of the turbo-supercharger by a mechanical linkage.

The stator assembly of the motor consists of eight pole pieces, each of which is wound with a coil of wire. Alternate coils are connected in

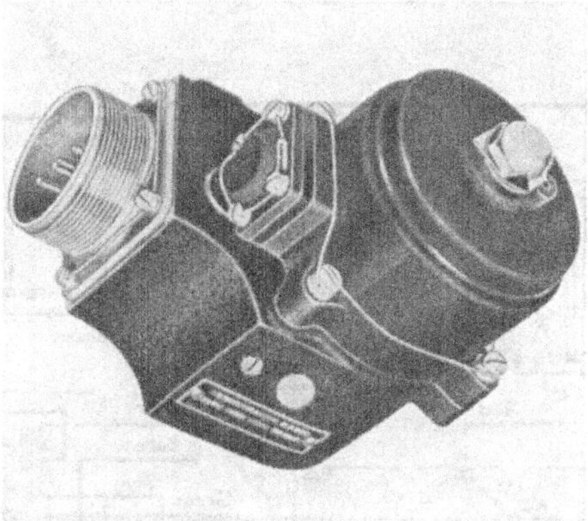

Fig. 394. View of the waste-gate motor showing stator end. (Courtesy Minneapolis-Honeywell Regulator Company)

series. Four of the coils form a fixed or line exciting winding, and four form an amplifying exciting winding. If the coils were numbered in rotation, 1, 3, 5, and 7 would form one winding, and 2, 4, 6, and 8 would form the other. Either set of coils may be used as the line exciting winding depending on the desired rotation of the motor.

The armature is of laminated steel with one brass lamination at each end. It is connected to the armature shaft by means of the clutch mechanism.

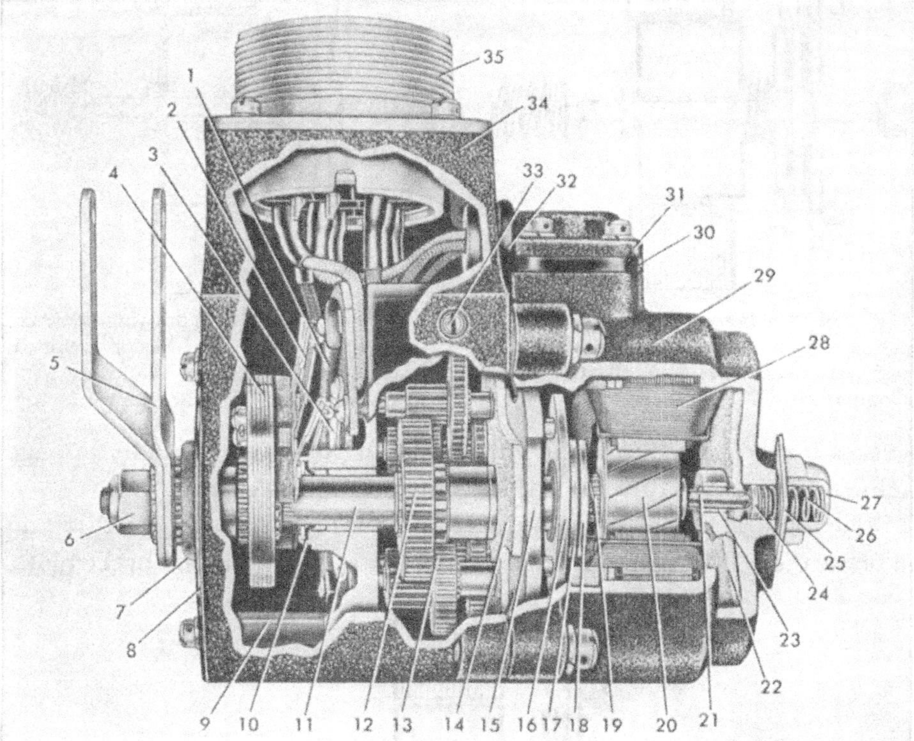

Fig. 395. A cutaway view of the waste-gate motor. (1) Potentiometer winding; (2) potentiometer wiper; (3) collector strip; (4) stop plate; (5) crank arm; (6) castellated nut; (7) dust washer; (8) cover plate; (9) mechanical stop; (10) Neoprene oil seal; (11) output shaft; (12) bull gear; (13) gear train; (14) felt oiler; (15) armature-assembly mounting plate; (16) brake plate; (17) brake disk; (18) clutch disk; (19) clutch spring; (20) armature; (21) armature shaft; (22) felt oiler; (23) sleeve bearing; (24) thrust plug; (25) brake-spring lock washer; (26) brake spring; (27) brake-spring cap; (28) stator coil; (29) stator housing; (30) terminal block; (31) terminal block cover; (32) rubber grommet; (33) oil plug; (34) gear housing; (35) connector. (Courtesy Minneapolis-Honeywell Regulator Company)

The clutch and brake mechanisms (Figure 396) are located together at the drive end of the motor assembly. The clutch is a safety device which prevents damage to the gear train when the potentiometer wiper assembly strikes a mechanical stop. The brake stops the shaft rotation whenever the amplifier phase winding is no longer energized. The armature is not directly connected to its shaft, but drives the shaft through the clutch mechanism. During normal motor operation, the armature drives the shaft with no slippage. When the wiper assembly

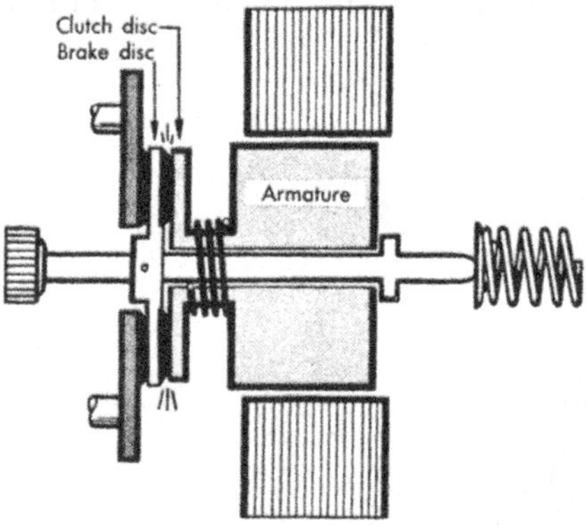

Fig. 396. A schematic drawing to show the clutch and brake mechanism. (Courtesy Minneapolis-Honeywell Regulator Company)

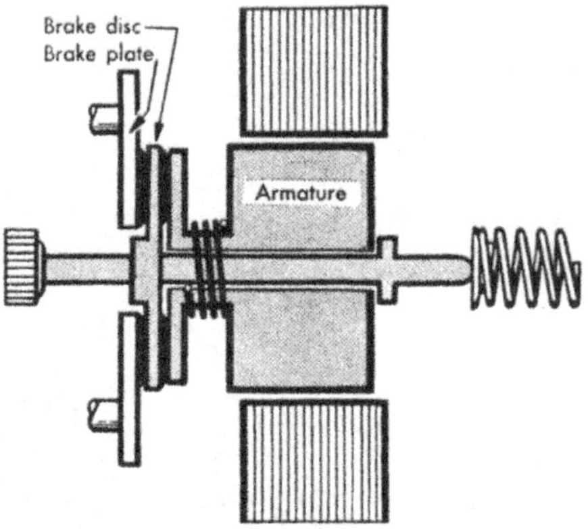

Fig. 397. A schematic drawing of brake operation: View A: Motor stopped, brake surfaces engaged. (Courtesy Minneapolis-Honeywell Regulator Company)

strikes a mechanical stop at either end of its travel, clutch slippage allows the rotor to turn for a short time around its shaft.

The braking surfaces (Figures 397 and 398) consist of a cork disk and a brake disk. The cork disk is inserted in the rigidly mounted brake plate.

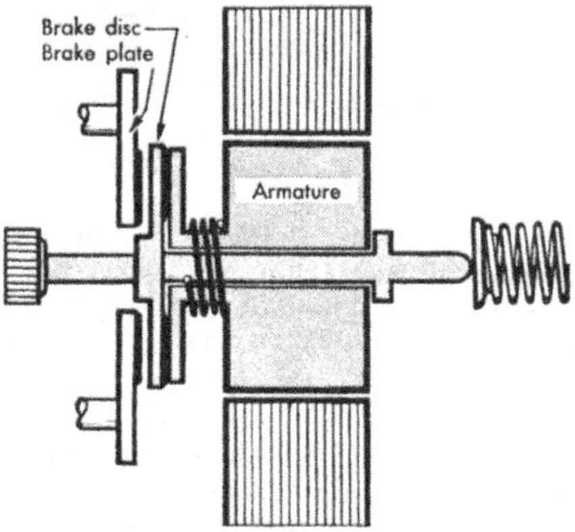

Fig. 398. A schematic drawing of brake operation. View B: Motor running, brake surfaces disengaged. (Courtesy Minneapolis-Honeywell Regulator Company)

The steel brake disk is part of the armature shaft. A brake spring under the cap housing on the end of the shaft holds the brake disk against the cork surface when the motor is not in operation. When the amplifier phase winding is energized, magnetic attraction on the armature pulls the shaft lengthwise, bringing the armature into alignment with both faces. This disengages the braking surfaces and allows shaft rotation.

Fig. 399. A view of the gear train in the waste-gate motor. (Courtesy Minneapolis-Honeywell Regulator Company)

When the amplifier phase winding ceases to be energized, the magnetic pull is decreased. The brake spring then displaces the armature shaft to re-engage the braking surfaces. The armature is then out of alignment with the pole faces.

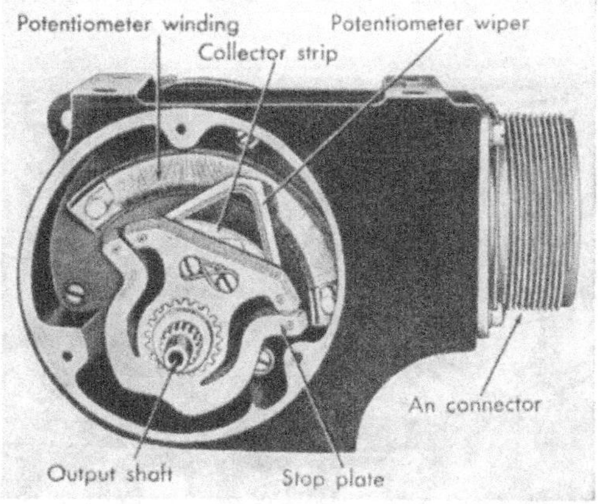

Fig. 400. A view of the balancing potentiometer in the waste-gate motor. (Courtesy Minneapolis-Honeywell Regulator Company)

The braking effect is greatly magnified through the gear train (Figure 399). The gear-train assembly is located in the gear housing and provides a speed reduction of 1689 to 1. The gear housing contains a quantity of oil which lubricates the gears.

Fig. 401. The waste-gate motor and linkage on a turbo-supercharger waste gate. (Courtesy Minneapolis-Honeywell Regulator Company)

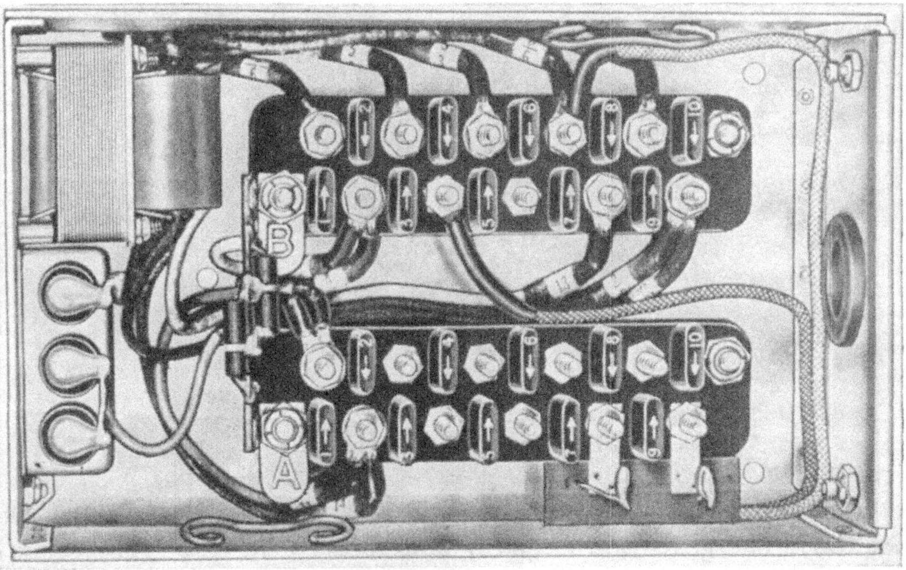

Fig. 402. The interior of a nacelle junction box. (Courtesy Minneapolis-Honeywell Regulator Company)

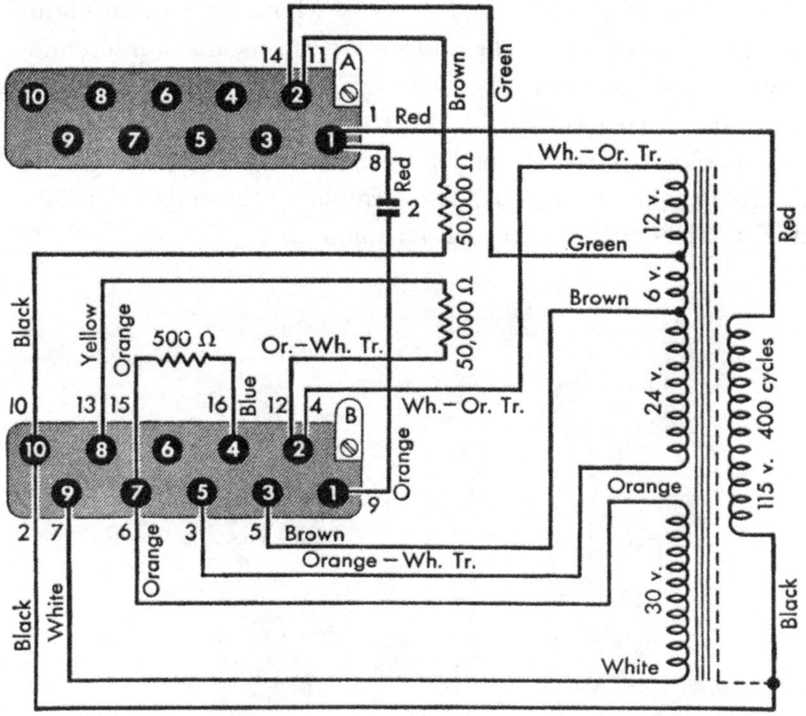

Fig. 403. Internal wiring of a nacelle junction box. (Courtesy Minneapolis-Honeywell Regulator Company)

The potentiometer (Figure 400) is located under the cover plate in the gear housing. This unit is the balancing potentiometer of the electric control circuit. It consists of a semicircular winding and a collector strip which is held in place by three mounting screws. Mechanical stops in the gear housing limit the rotation of the crank arm to 90°, and the stop plate is constructed so as to absorb the shock of striking the stops.

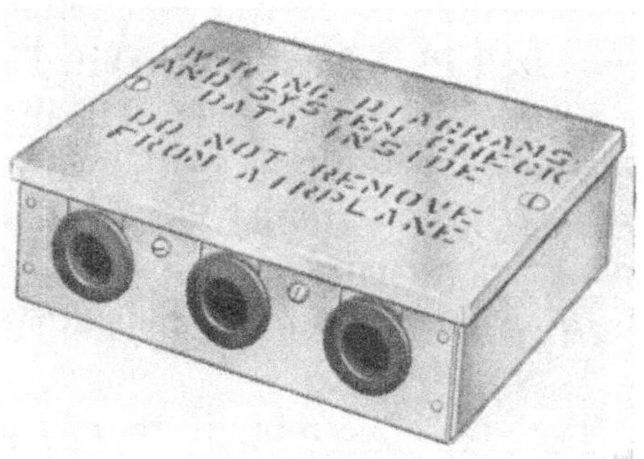

Fig. 404. An external view of the main junction box. (Courtesy Minneapolis-Honeywell Regulator Company)

The crank arm shown in Figure 401 is a double steel arm to which the waste-gate linkage is attached. The arm fits over the spline output shaft to which it is tightened by a nut. It is adjustable in steps of 22½°, permitting the crank arm to be placed in the best position relative to the waste-gate arm.

The nacelle junction box, the interior of which is shown in Figure 402, is located in the engine nacelle and provides a means of interconnecting the pressuretrol governor, the waste-gate-motor leads, and the main junction box. Two terminal blocks within the unit have ten terminals each. The blocks are identified by the letters, *A* and *B*, stamped on metal tabs attached to the box. The terminals are numbered 1–10 on the Bakelite dividers adjacent to each terminal post.

Fig. 405. The interior of the main junction box. (Courtesy Minneapolis-Honeywell Regulator Company)

Located in the nacelle junction box are a condenser and three resistors employed in the control circuit. Figure 403 shows the internal winding of the nacelle junction box. The main junction box (Figure 404), located in connection with the main instrument panel, is an aluminum box containing three terminal blocks of ten terminals each. These blocks are identified by tabs marked A, B, and C. The terminals are numbered 1–10 on the Bakelite dividers adjacent to each terminal post, as shown in Figure 405.

The entire assembly is entirely automatic in its operation and maintains equal manifold pressure at all times in one or more engines.

INDEX

The Aviation Collection by Sportsman's Vintage Press

www.SportsmansVintagePress.com

Aircraft Construction Handbook	by Thomas A. Dickinson
Aircraft Sheet Metal Work	by C. A. LeMaster
The Aircraft Apprentice	by Leslie MacGregor
Aircraft Woodwork	by Col. R. H. Drake
Aircraft Welding	by Col. R. H. Drake
Aircraft Sheet Metal	by Col. R. H. Drake
Aircraft Engines	by Col. R. H. Drake
Aircraft Electrical and Hydraulic Systems, and Aircraft Instruments	by Col. R. H. Drake
Aircraft Engine Maintenance and Service	by Col. R. H. Drake
Aircraft Maintenance and Service	by Col. R. H. Drake

Made in the USA
Monee, IL
07 March 2025

13624173R00223